INTRODUCTORY THEORY OF TOPOLOGICAL VECTOR SPACES

PURE AND APPLIED MATHEMATICS

A Program of Monographs, Textbooks, and Lecture Notes

MONOGRAPHS AND TEXTBOOKS IN PURE AND APPLIED MATHEMATICS

1. *K. Yano,* Integral Formulas in Riemannian Geometry (1970)
2. *S. Kobayashi,* Hyperbolic Manifolds and Holomorphic Mappings (1970)
3. *V. S. Vladimirov,* Equations of Mathematical Physics (A. Jeffrey, ed.; A. Littlewood, trans.) (1970)
4. *B. N. Pshenichnyi,* Necessary Conditions for an Extremum (L. Neustadt, translation ed.; K. Makowski, trans.) (1971)
5. *L. Narici et al.,* Functional Analysis and Valuation Theory (1971)
6. *S. S. Passman,* Infinite Group Rings (1971)
7. *L. Dornhoff,* Group Representation Theory. Part A: Ordinary Representation Theory. Part B: Modular Representation Theory (1971, 1972)
8. *W. Boothby and G. L. Weiss, eds.,* Symmetric Spaces (1972)
9. *Y. Matsushima,* Differentiable Manifolds (E. T. Kobayashi, trans.) (1972)
10. *L. E. Ward, Jr.,* Topology (1972)
11. *A. Babakhanian,* Cohomological Methods in Group Theory (1972)
12. *R. Gilmer,* Multiplicative Ideal Theory (1972)
13. *J. Yeh,* Stochastic Processes and the Wiener Integral (1973)
14. *J. Barros-Neto,* Introduction to the Theory of Distributions (1973)
15. *R. Larsen,* Functional Analysis (1973)
16. *K. Yano and S. Ishihara,* Tangent and Cotangent Bundles (1973)
17. *C. Procesi,* Rings with Polynomial Identities (1973)
18. *R. Hermann,* Geometry, Physics, and Systems (1973)
19. *N. R. Wallach,* Harmonic Analysis on Homogeneous Spaces (1973)
20. *J. Dieudonné,* Introduction to the Theory of Formal Groups (1973)
21. *I. Vaisman,* Cohomology and Differential Forms (1973)
22. *B.-Y. Chen,* Geometry of Submanifolds (1973)
23. *M. Marcus,* Finite Dimensional Multilinear Algebra (in two parts) (1973, 1975)
24. *R. Larsen,* Banach Algebras (1973)
25. *R. O. Kujala and A. L. Vitter, eds.,* Value Distribution Theory: Part A; Part B: Deficit and Bezout Estimates by Wilhelm Stoll (1973)
26. *K. B. Stolarsky,* Algebraic Numbers and Diophantine Approximation (1974)
27. *A. R. Magid,* The Separable Galois Theory of Commutative Rings (1974)
28. *B. R. McDonald,* Finite Rings with Identity (1974)
29. *J. Satake,* Linear Algebra (S. Koh et al., trans.) (1975)
30. *J. S. Golan,* Localization of Noncommutative Rings (1975)
31. *G. Klambauer,* Mathematical Analysis (1975)
32. *M. K. Agoston,* Algebraic Topology (1976)
33. *K. R. Goodearl,* Ring Theory (1976)
34. *L. E. Mansfield,* Linear Algebra with Geometric Applications (1976)
35. *N. J. Pullman,* Matrix Theory and Its Applications (1976)
36. *B. R. McDonald,* Geometric Algebra Over Local Rings (1976)
37. *C. W. Groetsch,* Generalized Inverses of Linear Operators (1977)
38. *J. E. Kuczkowski and J. L. Gersting,* Abstract Algebra (1977)
39. *C. O. Christenson and W. L. Voxman,* Aspects of Topology (1977)
40. *M. Nagata,* Field Theory (1977)
41. *R. L. Long,* Algebraic Number Theory (1977)
42. *W. F. Pfeffer,* Integrals and Measures (1977)
43. *R. L. Wheeden and A. Zygmund,* Measure and Integral (1977)
44. *J. H. Curtiss,* Introduction to Functions of a Complex Variable (1978)
45. *K. Hrbacek and T. Jech,* Introduction to Set Theory (1978)
46. *W. S. Massey,* Homology and Cohomology Theory (1978)
47. *M. Marcus,* Introduction to Modern Algebra (1978)
48. *E. C. Young,* Vector and Tensor Analysis (1978)
49. *S. B. Nadler, Jr.,* Hyperspaces of Sets (1978)
50. *S. K. Segal,* Topics in Group Kings (1978)
51. *A. C. M. van Rooij,* Non-Archimedean Functional Analysis (1978)
52. *L. Corwin and R. Szczarba,* Calculus in Vector Spaces (1979)

53. *C. Sadosky,* Interpolation of Operators and Singular Integrals (1979)
54. *J. Cronin,* Differential Equations (1980)
55. *C. W. Groetsch,* Elements of Applicable Functional Analysis (1980)
56. *I. Vaisman,* Foundations of Three-Dimensional Euclidean Geometry (1980)
57. *H. I. Freedan,* Deterministic Mathematical Models in Population Ecology (1980)
58. *S. B. Chae,* Lebesgue Integration (1980)
59. *C. S. Rees et al.,* Theory and Applications of Fourier Analysis (1981)
60. *L. Nachbin,* Introduction to Functional Analysis (R. M. Aron, trans.) (1981)
61. *G. Orzech and M. Orzech,* Plane Algebraic Curves (1981)
62. *R. Johnsonbaugh and W. E. Pfaffenberger,* Foundations of Mathematical Analysis (1981)
63. *W. L. Voxman and R. H. Goetschel,* Advanced Calculus (1981)
64. *L. J. Corwin and R. H. Szcarba,* Multivariable Calculus (1982)
65. *V. I. Istrăţescu,* Introduction to Linear Operator Theory (1981)
66. *R. D. Järvinen,* Finite and Infinite Dimensional Linear Spaces (1981)
67. *J. K. Beem and P. E. Ehrlich,* Global Lorentzian Geometry (1981)
68. *D. L. Armacost,* The Structure of Locally Compact Abelian Groups (1981)
69. *J. W. Brewer and M. K. Smith, eds.,* Emily Noether: A Tribute (1981)
70. *K. H. Kim,* Boolean Matrix Theory and Applications (1982)
71. *T. W. Wieting,* The Mathematical Theory of Chromatic Plane Ornaments (1982)
72. *D. B.Gauld,* Differential Topology (1982)
73. *R. L. Faber,* Foundations of Euclidean and Non-Euclidean Geometry (1983)
74. *M. Carmeli,* Statistical Theory and Random Matrices (1983)
75. *J. H. Carruth et al.,* The Theory of Topological Semigroups (1983)
76. *R. L. Faber,* Differential Geometry and Relativity Theory (1983)
77. *S. Barnett,* Polynomials and Linear Control Systems (1983)
78. *G. Karpilovsky,* Commutative Group Algebras (1983)
79. *F. Van Oystaeyen and A. Verschoren,* Relative Invariants of Rings (1983)
80. *I. Vaisman,* A First Course in Differential Geometry (1984)
81. *G. W. Swan,* Applications of Optimal Control Theory in Biomedicine (1984)
82. *T. Petrie and J. D. Randall,* Transformation Groups on Manifolds (1984)
83. *K. Goebel and S. Reich,* Uniform Convexity, Hyperbolic Geometry, and Nonexpansive Mappings (1984)
84. *T. Albu and C. Năstăsescu,* Relative Finiteness in Module Theory (1984)
85. *K. Hrbacek and T. Jech,* Introduction to Set Theory: Second Edition (1984)
86. *F. Van Oystaeyen and A. Verschoren,* Relative Invariants of Rings (1984)
87. *B. R. McDonald,* Linear Algebra Over Commutative Rings (1984)
88. *M. Namba,* Geometry of Projective Algebraic Curves (1984)
89. *G. F. Webb,* Theory of Nonlinear Age-Dependent Population Dynamics (1985)
90. *M. R. Bremner et al.,* Tables of Dominant Weight Multiplicities for Representations of Simple Lie Algebras (1985)
91. *A. E. Fekete,* Real Linear Algebra (1985)
92. *S. B. Chae,* Holomorphy and Calculus in Normed Spaces (1985)
93. *A. J. Jerri,* Introduction to Integral Equations with Applications (1985)
94. *G. Karpilovsky,* Projective Representations of Finite Groups (1985)
95. *L. Narici and E. Beckenstein,* Topological Vector Spaces (1985)
96. *J. Weeks,* The Shape of Space (1985)
97. *P. R. Gribik and K. O. Kortanek,* Extremal Methods of Operations Research (1985)
98. *J.-A. Chao and W. A. Woyczynski, eds.,* Probability Theory and Harmonic Analysis (1986)
99. *G. D. Crown et al.,* Abstract Algebra (1986)
100. *J. H. Carruth et al.,* The Theory of Topological Semigroups, Volume 2 (1986)
101. *R. S. Doran and V. A. Belfi,* Characterizations of C*-Algebras (1986)
102. *M. W. Jeter,* Mathematical Programming (1986)
103. *M. Altman,* A Unified Theory of Nonlinear Operator and Evolution Equations with Applications (1986)
104. *A. Verschoren,* Relative Invariants of Sheaves (1987)
105. *R. A. Usmani,* Applied Linear Algebra (1987)
106. *P. Blass and J. Lang,* Zariski Surfaces and Differential Equations in Characteristic $p > 0$ (1987)
107. *J. A. Reneke et al.,* Structured Hereditary Systems (1987)

108. *H. Busemann and B. B. Phadke,* Spaces with Distinguished Geodesics (1987)
109. *R. Harte,* Invertibility and Singularity for Bounded Linear Operators (1988)
110. *G. S. Ladde et al.,* Oscillation Theory of Differential Equations with Deviating Arguments (1987)
111. *L. Dudkin et al.,* Iterative Aggregation Theory (1987)
112. T. Okubo, *Differential Geometry* (1987)
113. *D. L. Stancl and M. L. Stancl,* Real Analysis with Point-Set Topology (1987)
114. *T. C. Gard,* Introduction to Stochastic Differential Equations (1988)
115. *S. S. Abhyankar,* Enumerative Combinatorics of Young Tableaux (1988)
116. *H. Strade and R. Farnsteiner,* Modular Lie Algebras and Their Representations (1988)
117. *J. A. Huckaba,* Commutative Rings with Zero Divisors (1988)
118. *W. D. Wallis,* Combinatorial Designs (1988)
119. *W. Więsław,* Topological Fields (1988)
120. *G. Karpilovsky,* Field Theory (1988)
121. *S. Caenepeel and F. Van Oystaeyen,* Brauer Groups and the Cohomology of Graded Rings (1989)
122. *W. Kozlowski,* Modular Function Spaces (1988)
123. *E. Lowen-Colebunders,* Function Classes of Cauchy Continuous Maps (1989)
124. *M. Pavel,* Fundamentals of Pattern Recognition (1989)
125. *V. Lakshmikantham et al.,* Stability Analysis of Nonlinear Systems (1989)
126. *R. Sivaramakrishnan,* The Classical Theory of Arithmetic Functions (1989)
127. *N. A. Watson,* Parabolic Equations on an Infinite Strip (1989)
128. *K. J. Hastings,* Introduction to the Mathematics of Operations Research (1989)
129. *B. Fine,* Algebraic Theory of the Bianchi Groups (1989)
130. *D. N. Dikranjan et al.,* Topological Groups (1989)
131. *J. C. Morgan II,* Point Set Theory (1990)
132. *P. Biler and A. Witkowski,* Problems in Mathematical Analysis (1990)
133. *H. J. Sussmann,* Nonlinear Controllability and Optimal Control (1990)
134. *J.-P. Florens et al.,* Elements of Bayesian Statistics (1990)
135. *N. Shell,* Topological Fields and Near Valuations (1990)
136. *B. F. Doolin and C. F. Martin,* Introduction to Differential Geometry for Engineers (1990)
137. *S. S. Holland, Jr.,* Applied Analysis by the Hilbert Space Method (1990)
138. *J. Okniński,* Semigroup Algebras (1990)
139. *K. Zhu,* Operator Theory in Function Spaces (1990)
140. *G. B. Price,* An Introduction to Multicomplex Spaces and Functions (1991)
141. *R. B. Darst,* Introduction to Linear Programming (1991)
142. *P. L. Sachdev,* Nonlinear Ordinary Differential Equations and Their Applications (1991)
143. *T. Husain,* Orthogonal Schauder Bases (1991)
144. *J. Foran,* Fundamentals of Real Analysis (1991)
145. *W. C. Brown,* Matrices and Vector Spaces (1991)
146. *M. M. Rao and Z. D. Ren,* Theory of Orlicz Spaces (1991)
147. *J. S. Golan and T. Head,* Modules and the Structures of Rings (1991)
148. *C. Small,* Arithmetic of Finite Fields (1991)
149. *K. Yang,* Complex Algebraic Geometry (1991)
150. *D. G. Hoffman et al.,* Coding Theory (1991)
151. *M. O. González,* Classical Complex Analysis (1992)
152. *M. O. González,* Complex Analysis (1992)
153. *L. W. Baggett,* Functional Analysis (1992)
154. *M. Sniedovich,* Dynamic Programming (1992)
155. *R. P. Agarwal,* Difference Equations and Inequalities (1992)
156. *C. Brezinski,* Biorthogonality and Its Applications to Numerical Analysis (1992)
157. *C. Swartz,* An Introduction to Functional Analysis (1992)
158. *S. B. Nadler, Jr.,* Continuum Theory (1992)
159. *M. A. Al-Gwaiz,* Theory of Distributions (1992)
160. *E. Perry,* Geometry: Axiomatic Developments with Problem Solving (1992)
161. *E. Castillo and M. R. Ruiz-Cobo,* Functional Equations and Modelling in Science and Engineering (1992)
162. *A. J. Jerri,* Integral and Discrete Transforms with Applications and Error Analysis (1992)
163. *A. Charlier et al.,* Tensors and the Clifford Algebra (1992)

164. *P. Biler and T. Nadzieja*, Problems and Examples in Differential Equations (1992)
165. *E. Hansen*, Global Optimization Using Interval Analysis (1992)
166. *S. Guerre-Delabrière*, Classical Sequences in Banach Spaces (1992)
167. *Y. C. Wong*, Introductory Theory of Topological Vector Spaces (1992)
168. *S. H. Kulkarni and B. V. Limaye*, Real Function Algebras (1992)

Additional Volumes in Preparation

INTRODUCTORY THEORY OF TOPOLOGICAL VECTOR SPACES

Yau-Chuen Wong
Department of Mathematics
The Chinese University of Hong Kong
Shantin NT, Hong Kong

CRC Press
Taylor & Francis Group
Boca Raton London New York

CRC Press is an imprint of the
Taylor & Francis Group, an **informa** business

CRC Press
Taylor & Francis Group
6000 Broken Sound Parkway NW, Suite 300
Boca Raton, FL 33487-2742

First issued in paperback 2019

ISBN-13: 978-0-8247-8779-0 (hbk)
ISBN-13: 978-0-367-40273-0 (pbk)

Library of Congress Cataloging-in-Publication Data

Wong, Yau-Chuen.
Introductory theory of topological vector spaces / Yau-Chuen Wong.
p. cm. -- (Monographs and textbooks in pure and applied mathematics ; 167)
Includes bibliographical references (p.) and index.
ISBN 0-8247-8779-X (acid-free paper)
1. Linear topological spaces. I. Title. II. Series.
QA322.W66 1992
515'.73--dc20 92-20435
CIP

Visit the Taylor & Francis Web site at
http://www.taylorandfrancis.com

and the CRC Press Web site at
http://www.crcpress.com

To My Children

KAREN KA–WING and JASON CHIT–KAY

Preface

This elementary text aims to present concisely the basic concepts and principles of functional analysis, as well as its applications, to first–year graduate and senior undergraduate students in mathematics. It includes some recent results that can be taught at an elementary level and enable students to reach quickly the frontiers of current mathematical research. Numerous helpful exercises are also included.

The first six chapters of this text are concerned with three fundamental results (the Hahn–Banach extension theorem, Banach's open mapping theorem and the uniform boundedness theorem) on Banach spaces, together with Grothendieck's structure theorem for compact sets in Banach spaces, and Helley's selection theorem as well. As the unit ball in Banach spaces is a neighborhood of 0 as well as a bounded set, to extend the concept of the unit ball to more general topological vector spaces, it is natural to study parallelly vector topologies and vector bornologies, which are included respectively in Chapters 7~11 and Chapters 12~15. Chapters 16~18 concentrate on the internal and external duality between bornologies and locally convex topologies. Chapters 19~22 are devoted to a study of some recent results on compact and weakly compact operators, operator ideals, and their applications to the important class of Schwartz spaces. The two final chapters deal respectively with some applications to fixed point theory, starting from Ky Fan's KKM coverings (Chapter 23 was written by Professor Shih Mau–Hsiang, Chung–Yuan Christian University at Taiwan), and duality theory on ordered vector spaces together with the continuity of positive linear mappings.

With regard to the selection of the materials for this book, I cannot agree more with Professor W. Rudin : "In order to write a book of moderate size, it was therefore necessary to select certain areas and to ignore others. I fully realize that almost any expert who looks at the table of contents will find that some of his (and my) favorite topics are missing, but this seems unavoidable. It was not my intention to write an encyclopedic treatise. I wanted to write a book that would open the way to further exploration" (quoted from his book Functional Analysis (Tata McGraw–Hill,1973)).

The list of references at the end of this book is not a bibliography; it contains only books and articles to which we have made reference in the text. We have made no systematic attempt to attribute theorems to their authors, nor have given any references to the original research paper in which they appeared.

This book has been used successfully at the Chinese University of Hong Kong since 1978, and was also used as a text for a one–semester course in functional analysis, at the graduate level, at Chung–Yuan Christian University during the academic year 1987–88. The material of this book will be useful to graduate students in analysis, as well as research workers in the field. It is hoped that students in mathematical physics, mathematical economics, or engineering will also profit from the presentation.

I would like to thank the National Science Council of the Republic of China for financial support during my stay in Taiwan. I am especially grateful to Chung–Yuan Christian University, and United College of the Chinese University of Hong Kong for both financial and moral support. The friendly atmosphere and the working environment at Chung–Yuan were conducive to scientific research.

I wish to express my deep gratitude to Professors Ky Fan, who encouraged me to write this book and suggested many helpful hints in the text. Also, I would like to thank Professors Shih Mau–Hsiang and Tu Shih–Tong (Chung–Yuan), Chen Ming–Po and Chiang Tzuu–Shuh (Institute of Mathematics, Academic Sinica at Taiwan) for stimulating discussions and valuable suggestions; special thanks go to Professor A.J.Ellis (University of Hong Kong) for his valuable comments. My appreciation is extended to Professor Shih Mau–Hsiang for the lectures he gave there and for allowing me to include his notes in Chapter 23.

It is a pleasure to acknowledge my indebtedness to many friends, colleagues and students for their help. Special thanks are due to Miss Chiang Yueh–hwa, Mr. Wen Chyi–yau, Mr. Fan Cheng–San Andrew, and Mr. Cheung Ming–Wai for their typing of the manuscript. Finally, I wish to express my appreciation to the editors and staff of Marcel Dekker for their effective cooperation.

Yau–Chuen Wong

Contents

1

Hahn-Banach's Extension Theorem

Throughout these lecture notes, the letter K will denote either the real field $\mathbb{R}$ or the complex field $\mathbb{C}$. Vector spaces mean vector spaces over the field K; vector spaces over $\mathbb{R}$ (resp. $\mathbb{C}$) are called real (resp. complex) vector spaces. For any subsets A and B of a vector space E and $\lambda, \mu \in K$, we define

$$\lambda A + \mu B = \{ \lambda a + \mu b : a \in A , b \in B \}.$$

The expression $\{x\} + A$ will be abbreviated by $x + A$ ($x \in E$), $(-1)A$ by $-A$ and $A + (-B)$ by $A - B$.

Let E be a vector space and $B \subseteq E$. Recall that B is:

(a) convex if $\lambda B + (1 - \lambda)B \subseteq B$ whenever $\lambda \in [0, 1]$;

(b) symmetric if $B = -B$;

(c) circled (or balanced) if $\lambda B \subseteq B$ whenever $|\lambda| \leq 1$;

(d) absolutely convex (or disked, disk) if it is convex and circled;

(e) absorbing (or radical at 0) if for any $x \in E$, there exists an $\lambda > 0$ such that

$$x \in \mu B \quad \text{for all } \mu \in K \text{ with } |\mu| \geq \lambda.$$

The intersection of a family of convex (resp. circled, disked) sets is convex (resp. circled, disked), hence the convex hull (resp. circled hull, disked hull) of a given set B, denoted by co(B) (resp. ch(B), ΓB), is defined to be the intersection of all convex (resp. circled, disked) sets containing B. If $B = \bigcup_{j \in \Lambda} B_j$, then we write $\underset{j}{co}(B_j)$ (resp. $\underset{j}{ch}(B_j)$, $\underset{j}{\Gamma} B_j$) for co(B) (resp. ch(B), ΓB). It is easily seen that

$$co(B) = \{ \Sigma_{i=1}^{n} \lambda_i x_i : x_i \in B, \lambda_i \geq 0 \text{ and } \Sigma_{i=1}^{n} \lambda_i = 1 \text{ for any } n \}$$

$$ch(B) = \bigcup_{|\lambda| \leq 1} \lambda B ;$$

$$\Gamma B = \{ \Sigma_{i=1}^{n} \mu_i x_i : x_i \in B \text{ and } \Sigma_{i=1}^{n} |\mu_i| \leq 1 \text{ for any } n \}.$$

Let E be a vector space. Recall that a map

$$p : E \longrightarrow \mathbb{R}$$

is a <u>sublinear functional</u> on E if it satisfies the following two conditions:

(SN1) (Subadditivity) $p(x + y) \leq p(x) + p(y)$ (for all $x, y \in E$);

(SN2) (Positive homogeneity) $p(\lambda x) = \lambda p(x)$ (for all $x \in E$ and $\lambda \geq 0$).

A sublinear functional p on E is a <u>seminorm</u> if

(SN3) (Homogeneity) $p(\mu x) = |\mu| p(x)$ (for all $x \in E$ and $\mu \in K$).

Let p be a sublinear functional on E. For any $\lambda > 0$, let

$$O^{(\lambda)} = O_E^{(\lambda)} = \{ x \in E : p(x) < \lambda \} \text{ and}$$

$$U_E^{(\lambda)} = \{ x \in E : p(x) \leq \lambda \} = U^{(\lambda)}.$$

Then $O_E^{(\lambda)}$ and $U_E^{(\lambda)}$ are convex and absorbing; if in addition, p is a seminorm then $O_E^{(\lambda)}$ and $U_E^{(\lambda)}$ are circled.It is of interest to know whether given an absorbing, disked subset V of E, there is some seminorm p on E for which

$$\text{either } V = O_E^{(1)} \text{ or } V = U_E^{(1)}.$$

We shall see that this is a bit too much ask for. Nevertheness one can get reasonably closed to such a result. To do this, we require the following:

(1.1) **Definition.** Let V be an absorbing subset of E. For each $x \in E$, let

$$p_V(x) = \inf\{ \lambda > 0 : x \in \lambda V \}.$$

Then p_V is called the <u>gauge</u> (or the <u>Minkowski functional</u>) of V.

Remark. The fact that V is absorbing ensures that $p_V(x)$ is <u>finite</u> for any $x \in E$.

(1.2) **Lemma.** *Let V be an absorbing subset of E and p_V the gauge of V.*

(a) *If V is convex, then p_V is a sublinear functional.*

(b) *If V is a disk, then p_V is a seminorm with the property that*

(1) $$\{x \in E : p_V(x) < 1\} \subseteq V \subseteq \{x \in E : p_V(x) \leq 1\}.$$

Moreover, if q is a seminorm on E such that

(2) $$\{x \in E : q(x) < 1\} \subseteq V \subseteq \{x \in E : q(x) \leq 1\},$$

then $p_V = q$.

Proof. (a) The positive homogeneity of p_V is obvious. To prove the subadditivity, let $\lambda > 0$ and $\mu > 0$ be such that

$$x \in \lambda V \text{ and } y \in \mu V.$$

Then the convexity of V ensures that

$$\frac{x + y}{\lambda + \mu} = \frac{\lambda}{\lambda + \mu}\,\frac{x}{\lambda} + \frac{\mu}{\lambda + \mu}\,\frac{y}{\mu} \in V,$$

so that $p_V(x + y) \leq \lambda + \mu$; consequently,

$$p_V(x + y) \leq p_V(x) + p_V(y)$$

since λ and μ were arbitrary.

(b) If V is circled then we have

$$p_V(\mu x) = |\mu| p_V(x) \quad \text{(for all } \mu \in K \text{ and } x \in E).$$

To prove (1), let $x \in E$ be such that $p_V(x) < 1$.By the definition of infimum, there is an λ with $0 < \lambda < 1$ such that $x \in \lambda V$, hence $x \in V$ (since V is circled and $0 < \lambda < 1$). The inclusion

$$V \subseteq \{ x \in E : p_V(x) \leq 1 \}$$

is obvious.

Finally, suppose that q is a seminorm on E such that (2) holds. Then we obtain from (1) that

(3) $$\{ x \in E : q(x) < 1 \} \subset \{ x \in E : p_V(x) \leq 1 \};$$

(4) $$\{ x \in E : p_V(x) < 1 \} \subset \{ x \in E : q(x) \leq 1 \}.$$

It is easily seen that

$$(3) \Leftrightarrow p_V(x) \leq q(x) \quad \text{(for all } x \in E).$$

Thus we obtain $q = p_V$ by (4).

(1.a) Gauge functionals : Let E be a vector space, and let $\{ B_i \}_{i \in \Lambda}$ be a non–empty family of absolutely convex subsets of E. For any $i \in \Lambda$, let $E(B_i) = \bigcup_n nB_i$ (the vector subspace of E spanned by B_i), and let γ_{B_i} be the gauge of B_i defined on $E(B_i)$ (since B_i is absorbing in $E(B_i)$), that is

$$\gamma_{B_i}(x) = \inf\{ \lambda > 0 : x \in \lambda B_i \} \quad \text{(for all } x \in E(B_i)).$$

Suppose further that

$$\gamma_{\cap}(x) = \sup\{ \gamma_{B_i}(x) : i \in \Lambda \} \text{ and}$$

$$\gamma_{\cup}(x) = \inf\{ \sum_{i\in\alpha} \gamma_{B_i}(x_i) : x = \sum_{i\in\alpha} x_i \text{ with } x_i \in B_i \text{ and } \alpha \in \mathcal{F}(\Lambda)\},$$

where $\mathcal{F}(\Lambda)$ is the direct set of all non-empty finite subset of Λ ordered by set inclusion. Then $\gamma_{\cap}(\cdot)$ is the gauge of $\bigcap_{i\in\Lambda} B_i$, and $\gamma_{\cup}(\cdot)$ is the gauge of $\Gamma_i B_i = \Gamma(\bigcup_{i\in\Lambda} B_i)$ (Junek [1983, P.26]).

(1.3) **Theorem** (Real Hahn–Banach's extension theorem). Let E be a real vector space, let p be a sublinear functional on E, let M be a vector subspace of E, and let g be a linear functional on M such that

$$g(y) \leq p(y) \quad (\text{for all } y \in M).$$

Then there exists a linear functional f on E with $f(y) = g(y)$ (for all $y \in M$) (called an extension of g) such that

$$f(x) \leq p(x) \quad (\text{for all } x \in E).$$

Proof. Let $x_0 \in E \setminus M$ and

$$M_1 = \langle M \cup \{ x_0 \} \rangle \quad (\text{the linear hull of } M \cup \{ x_0 \}).$$

Then each $z \in M_1$ can be uniquely expressed by

$$z = y + \lambda x_0 \quad \text{with } y \in M \text{ and } \lambda \in \mathbb{R}.$$

Any linear extension h of g on M_1 which is dominated by p must satisfy

(1) $$h(z) = g(y) + \lambda h(x_0) \leq p(z)$$

whenever $z = y + \lambda x_0 \in M_1$. Therefore the existence of such an extension h is equivalent to the choice of the real value $h(x_0)$. In order to choose $h(x_0)$, let us make the following observation : If $\lambda > 0$, we obtain from (1) that

$$h(x_0) \leq p(\frac{y}{\lambda} + x_0) - g(\frac{y}{\lambda}),$$

so that

$$h(x_0) \leq \inf\{\ p(y + x_0) - g(y) : y \in M\ \}; \tag{2}$$

if $\lambda < 0$, we obtain from (1) that

$$h(x_0) \geq \lambda^{-1}p(y + \lambda x_0) - \lambda^{-1}g(y) = -p(-\frac{y}{\lambda} - x_0) - g(\frac{y}{\lambda}),$$

so that

$$h(x_0) \geq \sup\{\ -p(y - x_0) + g(y) : y \in M\ \}. \tag{3}$$

Thus, if we choose any real number δ satisfying

$$\sup\{\ -p(y-x_0) + g(y) : y \in M\ \} \leq \delta$$

$$\leq \inf\{\ p(y + x_0) - g(y) : y \in M\ \}$$

and if we define, for any $z = y + \lambda x_0 \in M_1$,

$$h(z) = g(y) + \lambda\delta,$$

then in view of the choice of δ, h is actually a linear functional on M_1 such that

$$h = g \text{ on } M \quad \text{and } h(z) \leq p(z) \ \ (\text{for all } z \in M_1).$$

Now we complete the proof in terms of Zorn's lemma. Let us call a pair (Ψ, H) an extension of g, if it satisfies the following two conditions:

(i) H is a vector subspace of E containing M ;

(ii) Ψ is a linear functional on H such that
$\Psi = g$ on M and $\Psi \leq p$ on H.

Let $\mathcal{M}$ be the family of all extensions of g. Then $\mathcal{M} \neq \phi$ (since $(g, M) \in \mathcal{M}$), and $\mathcal{M}$ becomes an inductive ordered set under the ordering $\leq$ defined by

$$(\Psi_1, H_1) \leq (\Psi_2, H_2) \text{ if } H_1 \subset H_2 \text{ and } \Psi_2 = \Psi_1 \text{ on } H_1.$$

By Zorn's lemma, $\mathcal{M}$ has a maximal element (f, F). The maximality of (f, F), together with the argument of the extension h of g on M_1, ensures that f has the required properties.

To study the general case of Hahn–Banach's extension theorem, we first notice the following interesting result:

(1.b) The relationship between real linear functionals and complex linear functionals: Let E be a complex vector space. Then E can be viewed as a vector space over $\mathbb{R}$, hence any linear map f from E into $\mathbb{R}$ is called a real linear functional or $\mathbb{R}$–linear functional, and denoted by real $f \in E^*$. For emphasis we then refer to linear functionals on E as complex linear functionals or $\mathbb{C}$–linear functionals on E and denoted by complex $g \in E^*$.

(i) For any complex $f \in E^*$, the functional, defined by

$$f_R(x) = \operatorname{Re} f(x) \quad (\text{for all } x \in E),$$

is a real linear functional on E such that

$$f(x) = f_R(x) - i f_R(ix) \quad (\text{for all } x \in E).$$

(ii) For any real $h \in E^*$, the functional, defined by

$$g(x) = h(x) - i h(ix) \quad (\text{for all } x \in E),$$

is a complex linear functional on E such that

$$h(x) = \text{Re } g(x) \quad (\text{for all } x \in E).$$

(1.4) **Theorem** (Hahn–Banach's extension theorem). Let E be a vector space over K (where $K = \mathbb{R}$ or $\mathbb{C}$), let p be a seminorm on E, let M be a vector subspace of E, and let g be a linear functional on M such that

$$|g(y)| \leq p(y) \quad (\text{for all } y \in M).$$

Then there exists an $f \in E^*$ with $f(y) = g(y)$ (for all $y \in M$) such that

$$|f(x)| \leq p(x) \quad (\text{for all } x \in E).$$

Proof. If E is a real vector space then $f(-x) \leq p(-x) = p(x)$ (since p is a seminorm), thus $|f(x)| \leq p(x)$; consequently, the result follows from (1.3).

Suppose now that E is a complex vector space. Let M_R be the real restriction of M and

$$g_R = \text{Re } g.$$

Then M_R is a real vector subspace of E and g_R is a linear functional on M_R such that

$$g(y) = g_R(y) - ig_R(iy) \quad (\text{for all } y \in M)$$

and

$$g_R(y) \leq |g(y)| \leq p(y) \quad (\text{for all } y \in M_R).$$

By (1.3), there is a real linear functional γ on E such that

$$\gamma(y) = g_R(y) \quad (\text{for all } y \in M_R) \text{ and}$$

$$\gamma(x) \le p(x) \quad (\text{for all } x \in E).$$

Let

$$f(x) = \gamma(x) - i\gamma(ix) \quad (\text{for all } x \in E).$$

Then f is a linear functional on E such that

$$\gamma(x) = \operatorname{Re} f(x) \quad (\text{for all } x \in E).$$

As two sets M and M_R are identical and $iy \in M$ (for all $y \in M$), it follows that

$$f(y) = \gamma(y) - i\gamma(iy)$$

$$= g_R(y) - ig_R(iy) = g(y) \quad (\text{for all } y \in M).$$

For any given $x \in E$, let $\theta = \arg f(x)$. As $|f(x)| \ge 0$, we conclude that

$$|f(x)| = e^{-i\theta}f(x) = f(e^{-i\theta}x)$$

$$= \gamma(e^{-i\theta}x) \le p(e^{-i\theta}x) = p(x).$$

Therefore f is the required linear functional.

(1.c) Linear functionals (I): (i) Let Λ be a non–empty set, let $E_i = K$ (for all $i \in \Lambda$), let

$$K^\Lambda = \prod_{i\in\Lambda} E_i \text{ and } K^{(\Lambda)} = \bigoplus_{i\in\Lambda} E_i \quad (\text{the algebraic direct sum}).$$

Then K^Λ and $(K^{(\Lambda)})^*$ are algebraically isomorphic, denoted by $K^\Lambda \sim (K^{(\Lambda)})^*$. In particular, if $\Lambda = \{1,2,\cdots,n\}$, then $(K^n)^* \sim K^n$.

(ii) Let E be a vector space and $g, f_i \in E^*$ $(i = 1,2,\cdots,n.)$. Then g is a linear combination of $f_1,\cdots,f_n$ if and only if $\bigcap_{i=1}^{n} f_i^{-1}(0) \subseteq g^{-1}(0)$. [Using (i) and the induced mapping theorem.].

Exercises

1–1. Let E and F be vector spaces and $V \subset E$.

(a) Show that V is absorbing if and only if for any $x \in E$ there is some $\epsilon > 0$ such that $\alpha x \in V$ whenever $0 < |\alpha| \leq \epsilon$, in this case, $E = \bigcup_{n=1}^{\infty} nV$.

(b) The circled kernel of V, denoted by cik(V), is defined to be the union of all circled sets contained in V. Show that $\text{cik}(V) \neq \phi$ if and only if $0 \in V$; in this case,

$$\text{cik}(V) = \{x \in V : \lambda x \in V \text{ for all } |\lambda| \leq 1\} = \bigcap_{|\mu| \geq 1} \mu V.$$

(c) Show that the convex hull of a circled set is circled, and that(by an example) the circled hull of a convex set need not be convex.

(d) Show that $\Gamma V = \text{co}(\text{ch}V) = \{\Sigma_{i=1}^{n} \lambda_i x_i : x_i \in V, \Sigma_{i=1}^{n} |\lambda_i| \leq 1\}$.

(e) If B_i (i=1..m) are convex sets in E, then

$$\text{co} \bigcup_{i=1}^{m} B_i = \{\Sigma_{i=1}^{m} \mu_i b_i : b_i \in B_i (i=1..m), 0 \leq \lambda_i \leq 1, \Sigma_{i=1}^{m} \lambda_i = 1\}.$$

(f) Let $T: E \rightarrow F$ be a linear map (i.e. $T(\lambda x + \mu y) = \lambda Tx + \mu Ty$ (whenever $\lambda, \mu \in K$ and $x, y \in E$)). Show that the image of a set $W \subset E$ that is resp. convex, circled, disked, under T has the same corresponding property.

1–2. Prove (1.a).

1–3. Prove (1.b).

1–4. Prove (1.c).

1–5. Let E and F be vector spaces, let $T: E \rightarrow F$ be linear and M a vector subspace of E. We say that T is compatible with M if

$$M \subset \text{Ker } T = \{x \in E : Tx = 0\} \text{ (the kernel of T).}$$

(a) Suppose that T is compatible with M. Show that there exists a unique linear map $\tilde{T} : {}^{E}/_{M} \rightarrow F$ such that $T = \tilde{T} Q_M$, where $Q_M : E \rightarrow {}^{E}/_{M}$ is the quotient map ($\tilde{T}$ is called the map obtained from T by passing to the quotient.)

(b) Show that there exists a unique bijective linear map $\check{T} : {}^{E}/_{\text{Ker } T} \longrightarrow \text{Im } T$ (called the bijection associated with T) such that $T = J_T \check{T} Q_T$, where $Q_T : E \longrightarrow {}^{E}/_{\text{Ker } T}$ is the quotient map and $J_T : \text{Im } T \longrightarrow F$ is the canonical embedding. (The map $J_T \check{T} = \hat{T}$ is called the injection associated with T.)

1–6. Let E, F and G be vector spaces and $T : E \longrightarrow F$ a linear map.

(a) (Induced map from the left). Given a linear map $S: E \longrightarrow G$ and suppose that

(a.1) S is surjective (i.e. onto) and $\text{Ker } S \subset \text{Ker } T$.

Show that there exists a unique linear map $L : G \longrightarrow F$ such that

(a.2) $$T = LS.$$

Moreover, L is injective (i.e. one–one) if and only if

(a.3) $$\text{Ker } T \subset \text{Ker } S. \text{ (i.e. Ker } T = \text{Ker } S).$$

(b) (Induced map from the right) Given a linear map $S : G \longrightarrow F$ and suppose that

(b.1) S is injective and $\text{Im } T \subset \text{Im } S$.

Show that there exists a unique linear map $R : E \longrightarrow G$ such that

(b.2) $$T = SR.$$

Moreover, R is surjective if and only if

$$\text{Im } S \subset \text{Im } T \text{ (i.e. Im } T = \text{Im } S).$$

Note. This part is still true for sets ; in this case, all linear maps are only mappings.

(1–7) (A generalization of Hahn–Banach's extension theorem (Andenaes[1970])). Let E be a real vector space, let p be a sublinear functional on E, let M be a vector subspace of E and let $g \in M^*$ (a linear functional on M) be such that $g(z) \leq p(z)$ for all $z \in M$.

Show that for any $B \subset E$, there exists an $f \in E^*$ with the following properties:

(i) $f(z) = g(z)$ (for all $z \in M$) and $f(x) \leq p(x)$ $(x \in E)$;

(ii) if $h \in E^*$, satisfied (i), is such that $f|_B \leq h|_B$, then $f = h$ on B; [in other words, f is a maximal elements in the set $r(g,p) = \{f \in E^* : f=g$ on M and $f \leq p$ on $E\}$ with respect to the ordering $\leq_B$ on E^*, defined by

$$f \leq_B h \text{ if and only if } f \leq h \text{ on } B \; (f,h \in E^*).]$$

2

Banach Spaces and Hilbert Spaces

We begin with the following:

(2.1) **Definition**. A real–valued function $\|\cdot\|$, defined on a vector space E, is called a norm if it satisfies the following conditions:

(N1) $\|x\| = 0$ if and only if $x = 0$.

(N2) (Subadditivity) $\|x + y\| \leq \|x\| + \|y\|$ (for all $x, y \in E$).

(N3) (Homogeneity) $\|\lambda x\| = |\lambda| \, \|x\|$ (for all $x \in E$ and $\lambda \in K$).

A vector space equipped with a norm is called a normed vector space (or normed space for short). Hereafter we shall use E (or $(E, \|\cdot\|)$) for a normed space, also the open unit ball in E is defined by

$$O_E = \{ x \in E : \|x\| < 1 \} \quad \text{(or simply } O),$$

and the closed unit ball in E is defined by

$$U_E = \{ x \in E : \|x\| \leq 1 \} \quad \text{(or simply } U).$$

Remark (i). If (N1) is replaced by the following

$$(N1)^* \qquad \|0\| = 0,$$

then the real–valued function $\|x\|$ on E is only a seminorm, which is usually denoted by p. A vector space, equipped with a seminorm, is called a seminormed space.

Remark (ii). It is easily seen from (N2) that any norm $\|\cdot\|$ on E defines a metric d in a natural way

$$d(x, y) = \|x - y\| \quad \text{(for all } x, y \in E),$$

this metric is translation–invariant in the sense that

$$d(x,y) = d(x + z, y + z) \quad \text{(for all } z \in E).$$

We shall always assume that a normed space carries this metric and its associated topology, which is called the norm–topology and denoted by $\|\cdot\|$–top (or $\|\cdot\|_E$–top or norm–top).

Remark (iii). Two norms p_1 and p_2 on E are said to be equivalent if they define the same norm–topology on E. As a consequence of a local base (see (2.a.1) of (2.a)), it follows that two norms p_1 and p_2 on E are equivalent if and only if there exist $\lambda > 0$ and $\mu > 0$ such that

$$\mu \leq \frac{p_1(x)}{p_2(x)} \leq \lambda \quad \text{(for all } 0 \neq x \in E). \tag{2.1.a}$$

Remark (iv). We say that two normed spaces (E,p) and (F,q) over K are :

(a) metrically isomorphic (or isometric), denoted by $(E, p) \equiv (F, q)$, if there is a bijective linear map $T : E \longrightarrow F$ such that

$$q(Tx) = p(x) \quad \text{(for all } x \in E)$$

(T is referred to as a metric isomorphism (or isometry));

(b) topologically isomorphic (or isomorphic), denoted by $(E, p) \cong (F, q)$, if there is a bijective linear map $T : E \longrightarrow F$ which is a homeomorphism for the norm–top (T is referred to as a topological isomorphism (or an isomorphism)).

It is easily seen that a bijective linear map $T : E \longrightarrow F$ is a topological isomorphism if and only if the norm $\|\cdot\|_F$ on F, defined by

$$\|Tx\|_F = p(x) \quad \text{(for all } x \in E),$$

is equivalent to q, and this is the case (by (2.1.a)) if and only if there exist $\lambda > 0$ and $\mu > 0$ such that

$$\mu p(x) \leq q(Tx) \leq \lambda p(x) \quad \text{(for all } x \in E).$$

(2.a) The norm–topology: Let $(E,\|\cdot\|)$ be a normed space. Then the family

$$(2.a.1) \qquad \{ \frac{1}{n} O_E : n \geq 1 \} \quad (\text{or } \{ \frac{1}{n} U_E : n \geq 1 \}),$$

consisting of absolutely convex and absorbing sets, is a local base at 0 for $\|\cdot\|_E$–top; moreover, the $\|\cdot\|_E$–top has the following remarkable properties:

(i) The $\|\cdot\|_E$–top is compatible with the vector space operation, i.e., the maps

$$(x,y) \longrightarrow x + y : E \times E \longrightarrow E$$

and

$$(\lambda,x) \longrightarrow \lambda x : K \times E \longrightarrow E$$

are continuous. (Of course, we consider the product topology on the product spaces.)

(ii) For any $x_0 \in E$ and $0 \neq \gamma_0 \in K$, the translation, defined by

$$y \longrightarrow x_0 + \gamma_0 y \quad (\text{for all } y \in E),$$

is a homeomorphism from E onto E ; consequently, the family

$$(2.a.2) \qquad \{ x_0 + \frac{1}{n} U_E : n \geq 1 \} \quad (\text{or } \{ x_0 + \frac{1}{n} O_E : n \geq 1 \})$$

is a local base at x_0 for the $\|\cdot\|_E$–top.

As the translation and multiplication by non–zero scalars are homeomorphisms (see (2.a) (ii)), it follows that

$$\overline{x + A} = x + \overline{A}\,, \ \overline{\lambda A} = \lambda \overline{A} \ \text{ and } \ \overline{A} + \overline{B} \subseteq \overline{(A + B)}$$

whenever $x \in E$, $\lambda \in K$ and $A, B \subset E$ (where $\overline{A}$ is the closure of A); moreover, we have the following :

(2.2) **Proposition.** Let E be a normed space and $A, B \subset E$.

(a) $\overline{A} = \bigcap_{n=1}^{\infty} (A + \frac{1}{n} O_E) = \bigcap_{n=1}^{\infty} (A + \frac{1}{n} U_E)$.

(b) $A + G$ is open whenever G is open, hence

$A + \text{Int}B \subset \text{Int}(A + B)$ (where Int B denotes the interior of B).

(c) Let $K \subset E$ be compact and let B be closed. If $K \cap B = \phi$, then there exists some $m \geq 1$ such that

$$(K + \frac{1}{m} U_E) \cap (B + \frac{1}{m} U_E) = \phi . \tag{2.2.1}$$

Consequently, if $C \subset E$ is compact and B is closed, then $C + B$ is closed in E.

Proof. (a) As $\frac{1}{n+1} U_E \subset \frac{1}{n} O_E \subset \frac{1}{n} U_E$, it follows that

$$\bigcap_{n=1}^{\infty} (A + \frac{1}{n} O_E) = \bigcap_{n=1}^{\infty} (A + \frac{1}{n} U_E).$$

Now part (a) follows from the following computation:

$$\begin{aligned} x \in \overline{A} \quad & \Leftrightarrow (x + \frac{1}{n} O_E) \cap A \neq \phi \quad \text{(for all } n \geq 1) \\ & \Leftrightarrow \text{for any } n \geq 1, \text{ there is some } a_n \in A \text{ such that} \\ & \qquad x \in a_n + \frac{1}{n} O_E \quad (\text{since } O_E = -O_E) \\ & \Leftrightarrow x \in A + \frac{1}{n} O_E \quad \text{(for all } n \geq 1). \end{aligned}$$

(b) For any $x_0 \in E$, the translation $y \longrightarrow x_0 + y$ $(y \in E)$ is a homeomorphism, it follows that $x_0 + G$ is open, and hence from $A + G = \bigcup_{a \in A} (a + G)$ that $A + G$ is open.

Finally, since $A + \text{Int } B$ is an open subset of $A + B$, it follows that $A + \text{Int } B \subset \text{Int } (A + B)$.

(c) For any $x \in K$, the closedness of B and $B \cap K = \phi$ ensure that there exists some integer $n(x) \geq 1$ such that

$$(x + \frac{1}{n(x)} U_E + \frac{1}{n(x)} U_E + \frac{1}{n(x)} U_E) \cap B = \phi\,,$$

and surely

$$(x + \frac{1}{n(x)} U_E + \frac{1}{n(x)} U_E) \cap (B + \frac{1}{n(x)} U_E) = \phi\,. \tag{2.2.2}$$

Clearly $\{x + \frac{1}{n(x)} U_E : x \in K\}$ (or more precisely $\{x + \frac{1}{n(x)} O_E : x \in K\}$) forms an open covering of K, hence the compactness of K ensures that there is $\{ x_1, \cdots, x_k \} \subset K$ such that

$$K \subset \bigcup_{i=1}^{k} (x_i + \frac{1}{n(x_i)} U_E)\,. \tag{2.2.3}$$

Let $m = \max\{ n(x_1), \cdots, n(x_k) \}$. Then $\frac{1}{m} \leq \frac{1}{n(x_i)}$ $(i = 1, \cdots, k)$, hence we conclude from (2.2.2) and (2.2.3) that

$$(K + \frac{1}{m} U_E) \cap (B + \frac{1}{m} U_E) = \phi.$$

Finally, if $x \notin C + B$, then $(x - C) \cap B = \phi$. As C is compact, it follows that $x - C$ is compact, and hence from (2.2.1) that there is $m \geq 1$ such that $((x - C) + \frac{1}{m} U_E) \cap (B + \frac{1}{m} U_E) = \phi$ and a fortiori

$$((x - C) + \frac{1}{m} U_E) \cap B = \phi\,.$$

But this implies that

$$(x + \frac{1}{m} U_E) \cap (B + C) = \phi\,;$$

in other words, x is an interior point of $E \setminus (B + C)$, thus $B + C$ is closed.

A Banach space (or more briefly, B–space) is a normed space that, regarded as a metric space, is complete.

(2.b) The completion and a charaterization of completeness:

(i) Any normed space $(E, \|\cdot\|)$ is isometric to a dense subspace of a Banach space $\tilde{E}$, called the completion of $(E, \|\cdot\|)$, and this is unique up to a metric isomorphism. (The closed unit ball $U_{\tilde{E}}$ in $\tilde{E}$ is the closure in $\tilde{E}$ of either U_E or O_E.)

(ii) A normed space $(E, \|\cdot\|)$ is separable if and only if its completion $\tilde{E}$ is separable.

(iii) A normed space $(E, \|\cdot\|)$ is complete if and only if for any sequence $\{x_n\}$ in E with $\Sigma_{n=1}^{\infty} \|x_n\| < \infty$ (called the formal series $\Sigma_n x_n$ is absolutely convergent) the sequence $\{s_n\}$, defined by $s_n = \Sigma_{i=1}^{n} x_i$, is convergent to $z \in E$; in this case, we have

$$\|z\| \leq \Sigma_{n=1}^{\infty} \|x_n\|.$$

(iv) Let E be a normed space, let $K \subset E$ be totally bounded and closed, and let $B \subset E$ be complete. If $K \cap B = \phi$, then there exists some $m \geq 1$ such that $(K + \frac{1}{m} U_E) \cap (B + \frac{1}{m} U_E) = \phi$. [Compared with (2.2) (c).]

(2.3) **Lemma.** Let E be a normed space and M a vector subspace of E. If M is dense in E, then for any $z \in E$ and $\epsilon > 0$ there exists a sequence $\{u_n\}$ in M such that

$$z = \Sigma_{n=1}^{\infty} u_n \quad \text{and} \quad \Sigma_{n=1}^{\infty} \|u_n\| \leq (1 + \epsilon)\|z\|.$$

In other words, for any $z \in E$ and $\epsilon > 0$ there exists an absolutely convergent series $\Sigma_n u_n$ of $u_n \in M$ (with $\Sigma_n \|u_n\| \leq (1 + \epsilon)\|z\|$) such that $\Sigma_{n=1}^{\infty} u_n$ converges to z.

Proof. For each $n \geq 1$, there is some $x_n \in M$ such that

$$\|z - x_n\| \le \frac{\epsilon}{2^{n+1}} \|z\| .$$

Define

$$u_1 = x_1 \quad \text{and} \quad u_n = x_n - x_{n-1} \quad \text{(for all } n \ge 2).$$

Then

$$\Sigma_{i=1}^{n} u_i = x_n ,$$
$$\|u_1\| = \|x_1\| \le \|x_1 - z\| + \|z\| \le (\frac{\epsilon}{4} + 1) \|z\|$$

and

$$\|u_n\| \le \|x_n - z\| + \|x_{n-1} - z\|$$
$$\le (\frac{1}{2^{n+1}} + \frac{1}{2^n})\epsilon\|z\| \quad \text{(for all } n \ge 2).$$

Consequently, $z = \|\cdot\|\text{–}\lim_n x_n = \|\cdot\|\text{–}\lim_n \Sigma_{i=1}^{n} u_i$ and

$$\Sigma_{n=1}^{\infty}\|u_n\| \le [1 + \frac{\epsilon}{2^2} + \Sigma_{n=2}^{\infty}(\frac{\epsilon}{2^{n+1}} + \frac{\epsilon}{2^n})] \|z\|$$
$$= [1 + \epsilon(\Sigma_{n=1}^{\infty} \frac{1}{2^{n+1}} + \Sigma_{n=2}^{\infty} \frac{1}{2^n}] \|z\|$$
$$= (1 + \epsilon) \|z\|.$$

The concept of convergence of a series can be used to define a basis as follows:

(2.4) **Definition.** Let $(E, \|\cdot\|)$ be a normed space. A sequence $\{e_n\}$ in E is called a Schauder basis for E (also we say that E has a countable basis) if for any $x \in E$ there exists a unique sequence $\{\lambda_n\}$ in K such that

$$x = \|\cdot\| - \lim_n \Sigma_{i=1}^n \lambda_i e_i.$$

(2.5) **Proposition.** A normed space $(E,\|\cdot\|)$ with a Schauder basis must be separable.

Proof. Let $\{e_n\}$ be a Schauder basis for E and

$$C = \{\ \Sigma_{j=1}^n r_j e_j : n \geq 1 \text{ and } r_j \text{ are rational } \}.$$

Then C is countable as well as dense in E, hence E is separable.

Remark. From the preceding result, it is natural to ask the following question :

(Q) Does every separable Banach space have a Schauder basis?

This is a famous question raised by Banach more than fifty years ago. Because of almost all known separable Banach spaces had been shown to possess a Schauder basis, a positive answer was expected for a long time. Grothendieck made a deep analysis of this problem; he found many equivalent formulations and consequences but no solution; he conjectured a negative answer. In 1973, Enflo [1973] succeeded in constructing a separable, reflexive Banach space which has no Schauder basis: his ingenious but highly complicated methods were simplified to some degree by Davie [1973].

By a subspace of a normed space E is meant a vector subspace M of E; while a subspace of a B–space E is meant a closed vector subspace of E.

Let E be a normed space and M a subspace of E. We denote by

$$J_M^E : M \longrightarrow E \quad (\text{or simply } \ J_M : M \longrightarrow E)$$

the embedding map. If M is closed in E, then E/M becomes a normed space with respect to the quotient norm

$$\|x(M)\| = \inf\{ \|x + m\| : m \in M \} ,$$

where $x(M) = x + M$ denotes the equivalence class containing x. The quotient map from E onto E/M is denoted by Q_M^E or simply Q_M or Q. Of course, Q_M^E is continuous, moreover it is open as shown by the following:

(2.6) **Proposition.** Let $(E, \|\cdot\|)$ be a normed space, let M be a closed vector subspace of E and $Q_M : E \longrightarrow E/M$ the quotient map. Then

$$O_{E/M} = Q_M(O_E) \subset Q_M(U_E) \subset U_{E/M} , \tag{2.6.1}$$

hence Q_M is an open operator.

Proof. It is easily seen that $Q_M(O_E) \subset O_{E/M}$ since

$$\|Q_M(x)\| = \inf\{ \|x + u\| : u \in M \} \leq \|x\|.$$

To prove the converse, i.e., $O_{E/M} \subseteq Q_M(O_E)$, let $Q_M(x) \in O_{E/M}$. Then

$$\inf\{ \|x + u\| : u \in M \} = \|Q_M(x)\| < 1,$$

hence the definition of infimum shows that there is a $u_0 \in M$ such that

$$\|Q_M(x)\| \leq \|x + u_0\| < 1. \tag{2.6.2}$$

Now let $x_0 = x + u_0$. Then $x_0 \in O_E$ (by (2.6.2)) is such that $Q_M(x) = Q_M(x_0) \in Q_M(O_E)$ (by (2.6.2)), hence $O_{E/M} \subset Q_M(O_E)$.

Finally, since Q_M is continuous and $U_E = \overline{O}_E$, we conclude that $Q_M(\overline{O}_E) \subset \overline{Q_M(O_E)}$, and hence that

$$Q_M(U_E) = Q_M(\overline{O}_E) \subset \overline{Q_M(O_E)} = \overline{O}_{E/M} = U_{E/M} \, .$$

If E is complete (resp. separable) then so is E/M as shown by the following:

(2.c) <u>Quotient spaces and product spaces</u>: Let $(E,\|\cdot\|)$ and $(E_i,\|\cdot\|_i)$ $(i = 1,2,\cdots,n)$ be normed spaces and M a closed vector subspace of E.

(i) If $(E,\|\cdot\|)$ is complete (resp. separable), then so is the quotient space E/M (equipped with the quotient norm).

(ii) Consider the cartesian product $\prod_{i=1}^{n} E_i$, and define, for any $x = (x_1,\cdots,x_n) \in \prod_{i=1}^{n} E_i$, that

$$\|x\|_{\ell^\infty} = \sup_{1 \le i \le n} \|x_i\|_i; \qquad \|x\|_{\ell^2} = (\Sigma_{i=1}^{n} \|x_i\|_i^2)^{1/2}$$

$$\|x\|_{\ell^1} = \Sigma_{i=1}^{n} \|x_i\|_i \, .$$

Then $\|\cdot\|_{\ell^\infty}$, $\|\cdot\|_{\ell^2}$ and $\|\cdot\|_{\ell^1}$ are norms on $\prod_{i=1}^{n} E_i$, and these norm topologies coincide with the product topology; moreover, $\prod_{i=1}^{n} E_i$ is complete (resp. separable) (for these norms) if and only if each $(E_i,\|\cdot\|_i)$ is complete (resp. separable). (Usually, the norm $\|\cdot\|_{\ell^\infty}$ is called the <u>product norm</u> of $\|\cdot\|_i$ $(i = 1,\cdots,n)$, and we also write

$$\ell^\infty_n(E_i) = (\prod_{i=1}^{n} E_i,\|\cdot\|_{\ell^\infty}); \quad \ell^2_n(E_i) = (\prod_{i=1}^{n} E_i,\|\cdot\|_{\ell^2}) \text{ and}$$
$$\ell^1_n(E_i) = (\prod_{i=1}^{n} E_i,\|\cdot\|_{\ell^1}).)$$

All norms on a finite-dimensional vector space are equivalent as shown by the following:

(2.7) **Theorem.** Let $(E,\|\cdot\|)$ be an n–dimensional normed space, and let $\{e_1,\cdots,e_n\}$ be a basis of E. Then there exist numbers α and $\beta > 0$ such that

(2.7.1) $$\alpha(\Sigma_{i=1}^{n}|\lambda_i|^2)^{\frac{1}{2}} \le \|x\| \le \beta(\Sigma_{i=1}^{n}|\lambda_i|^2)^{\frac{1}{2}} \quad (x = \Sigma_{i=1}^{n}\lambda_i e_i \in E).$$

Consequently, any two norms on a finite–dimensional vector space are equivalent.

Proof. Recall Cauchy–Schwarz inequality in K^n:

(2.7.2) $$\Sigma_{i=1}^{n}|\zeta_i\eta_i| \le (\Sigma_{i=1}^{n}|\zeta_i|^2)^{\frac{1}{2}}(\Sigma_{i=1}^{n}|\eta_i|^2)^{\frac{1}{2}}$$

Now for any $x = \Sigma_{i=1}^{n}\lambda_i e_i \in E$, we have from (2.7.2) that

$$\begin{aligned}\|x\| &\le \Sigma_{i=1}^{n}|\lambda_i|\,\|e_i\| \\ &\le (\Sigma_{i=1}^{n}\|e_i\|^2)^{\frac{1}{2}}(\Sigma_{i=1}^{n}|\lambda_i|^2)^{\frac{1}{2}} \\ &\le \beta(\Sigma_{i=1}^{n}|\lambda_i|^2)^{\frac{1}{2}} \quad \text{where } \beta = (\Sigma_{i=1}^{n}\|e_i\|^2)^{\frac{1}{2}},\end{aligned}$$

and hence that

(2.7.3) $$|\,\|x\| - \|y\|\,| \le \|x - y\| \le \beta(\Sigma_{i=1}^{n}|\lambda_i - \zeta_i|^2)^{\frac{1}{2}}$$
$$(\text{for any } y = \Sigma_{i=1}^{n}\zeta_i e_i \in E).$$

(Namely, the norm $\|\cdot\|$ is a continuous function on E with respect to ℓ_n^2–norm.) It is easily seen that K^n is a Banach space under the norm defined by

(2.7.4) $$\|[\zeta_i]\|_2 = (\Sigma_{i=1}^{n}|\zeta_i|^2)^{\frac{1}{2}} \quad (\text{for all } [\zeta_i] \in K^n).$$

Let S be the unit sphere in K^n, that is

$$S = \{ [\zeta_i] \in K^n : \|[\zeta_i]\|_2 = 1 \},$$

and consider a function f on S, defined by

$$f(\zeta_1,\cdots,\zeta_n) = \|\Sigma_{i=1}^n \zeta_i e_i\| \quad (\text{for all } [\zeta_i] \in S). \tag{2.7.5}$$

As $0 \notin S$ and $[\zeta_i] = 0$ if and only if all $\zeta_i = 0$, it follows that

$$f(\zeta_1,\cdots,\zeta_n) > 0 \quad (\text{for all } [\zeta_i] \in S). \tag{2.7.6}$$

Moreover, (2.7.3) shows that f is continuous on S[in view of (2.7.5)]. Clearly, S is bounded and closed, hence S is compact (by Heine–Borel's theorem), thus f attains its infimum; namely, there exists an $[\eta_i^{(0)}] \in S$ such that

$$\alpha = f(\eta_1^{(0)},\cdots,\eta_n^{(0)}) = \inf\{ f(\xi_1,\cdots,\xi_n) : f(\xi_1,\cdots,\xi_n) \in S \} > 0.$$

Now, for any $0 \neq x = \Sigma_{i=1}^n \lambda_i e_i \in E$, let

$$\xi_i = \frac{\lambda_i}{(\Sigma_{j=1}^n |\lambda_j|^2)^{1/2}} \quad (\text{for all } i = 1,\cdots,n)$$

Then $[\xi_i] \in S$, hence (2.7.3), (2.7.5) and (2.7.7) show that

$$\begin{aligned} f(\xi_1,\cdots,\xi_n) &= \|\Sigma_{i=1}^n \xi_i e_i\| = \|\Sigma_{i=1}^n \frac{\lambda_i}{(\Sigma_{j=1}^n |\lambda_j|^2)^{1/2}} e_i\| \\ &= \frac{1}{(\Sigma_{j=1}^n |\lambda_j|^2)^{1/2}} \left\|\Sigma_{i=1}^n \lambda_i e_i\right\| \geq \alpha, \end{aligned}$$

it then follows from $x = \Sigma_{i=1}^n \lambda_i e_i$ that

$$\|x\| \geq \alpha(\Sigma_{i=1}^n |\lambda_i|^2)^{\frac{1}{2}}.$$

(2.8) **Corollary.** (a) Every finite–dimensional normed space is complete.

(b) Every finite–dimensional vector subspace of a normed space is closed.

Proof. (a) Let E be an n–dimensional normed space and let $\{e_1, \cdots, e_n\}$ be a basis of E. Then $K^n \cong E$ under the topological isomorphism T defined by

$$T([\xi_i]) = \Sigma_{i=1}^{n} \xi_i e_i \quad (\text{for all } [\xi_i] \in K^n).$$

As K^n is complete, it follows from $K^n \cong E$ that E is complete.

(b) Let G be a normed space and let M be an n–dimensional vector subspace of G. Then M is an n–dimensional normed space under the relative norm, hence M is complete by (a), thus M is closed in G (since the norm–topology is Hausdorff).

As a consequence of (2.8), we obtain the following interesting result.

(2.9) **Corollary.** If E is an infinite–dimensional Banach space, then any Hamel basis of E is uncountable. (A subset B of E is a Hamel basis if B is linearly independent and $E = \langle B \rangle$.)

Proof. Suppose that E has a countable Hamel basis $\{e_1, e_2, \cdots\}$. For any $n \geq 1$, let

$$M_n = \text{span}(\{e_1, \cdots, e_n\}).$$

Then M_n is closed (since $\dim M_n = n$) and $E = \bigcup_n M_n$. By Baire's Category theorem, there is some M_K containing a non–empty open set; in particular, $N(x,r) \subset M_K$ for some $x \in M_K$ and $r > 0$. Consequently,

$$rO_E \subset M_K$$

(since $N(x,r) = x + rO_E$ and $-x \in M_K$), it then follows from $E = \bigcup_n nO_E$ that $E \subset M_K$, and hence that $\dim E \leq k$, which gives a contradiction.

In order to give a characterization of finite–dimensional normed spaces, we need the following crucial lemma:

(2.10) **Riesz's Lemma.** Let E be a normed space and M a proper, closed vector subspace of E (i.e., $\overline{M} = M \neq E$). For any $\epsilon \in (0,1)$ there exists an $x_\epsilon \in E$ such that

(2.10.1) $$\|x_\epsilon\| = 1 \text{ and } \operatorname{dist}(x_\epsilon, M) = \inf\{\ \|x_\epsilon - m\| : m \in M\} \geq \epsilon.$$

Proof. It is clear that $\|Q_M(u)\| = \operatorname{dist}(u,M)$ (for any $u \in E$). Now choose an $z \in E\backslash M$; since $Q_M(z) \neq O(M)$ and the quotient norm $\|\cdot\|$ is a norm, it follows that

(2.10.2) $$\|Q_M(z)\| = \operatorname{dist}(z,M) = \alpha > 0,$$

and hence from $\alpha < \frac{\alpha}{\epsilon}$ and the definition of infimum that there is some $y_0 \in M$ such that

(2.10.3) $$\alpha \leq \|z - y_0\| < \frac{\alpha}{\epsilon}.$$

Let us define

$$x_\epsilon = \frac{z - y_0}{\|z - y_0\|}$$

Then $\|x_\epsilon\| = 1$; moreover, we have, for any $u \in M$, that

$$\begin{aligned} \|x_\epsilon - u\| &= \left\| \frac{z - y_0}{\|z - y_0\|} - u \right\| \\ &= \frac{1}{\|z - y_0\|} \left\| z - y_0 - \|z - y_0\|\, u \right\| \\ &= \frac{1}{\|z - y_0\|} \left\| z - (y_0 + \|z - y_0\|\, u) \right\| \\ &\geq \frac{\epsilon}{\alpha}\, \alpha = \epsilon \end{aligned}$$

(by (2.10.2) and (2.10.3)) since $y_0 + \|z - y_0\|\, u \in M$, thus $\operatorname{dist}(x_\epsilon, M) \geq \epsilon$.

Remark. In Riesz's lemma, the proper vector subspace M has to be necessarily closed. For instance, let $E = C[0,1]$ be equipped with the sup–norm $\|\cdot\|_\infty$ (for definition, see (2.17) (f) below) and let M be the vector subspace of all polynomials on [0,1]. Then $\overline{M} = C[0,1]$ and therefore the result fails to work in this case.

Also, one cannot generally take $\epsilon = 1$ in Riesz's lemma; in fact, the existence of u in a B–space E such that $\|u\| = 1$ and $\text{dist}(u,M) \geq 1$ is a characterization of reflexivity of E (see (3.d) the next section or Diestel [1984, p.5–6]).

(2.d) Finite–dimensional Banach spaces: Let E be a finite–dimensional normed space and M a proper, closed vector subspace of E. Then there exists a $u \in M$ such that

$$\|u\| = 1 \quad \text{and} \quad \text{dist}\,(u,M) = 1.$$

[Using Riesz's lemma and Heine–Borel's theorem on finite–dimensional space.]

(2.11) Theorem (Riesz). A normed space E is finite–dimensional if and only if its closed unit ball U_E is compact.

Proof. Necessity. Follows from Heine–Borel's theorem.

Sufficiency. We assume that U_E is compact, but dim $E = \infty$, and then show that this leads to a contradiction.

Choose $x_1 \in E$ with $\|x_1\| = 1$; then $M_1 = < \{x_1\} >$ is an 1–dimensional subspace of E which is closed and proper, by Riesz's lemma, there is some $x_2 \in E$ such that

$$\|x_2\| = 1 \quad \text{and} \quad \|x_2 - x_1\| \geq \frac{1}{2}.$$

$M_2 = < \{x_1,x_2\} >$ is a 2–dimensional proper closed subspace of E, by Riesz's lemma, there is an $x_3 \in E$ such that

$$\|x_3\| = 1 \quad \text{and} \quad \text{dist}\,(x_3,M_2) \geq 1/2;$$

in particular,

$$\|x_3 - x_2\| \geq 1/2 \quad \text{and} \quad \|x_3 - x_1\| \geq 1/2.$$

Continue this process, we obtain a sequence $\{x_n\}$ in E such that

$$\|x_n\| = 1 \quad \text{and} \quad \|x_n - x_m\| \geq 1/2 \quad \text{whenever} \quad n \neq m.$$

Obviously $\{x_n\}$ cannot have any convergent subsequence. This contradicts the compactness of U_E, hence $\dim E < \infty$.

An inner product space is a complex vector space H together with a function $[\cdot,\cdot]$: $H \times H \to \mathbb{C}$ (called the inner product) satisfying the following

(I1) $[\lambda_1 x_1 + \lambda_2 x_2, y] = \lambda_1 [x_1, y] + \lambda_2 [x_2, y]$ for all $x_1, x_2, y \in H$ and $\lambda_1, \lambda_2 \in \mathbb{C}$,

(I2) $[x,y] = \overline{[y\, ,\, x]}$ (the bar denotes the complex conjugate),

(I3) $[x,x] \geq 0$ for all $x \in H$,

(I4) $[x,x] = 0$ if and only if $x = 0$.

For a fixed $y \in H$, (I1) says that $[\cdot,y]$ is a linear functional on H, while for a fixed $x \in H$, (I1) and (I2) indicate that $[x,\cdot]$ is a conjugate–linear functional on H in the following sense

$$[x, \mu_1 y_1 + \mu_2 y_2] = \bar{\mu}_1 [x, y_1] + \bar{\mu}_2 [x, y_2].$$

Furthermore, if we define

$$\|x\| = [x,x]^{\frac{1}{2}} \quad \text{for all } x \in H, \tag{2.12.a}$$

then $\|x\| \geq 0$, $\|x\| = 0$ if and only if $x = 0$,and

$$\|\lambda x\| = |\lambda|\,\|x\| \quad \text{for all } \lambda \in \mathbb{C} \text{ and } x \in H.$$

Therefore it is natural to ask whether $\|.\|$ is a norm. The answer is affirmative as shown by the following result.

(2.12) **Lemma**. Let H be an inner product space. Then

(1) $$|[x,y]| \leq \|x\|\|y\| \quad \text{for all } x,y \in H,$$

consequently, we have

(2) $$\|x+y\| \leq \|x\| + \|y\| \quad \text{for all } x,y \in H.$$

Proof. If $y = 0$, then (1) is trivial. Therefore we assume that $y \neq 0$. For any $\lambda \in \mathbb{C}$,

$$0 \leq [x+\lambda y,x+\lambda y] = \|x\|^2+\lambda[y,x]+\overline{\lambda}[x,y]+\lambda\overline{\lambda}\|y\|^2;$$

in particular, if $\lambda = -[x,y]\|y\|^{-2}$, the last formula is easily seen to become

$$0 \leq \|x\|^2-\frac{|[x,y]|^2}{\|y\|^2},$$

which obtains (1). To prove (2), we notice that

$$\begin{aligned}\|x+y\|^2 = [x+y,x+y] &= \|x\|^2 + [x,y] + [y,x] + \|y\|^2 \\ &= \|x\|^2 + 2\mathrm{Re}[x,y] + \|y\|^2 \leq \|x\|^2 + 2|[x,y]| + \|y\|^2.\end{aligned}$$

Formula (2) then follows from (1).

Formula (1) is usually called Cauchy–Schwarz's inequality.

Therefore every inner product space is a normed space under the norm defined by (2.12.a), which is called the associated norm.

A Hilbert space is an inner product space which is complete under the associated norm (2.12.a).

(2.e) Some characterizations of inner product spaces: (a) Let H be an inner product space. Then:

(i) (Polarization identity) For any $x,y \in H$,

$$[x, y] = \frac{1}{4}\{\|x + y\|^2 - \|x - y\|^2 + i\|x + iy\|^2 - i\|x - iy\|^2\}.$$

(ii) (Parallelogram law) For any $x,y \in H$,

$$\|x + y\|^2 + \|x - y\|^2 = 2\|x\|^2 + 2\|y\|^2.$$

(b) A normed space $(X,\|\cdot\|)$ is a inner product space if and only if the norm $\|\cdot\|$ satisfies the parallelogram law. [On $H\times H$, the function, defined by $[x,y] = \frac{1}{4}\{\|x+y\|^2 - \|x-y\|^2 + i\|x+iy\|^2 - i\|x-iy\|^2$, is an inner product.]

Let H be an inner product space. Two vectors x and y in H are said to be orthogonal, denoted by $x \perp y$, if $[x,y] = 0$. A subset B of H is said to be orthogonal (resp. orthonormal) if

$$x \perp y \text{ for all } x,y \in B \text{ (resp. } x \perp y \text{ for all } x,y \in B \text{ and } \|x\|=1).$$

The orthogonal complement of B, denoted by $B^\perp$, is defined by

$$B^\perp = \{u \in H : [x,u]=0 \text{ for all } x \in B\}.$$

(2.f) Some properties of orthogonal complements : Let H be an inner product space and $B \subset H$.

(a) (Pythagorean theorem) If $x \perp y$ then

$$\|x+y\|^2 = \|x\|^2 + \|y\|^2.$$

(b) $B^\perp$ is a closed vector subspace of H.

(c) If B is an orthogonal set of non-zero vectors, then B is linearly independent.

Let Λ be a non-empty set. Then the family $\mathfrak{F}(\Lambda)$ of all non-empty finite subsets of Λ is a directed set under the set inclusion. If $[x_i,\Lambda]$ is a family in H , for any $\alpha \in \mathfrak{F}(\Lambda)$ we write

$$x_\alpha = \Sigma_{i\in\alpha} x_i ,$$

hence $\{x_\alpha, \alpha \in \mathfrak{F}(\Lambda)\}$ becomes a net in H which is called the associated net with $[x_i,\Lambda]$. If $\{x_\alpha, \alpha \in \mathfrak{F}(\Lambda)\}$ is convergent under the norm, then we write $\Sigma_\Lambda x_i$ to be its limit, that is

$$\Sigma_\Lambda x_i = \lim_{\alpha\in\mathfrak{F}(\Lambda)} \Sigma_{i\in\alpha} x_i.$$

The following result is a generalization of the Pythagorean theorem.

(2.13) Proposition. Let H be a Hilbert space and $[x_i,\Lambda]$ an orthogonal family in H. Then the following statements are equivalent.

(a) The net $\{x_\alpha, \alpha \in \mathfrak{F}(\Lambda)\}$ associated with $[x_i,\Lambda]$ is convergent.

(b) $\Sigma_\Lambda \|x_i\|^2 < \infty$.

In this case, we have

$$\|\Sigma_\Lambda x_i\|^2 = \Sigma_\Lambda \|x_i\|^2$$

Proof. (a) ⇒ (b) : In view of the continuity of the norm and the Pythagorean theorem, it follows from the existence of $\Sigma_\Lambda x_i$ that

$$\begin{aligned}\|\Sigma_\Lambda x_i\|^2 &= \|\lim_{\alpha\in\mathfrak{F}(\Lambda)} \Sigma_{i\in\alpha} x_i\|^2 = \lim_{\alpha\in\mathfrak{F}(\Lambda)} \|\Sigma_{i\in\alpha} x_i\|^2 \\ &= \lim_{\alpha\in\mathfrak{F}(\Lambda)} \Sigma_{i\in\alpha}\|x_i\|^2 < \infty ,\end{aligned}$$

and hence that $\|\Sigma_\Lambda x_i\|^2 = \Sigma_\Lambda \|x_i\|^2 < \infty$.

(b) ⇒ (a) : For any $\epsilon > 0$ there exists an $\alpha_o \in \mathfrak{F}(\Lambda)$ such that

$$\Sigma_{i\in\alpha}\|x_i\|^2 - \Sigma_{i\in\alpha_o}\|x_i\|^2 < \epsilon \text{ for all } \alpha \geq \alpha_o .$$

If $\alpha_1, \alpha_2 \in \mathfrak{F}(\Lambda)$ are such that $\alpha_j \geq \alpha_o$ $(j = 1,2)$, then we have by the Pythagorean theorem that

$$\begin{aligned}\|\Sigma_{i\in\alpha_1} x_i - \Sigma_{i\in\alpha_2} x_i\|^2 &= \sum_{i\in\alpha_1\setminus\alpha_2} \|x_i\|^2 + \sum_{i\in\alpha_2\setminus\alpha_1} \|x_i\|^2 \\ &\leq \sum_{i\in\alpha_1\cup\alpha_2} \|x_i\|^2 - \sum_{i\in\alpha_o} \|x_i\|^2 \leq \epsilon ,\end{aligned}$$

hence $\{\Sigma_{i\in\alpha} x_i, \alpha \in \mathfrak{F}(\Lambda)\}$ is a Cauchy net in H, thus the completeness of H ensures that $\Sigma_\Lambda x_i$ exists.

(2.14) Corollary. <u>Let</u> H <u>be a Hilbert space</u>, <u>let</u> $[e_i,\Lambda]$ <u>be an orthonormal family in</u> H <u>and let</u> M <u>be the smallest closed vector subspace of</u> H <u>generated by the set</u> $\{e_i : i\in\Lambda\}$. <u>Then</u>

$$M = \{\Sigma_\Lambda \lambda_i e_i : \lambda_i \in \mathbb{C} \text{ and } \Sigma_\Lambda |\lambda_i|^2 < \infty\} .$$

Proof. We first notice that if $\Sigma_\Lambda |\lambda_i|^2 < \infty$ then the set $\{\lambda_i e_i : i \in \Lambda\}$ is orthogonal and

$$\lim_{\alpha\in\mathfrak{F}(\Lambda)} \Sigma_{i\in\alpha}\|\lambda_i e_i\|^2 = \lim_{\alpha\in\mathfrak{F}(\Lambda)} \Sigma_{i\in\alpha}|\lambda_i|^2 = \Sigma_\Lambda |\lambda_i|^2 < \infty ,$$

hence $\lim_{\alpha\in\mathfrak{F}(\Lambda)} \Sigma_{i\in\alpha}\lambda_i e_i = \Sigma_\Lambda \lambda_i e_i$ exists in H by (2.13). Thus the set

$$N = \{\Sigma_\Lambda \lambda_i e_i : \lambda_i \in \mathbb{C}, \Sigma_\Lambda |\lambda_i|^2 < \infty\}$$

is a subset of H containing all e_i ($i \in \Lambda$). Further if $\Sigma_\Lambda \lambda_i e_i$ and $\Sigma_\Lambda \mu_i e_i$ belong to N, then

$$\Sigma_\Lambda |\lambda_i + \mu_i|^2 \leq 2(\Sigma_\Lambda |\lambda_i|^2 + \Sigma_\Lambda |\mu_i|^2) < \infty,$$

hence $\Sigma_\Lambda \lambda_i e_i + \Sigma_\Lambda \mu_i e_i \in N$. Consequently N is a vector subspace of H. This show that $N \subset M$.

In order to verify this result, it suffices to show that N is closed. Indeed, let $\{x^{(m)}\}$ be a Cauchy sequence in N and

$$x^{(n)} = \Sigma_\Lambda \lambda_i^{(n)} e_i \qquad \text{(for all n)}.$$

For any $i \in \Lambda$,

$$|\lambda_i^{(n)} - \lambda_i^{(m)}| \leq (\Sigma_\Lambda |\lambda_i^{(n)} - \lambda_i^{(m)}|^2)^{\frac{1}{2}} = \|x^{(n)} - x^{(m)}\|;$$

this implies that $\{\lambda_i^{(m)}, m \geq 1\}$ is a Cauchy sequence in $\mathbb{C}$, thus for any $i \in \Lambda$, we have $\lambda_i = \lim_m \lambda_i^{(m)}$. We claim that $\Sigma_\Lambda |\lambda_i|^2 < \infty$. Indeed, for any $\alpha \in \mathfrak{F}(\Lambda)$,

$$\Sigma_{i \in \alpha} |\lambda_i|^2 = \lim_n \Sigma_{i \in \alpha} |\lambda_i^{(n)}|^2 \leq \lim_n \Sigma_\Lambda |\lambda_i^{(n)}|^2 = \lim_n \|x^{(n)}\|^2 < \infty$$

in view of the completeness of H. Therefore $x = \Sigma_\Lambda \lambda_i e_i$ is well–defined and belongs to N. Now for any $\epsilon > 0$, we choose $n_0 > 0$ satisfying

$$\|x^{(n)} - x^{(m)}\| < \epsilon \qquad \text{(for all } n, m \geq n_0).$$

For any $\alpha \in \mathfrak{F}(\Lambda)$,

$$\Sigma_{i \in \alpha} |\lambda_i - \lambda_i^{(n)}|^2 = \lim_m \Sigma_{i \in \alpha} |\lambda_i^{(m)} - \lambda_i^{(n)}|^2$$

$$\le \limsup_n \|x^{(m)}-x^{(n)}\|^2 \le \epsilon^2 \quad \text{whenever } n \ge n_o.$$

Thus for any $n \ge n_o$, we have

$$\|x - x^{(n)}\|^2 = \Sigma_\Lambda |\lambda_i - \lambda_i^{(n)}|^2 \le \epsilon^2$$

which obtains our assertion.

A family $[e_i,\Lambda]$ in a Hilbert space H is called an orthonormal basis if the set $\{e_i: i\in\Lambda\}$ is orthonormal and the smallest closed vector subspace containing all e_i $(i\in\Lambda)$ is H.

(2.15) Corollary. If $\{e_i: i\in\Lambda\}$ is an orthonormal basis for a Hilbert space H, then for any $x \in H$ there exists uniquely a family $\{\lambda_i : i \in \Lambda\}$ (called the Fourier coefficients) contained in $\mathbb{C}$ such that $x = \Sigma_\Lambda \lambda_i e_i$; moreover

$$\lambda_i = [x,e_i] \text{ for all } i \in \Lambda .$$

Proof. By (2.14), there exist uniquely $\{\lambda_i : i \in \Lambda\}$ with $\Sigma_\Lambda |\lambda_i|^2 < \infty$ such that $x = \Sigma_\Lambda \lambda_i e_i$. For any $i\in\Lambda$, the continuity of the inner product ensures that

$$[x,e_i] = [\lim_{\alpha\in\mathfrak{F}(\Lambda)} \Sigma_{j\in\alpha}\lambda_j e_j, e_i] = \lim_{\alpha\in\mathfrak{F}(\Lambda)} \Sigma_{j\in\alpha} \lambda_j [e_j,e_i] = \lambda_i.$$

(2.16) Theorem. Every Hilbert space H possesses an orthonormal basis.

Proof. Let $\mathfrak{M}$ be the collection of all orthonormal subsets B of H with the partial ordering $B_1 \le B_2$. Let $\{B_j : j \in D\}$ be a chain in $\mathfrak{M}$. Then $\cup_{j\in D} B_j$ is obviously orthonormal, hence it is an upper bound of $\{B_j : j \in D\}$. Zorn's lemma ensures that $\mathfrak{M}$ has a maximal element B. To show that B is an orthonormal basis for H, let M be the smallest closed subspace of H containing B. If $M \neq H$, then there exists an $e \in M^{\perp}$ with $\|e\| = 1$, hence the set $B \cup \{e\}$ is an orthonormal subset of H containing B properly. This contradicts the maximality of B.

We conclude this section with some examples of Banach spaces.

(2.17) Examples. (a) The vector space, defined by

$$\ell^{\infty} = \{ \xi = [\xi_n] \in \mathbb{K}^{\mathbb{N}} : [\xi_n] \text{ is bounded } \},$$

is a Banach space under the sup–norm $\|\cdot\|_{\infty}$, defined by

$$\|\xi\|_{\infty} = \sup_{j \geq 1} |\xi_j| \quad (\text{for all } \xi = [\xi_n] \in \ell^{\infty}).$$

But ℓ^{∞} is not separable.

Proof. It is not hard to show that ℓ^{∞} is a Banach space. To see that ℓ^{∞} is not separable, let

$$B = \{ \xi = [\xi_j] \in \ell^{\infty} : \xi_j = 1 \text{ or } \xi_j = 0 \text{ for all } j \geq 1 \}.$$

Then it is easily seen that
$B = \{0,1\}^{\mathbb{N}}$ (the set of all maps from $\mathbb{N}$ into the 2–point set $\{ 0,1 \}$)

and that

$$\|\xi - \eta\|_{\infty} = 1 \quad \text{for all } \xi, \eta \in B \quad \text{with} \quad \xi \neq \eta,$$

hence B is not countable since

$$\text{card } B = 2^{\text{card } \mathbb{N}} = 2^{\aleph_0} = c \quad (\text{the power of the continuum}).$$

It then follows that B cannot contain a countable dense subset; in other words, B is not separable, thus ℓ^∞ is not separable. (Since it is easily shown that any subset of a separable normed space must be separable.)

(b) The subsets, defined by

$$c = \{ \xi = [\xi_i] \in \ell^\infty : \lim_j \xi_j \text{ exists } \}$$

and

$$c_0 = \{ \xi = [\xi_i] \in \ell^\infty : \lim_j \xi_j = 0 \},$$

are vector subspaces of ℓ^∞. Equipped with the sup–norm $\|\cdot\|_\infty$, c and c_0 are separable Banach spaces. Furthermore, for any $n \geq 1$, let

$$e^{(n)} = [\delta_j^{(n)}]_{j\geq 1} \quad \text{and} \quad e = (1,1,\cdots),$$

$$\text{where } \delta_j^{(n)} = \begin{cases} 1 & \text{if } j = n \\ 0 & \text{if } j \neq n . \end{cases}$$

Then the countable set $\{ e^{(n)} : n \geq 1 \} \cup \{ e \}$ is a Schauder basis for c, while $\{ e^{(n)} : n \geq 1 \}$ is a Schauder basis for c_0.

<u>**Remark**</u>. The space c contains c_0 as a subspace, and c_0 has codimension one, thus

$$c = c_0 \oplus < \{ e \} >$$

Every convergent sequence $[\xi_n] \in c$ can be represented in the form

$$[\xi_n] = [\eta_n] + \lambda e \text{ with } \lambda = \lim_j \xi_j \text{ and } [\eta_n] = [\xi_n - \lambda] \in c_0 .$$

(As $\lim_j \xi_j = \lambda$, it follows that $\lim_n \sup_{j\geq n} |\xi_j - \lambda| = 0$, and hence that

$$\|\Sigma_{k=1}^{\infty} \xi_k e^{(k)} - \Sigma_{j=1}^{n} (\xi_j - \lambda) e^{(j)} - \lambda e\|_\infty$$

$$= \|(\lambda, \cdots, \lambda, \xi_{n+1}, \cdots) - \lambda e\|_\infty$$

$$= \sup_{j\geq n+1} |\xi_j - \lambda| \longrightarrow 0 \quad (\text{as } n \longrightarrow \infty).)$$

(c) The vector space, defined by

$$\ell^1 = \{ \xi = [\xi_n] \in \mathbb{K}^{\mathbb{N}} : \Sigma_n |\xi_n| < \infty \},$$

is a separable Banach space under the norm

$$\|[\xi_n]\|_{\ell^1} = \Sigma_{n=1}^{\infty} |\xi_n| \quad (\text{for all } [\xi_n] \in \ell^1).$$

Furthermore, the sequence $\{ e^{(n)}, n \geq 1 \}$ is a Schauder basis for ℓ^1.

(d) For any p with $1 \leq p < \infty$, the set , defined by

$$\ell^p = \{ \xi = [\xi_j] \in \mathbb{K}^{\mathbb{N}} : \Sigma_{j=1}^{\infty} |\xi_j|^p < \infty \},$$

is a separable Banach space under the norm

$$\|\xi\|_p = (\Sigma_{j=1}^{\infty} |\xi_j|^p)^{1/p} \quad (\text{for all } \xi = [\xi_j] \in \ell^p).$$

Moreover, the sequence $\{ e^{(n)}, n \geq 1 \}$ is a Schauder basis for ℓ^p

(e) If $1 \leq p_1 < p_2 \leq \infty$, then $\ell^{p_1} \subsetneq \ell^{p_2}$, (e.1) $$\|[\xi_j]\|_{p_2} = (\Sigma_{k=1}^{\infty} |\xi_k|^{p_2})^{1/p_2} \le (\Sigma_{k=1}^{\infty} |\xi_k|^{p_1})^{1/p_1}$$

$$= \|[\xi_j]\|_{p_1} \quad (\text{for all } [\xi_i] \in \ell^{p_1}).$$

(e.2) $$\lim_{p\to\infty} \|[\eta_j]\|_p = \lim_{p\to\infty} (\Sigma_{k=1}^{\infty} |\eta_k|^p)^{1/p} = \inf_{p\ge 1} \|[\eta_j]\|_p$$
$$= \sup_{j\ge 1} |\eta_j| = \|[\eta_j]\|_{\infty} \quad (\text{for all } [\eta_j] \in \ell^p).$$

Proof. To prove the first part, let $0 \ne [\zeta_i] \in \ell^{p_1}$ and

(1) $$\eta_j = \frac{\zeta_j}{\| [\zeta_i] \|_p} \quad (\text{for all } j = 1,2,\cdots).$$

It then follows that $|\eta_j| \le 1$ (for all $j \ge 1$), and hence from $1 \le p_1 < p_2 \le \infty$ that

$$|\eta_j|^{p_2} \le |\eta_j|^{p_1} \quad (\text{for all } j \ge 1);$$

thus

$$\Sigma_{j=1}^{\infty} |\eta_j|^{p_2} \le \Sigma_{j=1}^{\infty} |\eta_j|^{p_1} = 1.$$

Consequently,

(2) $$(\Sigma_{j=1}^{\infty} |\zeta_j|^{p_2})^{1/p_2} \le \|[\zeta_j]\|_{p_1}$$

(by (1)); this implies that $[\xi_j] \in \ell^{p_2}$ and that (e.1) holds.

To prove (e.2), we first notice from (e.1) that $(\Sigma_{k=1}^{\infty} |\eta_k|^p)^{1/p} \downarrow$ (for all $p \ge 1$), hence

(3) $$\lim_{p\to\infty} \left(\Sigma_{k=1}^{\infty} |\eta_k|^p\right)^{1/p} = \inf_{p\geq 1} \left(\Sigma_{k=1}^{\infty} |\eta_k|^p\right)^{1/p}.$$

On the other hand, for any $\epsilon > 0$, $p_0 \geq 1$ and $[\eta_i] \in \ell^{p_0}$, there is some $N \geq 1$ such that

$$\left(\Sigma_{k>n}^{\infty} |\eta_k|^{p_0}\right)^{1/p_0} < \epsilon \quad \text{(for all } n \geq N).$$

For all p with $p > p_0$, we obtain from (e.1) that

$$\begin{aligned}
\left(\Sigma_{k=1}^{\infty} |\eta_k|^p\right)^{1/p} &\leq \left(\Sigma_{k=1}^{N} |\eta_k|^p\right)^{1/p} + \left(\Sigma_{k=N+1}^{\infty} |\eta_k|^p\right)^{1/p} \\
&\leq \left(\Sigma_{k=1}^{N} |\eta_k|^p\right)^{1/p} + \left(\Sigma_{k=N+1}^{\infty} |\eta_k|^{p_0}\right)^{1/p_0} \\
&\leq \left(\Sigma_{k=1}^{N} |\eta_k|^p\right)^{1/p} + \epsilon \\
&\leq N^{1/p} \max_{1\leq k\leq N} |\eta_k| + \epsilon
\end{aligned}$$

(since $\Sigma_{k=1}^{N} |\eta_k|^p \leq N \max_{1\leq k\leq N} |\eta_k|^p$), so that

(4) $$\lim_{p\to\infty} \left(\Sigma_{k=1}^{\infty} |\eta_k|^p\right)^{1/p} \leq \max_{1\leq k\leq N} |\eta_k| + \epsilon \leq \sup_{j\geq 1} |\eta_j| + \epsilon.$$

By the definition of supremum, there is some η_{k_0} such that

$$|\eta_{k_0}| > \sup_{j\geq 1} |\eta_j| - \epsilon.$$

Clearly, $|\eta_{k_0}|^p \leq \Sigma_{k=1}^{\infty} |\eta_k|^p$, it then follows that

$$(\Sigma_{k=1}^{\infty} |\eta_k|^p)^{1/p} \geq |\eta_{k_0}| > \sup_{j\geq 1} |\eta_j| - \epsilon,$$

and hence from (3) that

(5) $$\lim_{p\to\infty} (\Sigma_{k=1}^{\infty} |\eta_k|^p)^{1/p} \geq \sup_{j\geq 1} |\eta_j| - \epsilon.$$

Combining (4) and (5), we obtain (e.2).

Finally, to prove that $\ell^{p_1} \neq \ell^{p_2}$, let $p = \frac{p_1 + p_2}{2}$ and

$$\xi_k = k^{-1/p} \quad (\text{for all } k \geq 1).$$

Then $p_1 < p < p_2$, $\frac{p_2}{p} > 1$ and $\frac{p_1}{p} < 1$, it follows that

$$[k^{-1/p}]_{k\geq 1} \in \ell^{p_2} \quad \text{but} \quad [k^{-1/p}]_{k\geq 1} \notin \ell^{p_1}.$$

(f) <u>The space</u> $C(\Omega)$: Let Ω be a compact Hausdorff space, and let $C(\Omega)$ (resp. $C_{\mathbb{R}}(\Omega)$) be the set of all complex–valued (or real–valued) continuous functions on Ω. Then $C(\Omega)$ (resp. $C_{\mathbb{R}}(\Omega)$) is a Banach space under the sup–norm $\|\cdot\|_\infty$, defined by

$$\|f\|_\infty = \max_{t\in\Omega} |f(t)| \quad (\text{for all } f \in C(\Omega)).$$

(g) For $1 \leq p < \infty$, let $L^{(p)}[a,b]$ (where $-\infty < a < b < \infty$) be the vector space consisting of all complex–valued (or real–valued) Lebesgue measurable functions $x(\cdot)$ on $[a,b]$, which are p–th power integrable in the sense that

$$\|x\|_p = (\int_a^b |x(t)|^p dm(t))^{1/p} \quad (\text{for all } x(\cdot) \in L^{(p)}[a,b]),$$

(where m always denote the Lebesgue measure on $[a,b]$), and let

$$N = \{ x(\cdot) \in L^{(p)}[a,b]: x(t) = 0 \text{ a.e. } \}.$$

Then N is a closed vector subspace of $L^{(p)}[a,b]$, hence the quotient space $L^{(p)}[a,b]/N$, equipped with the quotient norm $\|\cdot\|_p$, is a Banach space, which is usually denoted by $L^p[a,b]$.

(h)<u>The space</u> $L^\infty[a,b]$: Let $L^\infty[a,b]$ be the set of all essentially bounded, Lebesgue measurable functions on $[a,b]$, more precisely the set of equivalence classes of all essentially bounded, Lebesgue measurable functions on $[a,b]$. (A Lebesgue measurable function $x(\cdot)$ on $[a,b]$, except perhaps a set of measure zero, is said to be <u>essentially bounded</u> if there is a $C \geq 0$ such that $|x(\cdot)| \leq C$ a.e. on $[a,b]$). Then $L^\infty[a,b]$ is a Banach space under the essential sup–norm, defined by

$$\|x\|_\infty = \operatorname*{ess\,sup}_{[a,b]} |x(\cdot)| = \inf \{ \sup_{t\in[a,b]\setminus E_0} |x(t)| : m(E_0) = 0 \}.$$

The following example should be compared with Example (e).

(i) <u>If</u> $1 \leq p_1 < p_2 < \infty$, <u>then</u> $L^{p_2}[a,b] \subsetneq L^{p_1}[a,b]$,

$$\text{(i.1)} \qquad \|x\|_{p_1} \leq \|x\|_{p_2}(b-a)^{\frac{1}{p_1}-\frac{1}{p_2}} \quad (\text{for all } x(\cdot) \in L^{p_2}[a,b])$$

and

$$\text{(i.2)} \qquad \|x\|_\infty = \lim_{\substack{p\to\infty \\ p\geq 1}} \|x\|_p \quad (\text{for all } x(\cdot) \in L^\infty[a,b]).$$

<u>Note</u>. The preceding example holds only for a <u>finite measure space</u>; for a general measure space no such inclusion holds.

Proof. For any $x(\cdot) \in L^{p_2}[a,b]$, we have $|x(\cdot)|^{p_1} \in L^{\frac{p_2}{p_1}}[a,b]$. Now let q be the conjugate index of $\frac{p_2}{p_1}$ (i.e., $\frac{1}{q} + \frac{1}{\frac{p_2}{p_1}} = 1$). Then

$$q = \frac{p_2}{p_2 - p_1}.$$

Hence we apply Hölder's ineqality to $|x(\cdot)|^{p_1}$ and $|y(\cdot)| = 1$, we obtain

$$\begin{aligned}\int_a^b |x(t)|^{p_1}\, dm(t) &\le \left(\int_a^b (|x(t)|^{p_1})^{\frac{p_2}{p_1}}\, dm(t)\right)^{\frac{p_1}{p_2}} (b-a)^{\frac{p_2-p_1}{p_2}} \\ &= \left(\int_a^b (|x(t)|^{p_2}\, dm(t)\right)^{\frac{p_1}{p_2}} (b-a)^{\frac{p_2-p_1}{p_2}},\end{aligned}$$

thus

$$\begin{aligned}\|x\|_{p_1} &= \left(\int_a^b |x(t)|^{p_1}\, dm(t)\right)^{1/p_1} \le \|x\|_{p_2} (b-a)^{\frac{p_2-p_1}{p_1 p_2}} \\ &= \|x\|_{p_2} (b-a)^{\frac{1}{p_1} - \frac{1}{p_2}}\end{aligned}$$

This proves (i.1) and $L^{p_2}[a,b] \subset L^{p_1}[a,b]$.

To prove (i.2), let $x(\cdot) \in L^{\infty}[a,b]$ and

$$\lambda = \|x\|_{\infty} = \operatorname*{ess\,sup}_{[a,b]} |x(\cdot)|.$$

Then $|x(\cdot)| \le \lambda$ on $[a,b]$ (a.e.), hence

$$\|x\|_p \le \left(\int_a^b \lambda^p \, dm(t)\right)^{\frac{1}{p}} = \lambda\,(b-a)^{\frac{1}{p}}.$$

The inequality (i.1) shows that $\|x\|_p \uparrow$ (for all $p \ge 1$), so that

$$\overline{\lim_{p\to\infty}} \|x\|_p = \sup_{p\ge 1} \|x\|_p \le \lambda.$$

On the other hand, for any $\epsilon > 0$, there exists a Lebesgue measurable set B_0 in $[a,b]$ such that

$$m(B_0) = 0 \quad \text{and} \quad \sup_{t\in[a,b]\setminus B_0} |x(t)| > \lambda - \epsilon.$$

Now the set $B = [a,b]\setminus B_0$ is measurable with $0 < m(B) \le b - a$ such that

$$\|x\|_p \ge \left(\int_B |x(t)|^p \, dm(t)\right)^{\frac{1}{p}} > (\lambda - \epsilon)\,(m(B))^{\frac{1}{p}},$$

so that

$$\underline{\lim_{p\to\infty}} \|x\|_p \ge \lambda - \epsilon.$$

Thus

$$\lim_{p\to\infty} \|x\|_p = \lambda = \|x\|_\infty.$$

By a similar argument give in the proof of (a) —— (e) of (2.17), one can verify the following more general Banach (sequence) spaces.

(2.g) <u>The spaces</u> $\ell^p(\Lambda)$ (I): Let Λ be a non–empty index set. Then the collection of all non–empty finite subsets of Λ, denoted by $\mathcal{F}(\Lambda)$, is a directed set ordered by the set inclusion. Elements in $\mathcal{F}(\Lambda)$ will be denoted by α, β, γ etc. A family $[\lambda_i,\Lambda]$ of numbers is said to be <u>summable</u> if the net $\{ \Sigma_{i\in\alpha}\lambda_i, \mathcal{F}(\Lambda) \}$ converges. The uniquely determined limit λ is called the <u>sum of</u> $[\lambda_i,\Lambda]$, and denoted by

$$\lambda = \Sigma_\Lambda \lambda_i \quad \text{or} \quad \lambda = \Sigma_i \lambda_i.$$

(i) A family $[\lambda_i, \Lambda]$ of numbers is summable if and only if it is absolutely summable in the sense that the family $[|\lambda_i|, \Lambda]$ of positive numbers is summable, and this is the case if and only if $\sup_{\alpha \in \mathcal{F}(\Lambda)} \Sigma_{i \in \alpha} |\lambda_i| < \infty$.

(ii) If $[\lambda_i, \Lambda]$ is a summable family of numbers,then it contains at most countably many non–zero terms.

(iii) For $1 \leq p < \infty$, the collection $\ell^p(\Lambda)$ of all families $[\lambda_i, \Lambda]$ of numbers for which $[|\lambda_i|^p, \Lambda]$ is summable forms a Banach space with respect to the operation

(2.g.1) $$a[\lambda_i, \Lambda] + b[\mu_i, \Lambda] = [a\lambda_i + b\mu_i, \Lambda] \quad (a, b \in K, [\lambda_i], [\mu_i] \in \ell^p(\Lambda)),$$

and with the norm

$$\|[\lambda_i, \Lambda]\|_p = (\Sigma_i |\lambda_i|^p)^{\frac{1}{p}}.$$

In particular, if $\Lambda = \mathbb{N}$ then $\ell^p(\mathbb{N})$ is the usual Banach spaces ℓ^p.

(iv) The collection $\ell^\infty(\Lambda)$ of all bounded families $[\lambda_i, \Lambda]$ of numbers forms a Banach space with respect to the operation (2.g.1) and the norm

(2.g.2) $$\|[\lambda_i, \Lambda]\|_\infty = \sup_{i \in \Lambda} |\lambda_i| \quad (\text{for all } [\lambda_i] \in \ell^\infty(\Lambda)).$$

(v)A family $[\lambda_i, \Lambda]$ of numbers is called a null family (or converge to 0) if for any $\epsilon > 0$ there exists an $\alpha \in \mathcal{F}(\Lambda)$ such that

$$|\lambda_i| < \epsilon \qquad \text{for all } i \notin \alpha.$$

The collection $c_0(\Lambda)$ of all null families of numbers forms a Banach space with respect to the operation (2.g.1) and with the sup–norm (2.g.2). In particular, if $\Lambda = \mathbb{N}$ then $\ell^\infty(\mathbb{N})$ and $c_0(\mathbb{N})$ are the usual Banach spaces ℓ^∞ and c_0 respectively.

(vi) As usual, we denote the j–th unit family by $e^{(j)} = [\delta_{ji}, i \in \Lambda]$, where δ_{ji} is Kronecker's symbol. Therefore each element $[\lambda_i,\Lambda]$ in $\ell^1(\Lambda)$ (or in $c_0(\Lambda)$) has the following representation :

(2.g.3) $$[\lambda_i,\Lambda] = \Sigma_\Lambda \lambda_j e^{(j)} \text{ and}$$

(2.g.4) $$\lambda_j = < [\lambda_i, \Lambda], e^{(j)} > \quad (\text{for all } j \in \Lambda).$$

Of course, the convergence of (2.g.3) is understood for the $\ell^1(\Lambda)$–norm or the $\|\cdot\|_\infty$–norm dependent upon on the element $[\lambda_i,\Lambda]$ belonging to $\ell^1(\Lambda)$ or $c_0(\Lambda)$. It should be noted that elements $[\lambda_i,\Lambda]$ in $\ell^\infty(\Lambda)$ cannot have any representation analogues to (2.g.3), but we still have

$$\lambda_j = <[\lambda_i,\Lambda], e^{(j)}> \quad (\text{for all } j \in \Lambda).$$

Exercises

2–1. Prove (2.a).

2–2. Prove (2.b).

2–3. Prove (2.c).

2–4. Let E be a normed space, let M be a closed vector subspace of E, let $Q_M : E \longrightarrow {}^E/_M$ be the quotient map and $B \subset E$.

(a) Show that $Q_M(B)$ is closed in ${}^E/_M$ if and only if $B + M$ is closed in E.

(b) $Q_M^{-1}(Q_M(B)) = B + M$.

(c) If, in addition, B is a finite–dimensional subspace of E, then B+M is closed in E.

2–5. Let E be a normed space, let $\{x_n\}$ be a Cauchy sequence in E and $u \in E$. Show that one of the following two condition ensures that $\lim_n x_n = u$:

(i) $\{x_n\}$ has a subsequence $\{x_{n_k}\}$ with $u = \lim_k x_{n_k}$.

(ii) u is a cluster point of $\{x_n\}$ in the sense that for any $\epsilon > 0$ there exist infinite indices k such that

$$\|u - x_k\| < \epsilon.$$

2–6. Prove (2.d).

2–7. Show that:

(a) Any Schauder basis in a B–space must be linearly independent.

(b) In a finite–dimensional B–space, Schauder basis and Hamel basis coincide.

(c) Any separable Hilbert space possesses a Schauder basis.

2–8. Prove (2.e).

2–9. Prove (2.f).

2–10. Prove (2.g).

3

Operators and Some Consequences of Hahn-Banach's Extension Theorem

Let $(E, \|\cdot\|)$ and $(F, \|\cdot\|)$ be normed spaces over K. We denote by $L^*(E, F)$ the vector space of all linear maps from E into F, while the set consisting of all continuous linear maps (called <u>operators</u>) from E into F is a vector subspace of $L^*(E, F)$, which is denoted by $L(E, F)$. In particular, we write

$$E^* = L^*(E, K) \text{ and } E' = L(E, K).$$

E^* is called the <u>algebraic dual</u> of E, while E' is called the <u>topological dual</u> of E. For any $f \in E^*$ and $x \in E$, we write

$$\langle x, f \rangle = f(x).$$

(3.a) A characterization of continuity of linear maps: Let E and F be normed spaces. The following conditions are equivalent for an $T \in L^*(E, F)$:

(i) T is continuous at 0.

(ii) T is continuous on E.

(iii) T is bounded in the sense that there exists an $\lambda > 0$ such that

$$\| Tx \| \leq \lambda \| x \| \quad \text{(for all } x \in E).$$

The following result shows that $L(E, F)$ is a normed space under the operator norm:

(3.1) **Theorem**. <u>Let</u> E <u>and</u> F <u>be normed spaces over</u> K. <u>Then</u> $L(E, F)$ <u>becomes a normed space over</u> K, <u>under the operator norm</u> $\|\cdot\|$, <u>defined by</u>

$$\| T \| = \sup \{ \| Tx \| : x \in U_E \} \quad \text{(for all } T \in L(E, F)).$$

Moreover, if F is complete, then so is L(E, F). In particular, the topological dual E' of E is always a Banach space (called the Banach dual of E).

The norm–topology on L(E, F), induced by the operator norm, is called the uniform norm–topology.

Remark. The operator norm can be represented by

$$(3.1.1) \qquad \|T\| = \inf\{\, \alpha \geq 0 : \|Tx\| \leq \alpha \|x\| \text{ for all } x \in E \,\}$$

$$= \sup\{\, \|Tx\| : x \in O_E \,\}$$

$$= \sup\{\, \|Tx\| : \|x\| = 1 \,\}$$

$$= \sup\{\, \frac{\|Tx\|}{\|x\|} : 0 \neq x \in E \,\} \quad (\text{for all } T \in L(E, F)).$$

The proof is routine, hence will be omitted.

(3.b) Isomorphic embedding: Let E and F be normed spaces and $T \in L(E, F)$. We say that T is isomorphic embedding if T has a bounded inverse $T^{-1} : T(E) \longrightarrow E$. It is not hard to show that T is isomorphic embedding if and only if

$$\inf\{\, \|Tx\| : \|x\| = 1 \,\} > 0$$

(compared with (3.1.1)) or equivalently, there exists an $r > 0$ such that

$$\|Tx\| \geq r\|x\| \quad (\text{for all } x \in E).$$

(3.c) Linear mappings between finite–dimensional Banach spaces: Let E and F be finite dimensional Banach spaces. Then any linear map $T : E \longrightarrow F$ is continuous. Consequently,

$$\dim E' = \dim E^* = \dim E < \infty.$$

(3.2) **Theorem** (Induced mapping theorem). *Let* E, F *and* G *be normed spaces over* **K** . *Given the following operators*

$$\begin{array}{ccc} E & \xrightarrow{\ T\ } & F \\ {\scriptstyle S}\big\downarrow & & \\ G & & \end{array}$$

Suppose that S *is surjective and that*

(3.2.1) $$\| Tx \| \le \lambda \| Sx \| \quad (\textit{for all } x \in E)$$

for some $\lambda > 0$. *Then there exists a unique* $R \in L(G,F)$ *such that*

(3.2.2) $$T = RS \quad \textit{and} \quad \| R \| \le \lambda.$$

Proof. For any $y \in G$, the surjectivity of S ensures that there is an $x \in E$ such that $y = Sx$, hence we define R on G by setting

(3.2.3) $$Ry = Tx.$$

(Remember that $x \in E$ is such that $y = Sx$.) (3.2.1) implies that R is well–defined since Ker S $\subset$ Ker T. [Let $y = 0$. Then $0 = y = Sx$ implies that $x \in$ Ker S $\subset$ Ker T, hence $0 = Tx = Ry$.] Clearly R is linear and satisfies RS = T (by (3.2.3)). To prove the continuity of R, let $y \in G$ and let $x \in E$ be such that $y = Sx$. Then (3.2.1) shows that

$$\| Ry \| = \| Tx \| \le \lambda \| Sx \| = \lambda \| y \|,$$

so that $\| R \| \le \lambda$.

Finally, the uniqueness follows from the surjectivity of S.

For a dual result of (3.2), see Ex. 3–17.

(3.3) **Theorem** (Hahn–Banach). *Let* E *be a normed space and* M *a vector subspace of* E. *For any continuous linear functional* f_0 *on* M, *there is an* $f \in E'$ *with* $f(m) = f_0(m)$ *for all* $m \in M$ (called an *extension* of f_0) such that

$$\| f \| = \| f_0 \|_M (= \sup \{ |f_0(m)| : m \in M , \|m\| \leq 1 \}).$$

Proof. For any $x \in E$, let

(3.3.1) $$p(x) = \| f_0 \|_M \| x \|.$$

Then p is a norm on E such that

$$|f_0(m)| \leq \|f_0\|_M \|m\| = p(m) \quad \text{(for all } m \in M).$$

By the Hahn–Banach extension theorem, there exists an $f \in E^*$ extending f_0 such that

$$| f(x) | \leq p(x) \quad \text{(for all } x \in E).$$

It then follows from (3.3.1) that $\| f \| \leq \| f_0 \|_M$, and hence that $f \in E'$. On the other hand, if $m \in M$, then

$$|f_0(m)| = |f(m)| \leq \|f\| \|m\|,$$

hence $\|f_0\|_M \leq \|f\|$; consequently, $\|f\| = \|f_0\|_M$.

(3.4) **Corollary.** Let E be a normed space. For any $0 \neq x_0 \in E$, there exists an $f \in E'$ such that

(3.4.1) $$\| f \| = 1 \quad \text{and} \quad f(x_0) = \| x_0 \|.$$

Consequently,

(3.4.2) $$\| x \| = \sup \{ | g(x) | : g \in E', \| g \| = 1 \}$$
$$= \sup \{ | g(x) | : g \in E', \| g \| \leq 1 \}.$$

Proof. Let $M = \{ \lambda x_0 : \lambda \in K\}$ and define

$$f_0(\lambda x_0) = \lambda \, \| x_0 \| \quad (\text{for all } \lambda \in K).$$

Then M is a subspace of E and f_0 is a continuous linear functional on M with $\| f_0 \|_M = 1$ hence (3.4.1) follows from (3.3).

To prove (3.4.2), we first notice that it is trivial for $x = 0$, hence we assume that $x \neq 0$. As

$$|h(x)| \leq \|h\| \, \|x\| \quad (\text{for all } h \in E'),$$

it follows that

$$\sup \{ |g(x)| : g \in E', \|g\| = 1 \}$$

$$\leq \sup \{ |h(x)| : h \in E', \|h\| \leq 1 \} \leq \|x\|.$$

On the other hand, for any $0 \neq x \in E$, the first part of (3.4) shows that there exists an $f \in E'$ such that

$$\|f\| = 1 \text{ and } f(x) = \|x\|,$$

so that

$$\|x\| = f(x) \leq \sup \{ |g(x)| : g \in E', \|g\| = 1 \}.$$

Thus (3.4.2) holds.

Let E be a normed space and let E' be the Banach dual of E. Then E' is a Banach space under the norm

$$\|f\| = \sup \{ |f(x)| : x \in U_E \} \quad (\text{for all } f \in E').$$

The Banach dual of E', denoted by E'', is called the Banach bidual of E. We write K_E for the evaluate map from E into E'', defined by

$$\langle f, K_E x\rangle = \langle x, f\rangle = f(x) \quad (\text{for all } f \in E'),$$

then (3.4.2) shows that

$$\|K_E x\| = \|x\|.$$

A Banach space E is said to be reflexive if K_E is surjective.

From (3.c), any finite–dimensional Banach space must be reflexive. It can be shown (see (a) and (b) of (3.9) below) that there exists an infinite–dimensional Banach space which is not reflexive.

As an application of (3.4),we verify the following result which is very useful for studying approximation theory.

(3.5) **Theorem.** Let E be a normed space, let N be a vector subspace of E , and let $u_0 \in E$ be such that

(3.5.1) $$\rho = \text{dist}\,(u_0, N) = \inf\{\, \|u_0 - w\| : w \in N \,\} > 0.$$

Then there exists an $f \in E'$ such that

(i) $\|f\| = 1$;
(ii) $f(u_0) = \text{dist}\,(u_0,N)$;
(iii) $N \subset f^{-1}(0)$.

Proof. It is easily seen that

$$\text{dist}\,(x,N) = \text{dist}\,(x,\overline{N}) \quad (\text{for any } x \in E);$$

and that

$$\|\, Q_{\overline{N}}(x)\, \| = \text{dist}\,(x,\overline{N}) \quad (\text{for all } x \in E),$$

where $Q_{\overline{N}} : E \longrightarrow E/_{\overline{N}}$ is the quotient map. It then follows from (3.5.1) that $Q_{\overline{N}}(u_0) \neq 0(\overline{N})$ (since $\| Q_{\overline{N}}(u_0) \| = \text{dist}(u_0, \overline{N}) = \rho > 0$), and hence from (3.4) that there exists an $\tilde{f} \in (E/_{\overline{N}})'$ such that

$$\| \tilde{f} \| = 1 \text{ and } \tilde{f}(Q_{\overline{N}}(u_0)) = \| Q_{\overline{N}}(u_0) \| = \rho.$$

Thus $f = \tilde{f} \circ Q_{\overline{N}} \in E'$ has the required properties (since $N \subset \overline{N} = \text{Ker } Q_{\overline{N}}$ and $O_{E/_{\overline{N}}} = Q_{\overline{N}}(O_E)$).

(3.d) Let E be a Banach space.

(i) If $f \in E'$ is such that $\|f\| = 1$, then

$$\text{dist}(x, f^{-1}(o)) = |f(x)| \text{ (for any } x \in E).$$

(ii) E is reflexive if and only if for any proper closed vector subspace M of E there exists an $0 \neq u \in E$ such that

$$\text{dist}(u, M) = \|u\|.$$

[The necessity follows from (3.5) and (3.3), while the sufficiency follows from (i) and James′ theorem (see James [1964]), which states that a Banach space G is reflexive if and only if any $f \in G'$ attains its supremum on U_G.]

(3.e) Separability: A normed space E for which its Banach dual E' is separable must be separable; there exists a separable Banach space whose Banach dual is not separable. [Let $\{ f_n : n \geq 1 \}$ be a countable dense subset of E' and let $x_n \in U_E$ be such that $|f(x_n)| > \frac{1}{2} \|f_n\|$. Then

$$M = \{ \Sigma_{j=1}^{k} r_j a_j : a_j \in \{ x_n : n \geq 1 \} \text{ and } r_j \text{ are rational } \}$$

is countable, and use (3.5) to show that M is dense in E.]

Let E and F be normed spaces and $T \in L(E, F)$. We denote by T' the restriction on F' of the algebraic adjoint T^* of T; that is,

$$\langle x, T'y' \rangle = \langle x, T^* y' \rangle = \langle Tx, y' \rangle \quad \text{(for all } x \in E \text{ and } y' \in F'\text{)}.$$

It then follows that $T'y' \in E'$ is such that

$$\|T'y'\| \leq \|T\| \, \|y'\| ,$$

and hence that $T' \in L(F', E')$ and $\|T'\| \leq \|T\|$ (it can be shown that $\|T\| = \|T'\|$). T' is called the dual operator of T. The dual operator of T', denoted by T'', is referred to as the bidual operator of T. Clearly, $T'' : E'' \longrightarrow F''$ is an operator such that

$$T''K_E = K_F T.$$

Moreover, it is easily seen that

$$I_{E'} = (K_E)' K_{E'} .$$

Let E be a normed space. For any subset B of E, the set, defined by

$$B^{\perp} = \{ x' \in E' : \langle b, x' \rangle = 0 \text{ (for all } b \in B) \},$$

is called the annihilator of B. Dually, if $D \subset E'$ then the annihilator of D is defined by

$$D^{\top} = \{ x \in E : \langle x, d' \rangle = 0 \text{ (for all } d' \in D) \}.$$

(3.f) Dual operators and annihilators: Let E, F be normed spaces, let $T \in L(E, F)$, let $B \subset E$ and $D \subset E'$.

(i) $\|T'\| = \|T\|$

(ii) $B^{\perp}$ (resp. $D^{\top}$) is a closed vector subspace of E' (resp. E) ; moreover, $(B^{\perp})^{\top}$ is the smallest closed vector subspace of E containing B. [Using (3.5).]

As an application of (3.3). We verify the following:

(3.6) **Theorem.** Let $(E, \|.\|)$ be a normed space and M a closed vector subspace

<u>of</u> E. (The quotient norm on E/M (resp. $E'/M^{\perp}$) is still denoted by $\|.\|$).

(a) <u>The Hahn–Banach extension theorem</u> (see (3.3)) <u>extends each</u> $m' \in M'$ <u>to an</u> $x' \in E'$, <u>hence we define</u>

$$\psi(m') = x' + M^{\perp} \tag{3.6.1}$$

<u>Then</u> $M' \equiv E'/M^{\perp}$ <u>under the metric isomorphism</u> (i.e., <u>isometry</u>) ψ.

(b) <u>Let</u> $Q_M : E \longrightarrow E/M$ <u>be the quotient map, and let</u> $Q'_M : (E/M)' \longrightarrow E'$ <u>be its dual map</u>. <u>Then</u> $(E/M)' \equiv M^{\perp}$ <u>under the metric isomorphism</u> Q'_M.

<u>Proof.</u> (a) If x' and x_1', in E', are extensions of $m' \in M'$, then $x' - x_1' \in M^{\perp}$, hence $x' + M^{\perp} = x_1' + M^{\perp}$, thus ψ is well–defined. As the restriction of every $x' \in E'$ to M is continuous on M, it follows that $\psi : M' \longrightarrow E'/M^{\perp}$ is surjective. It is easy to check that ψ is linear, it remains to show that ψ is a metric isomorphism. To do this, let $m' \in M'$. By (3.3), m' has an extension $x' \in E'$ with

$$\|x'\| = \|m'\|_M = \sup \{ |\langle m,m'\rangle| : m \in U_M \}. \tag{3.6.2}$$

Thus $x' + M^{\perp}$ is the set consisting of all (continuous and linear) extensions of m'. Clearly if $\omega' \in E'$ is an extension of m', then $\|m'\|_M \leq \|\omega'\|$, hence

$$\|m'\|_M \leq \inf \{ \|x' + u'\| : u' \in M^{\perp} \} = \| x' + M^{\perp} \| \tag{3.6.3}$$

(by the definition of quotient norm and $x' + M^{\perp}$). On the other hand,

$$\|x' + M^{\perp}\| = \inf \{ \|x' + u'\| : u' \in M^{\perp} \} \leq \|x'\|.$$

We conclude from (3.6.2) that

$$\|x' + M^{\perp}\| \leq \|x'\| = \|m'\|_M ,$$

and hence from (3.6.3) that

$$\|m'\|_M = \|x' + M^{\perp}\| = \|\psi(m')\| .$$

(b) The continuity of Q_M implies $Q_M' \in L((E/M)', E')$. We first claim that Q_M' is an isometry (into), that is

$$\|Q_M'(y')\| = \|y'\| \quad \text{(for all } y' \in (E/M)'). \tag{3.6.4}$$

In fact, we first notice from (2.6) that

$$O_{E/M} = Q_M(O_E) \subset Q_M(U_E) \subset U_{E/M} .$$

Now for any $y' \in (E/M)'$, we have

$$\begin{aligned} \|Q_M'(y')\| &= \sup_{x \in O_E} |\langle x, Q_M'(y')\rangle| = \sup_{x \in O_E} |\langle Q_M x, y'\rangle| \\ &= \sup_{\hat{x} \in O_{E/M}} |\langle \hat{x}, y'\rangle| = \|y'\| \end{aligned}$$

which proves our assertion .

We complete the proof by showing that

$$Q_M'((E/M)') = M^{\perp} . \tag{3.6.5}$$

Indeed, let $u' \in M^{\perp}$. Then we have the following operators

$$\begin{array}{ccc} E & \xrightarrow{\;u'\;} & K \\ {\scriptstyle Q_M}\downarrow & & \\ E/M & & \end{array}$$

such that Q_M is surjective and

$$\text{Ker } Q_M = M \subset \text{Ker } u'.$$

Hence, there exists a linear functional y' on E/M such that

$$y' \circ Q_M = u' .$$

As $O_{E/M} = Q_M(O_E)$, we obtain

$$\|u'\| = \sup\{ |\langle x,u'\rangle| : x \in O_E \} = \sup \{ |\langle x, y' \circ Q_M\rangle| : x \in O_E \}$$

$$= \sup \{ |\langle Q_M(x), y'\rangle| : x \in O_E \} = \|y'\| ,$$

hence $y' \in (E/M)'$ is such that $u' = y' \circ Q_M = Q_M'(y')$, thus (3.6.5) holds.

Combining (3.6.4) and (3.6.5), the result follows.

It is known (see (3.c)) that every finite–dimensional Banach space is reflexive, and that there are infinite dimensional Banach spaces which are, in general, not reflexive. However, every point in E'' is related to a point in E by means of finite dimensional subspace of E' as shown by the following important and useful result:

(3.7) **Theorem** (Helley's selection theorem). Let E be a normed space, let $\Psi'' \in E''$ and let N be a finite–dimensional vector subspace of E' . For any $\epsilon > 0$ there exists an $u_0 \in E$ such that

$$\|u_0\| < \|\Psi''\| + \epsilon \quad \text{and} \quad K_E u_0 = \Psi'' \quad \text{on } N.$$

Proof. The annihilator of N (in E) i.e.,

$$M = N^{\top} = \{ x \in E : \langle x, n'\rangle = 0 \ \ (\text{for all } n' \in N) \}$$

is a closed vector subspace of E such that

$$M^{\perp}=(N^{\top})^{\perp}=N$$

(since $\dim N < \infty$ and surely N is closed in E'). Let

$$M^{\perp\perp} = N^{\perp} = \{\, x'' \in E'' : \langle n', x''\rangle = 0 \ \ (\text{for all } \ n' \in N\ ) \,\} \tag{3.7.1}$$

(i.e., the annihilator of N in E"). Then (3.6) shows that

$$(E/M)' \equiv M^{\perp} \equiv N \ \text{ and } \ N' \equiv E''/M^{\perp\perp}, \tag{3.7.2}$$

so that

$$(E/M)'' \equiv N' \equiv E''/M^{\perp\perp}\,. \tag{3.7.3}$$

It then follows that the following diagram commutes

$$\begin{array}{ccc} E & \xrightarrow{K_E} & E'' \\ {\scriptstyle Q_M}\downarrow & & \downarrow{\scriptstyle Q_{M^{\perp\perp}}} \\ E/M & \xrightarrow[K_{E/M}]{} & E''/M^{\perp\perp} \end{array} \tag{3.7.4}$$

[For any $x \in E$, since $N' \equiv E''/_{M^{\perp\perp}}$ (by (3.7.2)), it is required to show that $\langle n', (Q_{M^{\perp\perp}} \circ K_E)x\rangle = \langle n', (K_{E/M} \circ Q_M)x\rangle$ (for all $n' \in N$). Indeed, we have from (3.7.1) that

$\langle n', Q_{M^{\perp\perp}}(K_E x)\rangle = \langle n', K_E x + M^{\perp\perp}\rangle = \langle n', K_E x\rangle = \langle x, n'\rangle$, and $\langle n', (K_{E/M} \circ Q_M)x\rangle = \langle Q_M x, n'\rangle = \langle x + M, n'\rangle = \langle x, n'\rangle$ (since $M = N^{\top}$), thus (3.7.4) holds.] Moreover, we claim that $K_{E/M}$ is surjective, hence

$$E/M \equiv E''/_{M^{\perp\perp}} \ \text{ under } \ K_{E/M}\,. \tag{3.7.5}$$

In fact,since $\dim N < \infty$, it follows from (3.7.2) that $\dim E/M < \infty$, and hence that

E/M is reflexive, thus $K_{E/M}$ is surjective (since $K_{E/M} : E/M \longrightarrow (E/M)'' \equiv E''/_{M^{\perp\perp}}$ is the evaluation map). This proves our assertion.

Now, as $Q_{M^{\perp\perp}}(\Psi'') \in E''/_{M^{\perp\perp}}$ (notice that $\Psi'' \in E''$), it follows from the surjectivity of $K_{E/M} \circ Q_M$ (see (3.7.5)) that there is an $x_0 \in E$ such that

$$Q_{M^{\perp\perp}}(\Psi'') = K_{E/M}(Q_M x_0) , \tag{3.7.6}$$

and hence from $\|K_{E/M} z\| = \|z\|$ (for all $z \in E/M$) and $\|Q_{M^{\perp\perp}}\| \le 1$ that

$$\|Q_M(x_0)\| = \|K_{E/M}(Q_M x_0)\| = \|Q_{M^{\perp\perp}}(\Psi'')\| \le \|\Psi''\|. \tag{3.7.7}$$

By the definition of quotient norms, there is an $u_0 \in E$ such that

$$Q_M(u_0) = Q_M(x_0) \text{ and } \|u_0\| < \|Q_M(x_0)\| + \epsilon .$$

It then follows from (3.7.6) and (3.7.7) that

$$K_{E/M}(Q_M u_0) = K_{E/M}(Q_M x_0) = Q_{M^{\perp\perp}}(\Psi'') \text{ and} \tag{3.7.8}$$

$$\|u_0\| < \|Q_M(x_0)\| + \epsilon < \|\Psi''\| + \epsilon ,$$

and hence from $K_{E/M} \circ Q_M = Q_{M^{\perp\perp}} \circ K_E$ (see (3.7.4)) that $Q_{M^{\perp\perp}}(\Psi'' - K_E u_0) = 0$, so that

$$\Psi'' - K_E u_0 \in \text{Ker } Q_{M^{\perp\perp}} = M^{\perp\perp} = N^{\perp}$$

or, equivalently

$$<n', \Psi''> = <n', K_E u_0> \text{ (for all } n' \in N). \tag{3.7.9}$$

From (3.7.8) and (3.7.9), we obtain our assertions.

Some author also call the following result as Helley's selection theorem.

(3.7)* **Theorem** (Helley's selection theorem). Let E be a normed space and $\lambda > 0$. Suppose that $\{x_1', \cdots, x_n'\} \subset E'$ and $\{r_1, \cdots, r_n\} \subset K$ are given and satisfy

(1) $$|\Sigma_{i=1}^n r_i \mu_i| \le \lambda \|\Sigma_{i=1}^n \mu_i x_i'\| \quad \text{for all choices } \{\mu_1, \cdots, \mu_n\} \subset K.$$

Then for any $\epsilon > 0$, there exists a $u_0 \in E$ such that

$$\|u_0\| < \lambda + \epsilon \text{ and } \langle u_0, x_i'\rangle = r_i \quad (i = 1, \cdots, n).$$

Proof. We employ (3.7) to verify this theorem. Clearly

$$N = \langle\{x_1', \cdots, x_n'\}\rangle$$

is a finite-dimensional vector subspace of E'. Let us define φ on N by

(2) $$\varphi(\Sigma_{i=1}^n \beta_i x_i') = \Sigma_{i=1}^n \beta_i r_i \quad (\text{for all } \beta_i \in K).$$

Then φ is clearly a linear functional on N which is continuous since

(3) $$|\varphi(\Sigma_{i=1}^n \beta_i x_i')| = |\Sigma_{i=1}^n \beta_i r_i| \le \lambda \|\Sigma_{i=1}^n \beta_i x_i'\|$$

(by (1)), hence (3.3) ensures that there exists $\Psi'' \in E''$ such that

(4) $$\Psi'' = \varphi \text{ (on N) and } \|\Psi''\| = \|\varphi\|_N .$$

By (3.7), there is a $u_0 \in E$ such that

$$\|u_0\| < \|\Psi''\| + \epsilon \text{ and } K_E u_0 = \Psi'' \text{ on N.}$$

By (3) and (4)

$$\|u_0\| < \|\Psi''\| + \epsilon = \|\varphi\|_N + \epsilon \leq \lambda + \epsilon ,$$

also we obtain from (2) that

$$<u_0, x_i'> = <x_i', K_E u_0> = <x_i', \Psi''> = <x_i', \varphi> = r_i$$

(for all $i = 1,\cdots,n$).

<u>**Remark**</u>. (3.7) can be proved by using $(3.7)^*$.

In $(3.7)^*$, we consider the problem of the existence of $u_0 \in E$ such that

$$<u_0, x_i'> = r_i \quad (\text{for all } i = 1, \cdots,n).$$

The dual result is given by the following:

(3.8) **Theorem** (Riesz). <u>Let</u> E <u>be a normed space and let</u>

$$\{ x_i : i \in \Lambda \} \subset E \quad \underline{\text{and}} \quad \{ r_i : i \in \Lambda \} \subset K.$$

<u>Then the following two statements are equivalent</u>.

(a) <u>There exists a</u> $u' \in E'$ <u>such that</u>

(3.8.1) $$< x_i , u' > = r_i \quad (\text{for all } i \in \Lambda).$$

(b) <u>There exists a</u> $\lambda > 0$ <u>such that</u>

(3.8.2) $$| \Sigma_{i\in\Lambda} \mu_i r_i | \leq \lambda \| \Sigma_{i\in\Lambda} \mu_i x_i \|$$

<u>for any family</u> $\{ \mu_i : i \in \Lambda \}$ <u>in</u> K <u>with only finitely many non–zero</u> μ_i .

Proof. (a) ⇒ (b) : If $u' \in E'$ is such that (3.8.1) holds, then for any family $\{ \mu_i : i \in \Lambda \}$ in K with only finitely many non–zero μ_i, we have

$$|\Sigma_{i\in\Lambda} \mu_i r_i| = |\Sigma_{i\in\Lambda} \mu_i <x_i, u'>| = |<\Sigma_{i\in\Lambda} \mu_i x_i, u'>|$$

$$\leq \|u'\| \, \|\Sigma_{i\in\Lambda} \mu_i x_i\| ,$$

thus (b) follows if we take $\lambda = \|u'\|$.

(b) ⇒ (a) : Suppose that $M = <\{ x_i : i \in \Lambda \}>$, and let us define h on M by setting

(3.8.3) $$h(\Sigma_{i\in\Lambda} \mu_i x_i) = \Sigma_{i\in\Lambda} \mu_i r_i \quad \text{for all } \{ \mu_i : i \in \Lambda \} \subseteq K.$$

Then h is well–defined [if $\Sigma_{i\in\Lambda} \mu_i x_i = 0$ then $\Sigma_{i\in\Lambda} \mu_i r_i = 0$ (by (3.8.2)), hence $h(\Sigma_{i\in\Lambda} \mu_i x_i) = 0$ by (3.8.3)], linear and continuous with

(3.8.4) $$|h(\Sigma_{i\in\Lambda} \mu_i x_i)| = |\Sigma_{i\in\Lambda} \mu_i r_i| \leq \lambda \| \Sigma_{i\in\Lambda} \mu_i x_i \| .$$

By (3.3), there exists a $u' \in E'$ such that

$$\| u' \| = \| h \|_M \leq \lambda$$

and $u' = h$ on M, in particular, we obtain from (3.8.3) that

$$<x_i, u'> = h(x_i) = r_i \quad (\text{for all } i \in \Lambda).$$

(3.9) **Examples.** (a) $\ell^\infty \equiv (\ell^1)'$; more precisely, every $f \in (\ell^1)'$ can be represented as the form

(a.1) $$<[\xi_i], f> = \Sigma_{n=1}^{\infty} \xi_n \eta_n \qquad \text{for all } \xi = [\xi_i] \in \ell^1,$$

where $\eta = [\ \eta_n\ ] \in \ell^\infty$ and

(a.2) $$\|f\| = \|\eta\|_\infty.$$

Proof. Let $\eta = [\eta_n] \in \ell^\infty$. For any $\xi = [\zeta_i] \in \ell^1$,

(a.3) $$|\Sigma_{n=1}^\infty \xi_n\eta_n| \leq \Sigma_{n=1}^\infty |\xi_n\eta_n| \leq (\sup_n|\eta_n|)\ (\Sigma_{n=1}^\infty|\xi_n|) < \infty,$$

hence (a.1) defines a functional on ℓ^1, which is obviously linear; moreover, we see from (a.3) that $f \in (\ell^1)'$ and

(a.4) $$\|f\| \leq \sup_n |\eta_n| = \|\eta\|_\infty .$$

Conversely, let $f \in (\ell^1)'$ and

$$<e^{(n)}, f> = \eta_n \qquad \text{(for all } n \geq 1)$$

where $\{ e^{(i)} : i \geq 1 \}$ is a Schauder basis for ℓ^1. Then

$$|\eta_n| = |<e^{(n)}, f>| \leq \|f\|\ \|e^{(n)}\|_1 = \|f\| \qquad \text{(for all } n \geq 1)$$

hence $\eta = [\eta_n] \in \ell^\infty$ is such that

(a.5) $$\|\eta\|_\infty \leq \|f\|.$$

Moreover, for any $\xi = [\xi_i] \in \ell^1$, since $\xi = \|.\|\text{-}\lim_n \Sigma_{i=1}^n \xi_i e^{(i)}$, it follows from the continuity of f that

$$<\xi, f> = \lim_n \Sigma_{i=1}^n \xi_i < e^{(i)}, f > = \Sigma_{i=1}^\infty \xi_i\eta_i ,$$

that is, (a.1) holds. Combining (a.4) and (a.5), we obtain (a.2).

(b) $\ell^1 \equiv c_0'$; more precisely, every $f \in c_0'$ can be represented as the form (b.1) $$<[\zeta_i], f> = \Sigma_{i=1}^{\infty} \zeta_i \eta_i \quad (\text{ for all } \zeta = [\zeta_i] \in c_0),$$

where $\eta = [\eta_i] \in \ell^1$ and

(b.2) $$\|f\| = \|\eta\|_1 = \Sigma_{n=1}^{\infty} |\eta_n|.$$

Proof. Let $\eta = [\eta_n] \in \ell^1$. For any $\zeta = [\zeta_n] \in c_0$,

(b.3) $$|\Sigma_{n=1}^{\infty} \zeta_n \eta_n| \leq \Sigma_{n=1}^{\infty} |\eta_n||\zeta_n| \leq (\sup_{j\geq 1} |\zeta_j|)(\Sigma_{n=1}^{\infty} |\eta_n|) < \infty,$$

hence (b.1) defines a functional on c_0, which is linear and continuous (by(b.3)),hence $f \in c_0'$ and

(b.4) $$\|f\| \leq \|\eta\|_1 = \Sigma_{n=1}^{\infty} |\eta_n| .$$

Conversely, let $f \in c_0'$ and

$$\eta_n = <e^{(n)}, f> \text{ for all } n \geq 1 ,$$

where $\{ e^{(n)} : n > 1 \}$ is a Schauder basis for c_0. For any $\zeta = [\zeta_n] \in c_0$, since $\xi = \|.\|_\infty\text{-}\lim_n \Sigma_{i=1}^{n} \xi_i e^{(i)}$, it follows from the continuity of f that

$$<\xi, f> = \lim_n \Sigma_{i=1}^{n} \xi_i <e^{(i)}, f> = \Sigma_{i=1}^{\infty} \xi_i \eta_i ,$$

namely, (b.1) holds. To prove that

(b.5) $$\eta = [\eta_n] \in \ell^1 \text{ and } \|\eta\|_1 \leq \|f\| ,$$

let us define x_j by setting

$$x_j = \begin{cases} \dfrac{|\eta_j|}{\eta_j} & \text{if } \eta_j \neq 0 \\ 0 & \text{if } \eta_j = 0. \end{cases}$$

Then $x^{(n)} = \Sigma_{i=1}^{n} x_i e^{(i)} \in c_0$ is such that $\|x^{(n)}\|_\infty = 1$ and

$$<x^{(n)}, f> = \Sigma_{i=1}^{n} x_i < e^{(i)}, f> = \Sigma_{i=1}^{n} x_i \eta_i = \Sigma_{i=1}^{n} |\eta_i| \leq \|f\| .$$

This shows that $\Sigma_j |\eta_j|$ converges and that $\Sigma_{i=1}^{\infty} |\eta_i| \leq \|f\|$, thus (b.5) holds.

Now we are going to study the Banach dual of c, we also find that $c' \equiv \ell^1$ but the representation is quite different from (b.1).

(c) $\ell^1 \equiv c'$; more precisely, every $f \in c'$ can be represented as the form

(c.1) $$<[\xi_i], f> = \mu_1(\lim_n \xi_n) + \Sigma_{i=2}^{\infty} \mu_i \xi_{i-1} \text{ (for all } \xi = [\xi_i] \in c),$$

where $\mu = [\mu_i]_{i \geq 1} \in \ell^1$ and

(c.2) $$\|f\| = \|\mu\|_1 = \Sigma_{i=1}^{\infty} |\mu_i|.$$

Proof. Suppose that $\mu = [\mu_i] \in \ell^1$. For any $\xi = [\xi_i] \in c$, since $|\lim_n \xi_n| \leq \sup_n |\xi_n|$, it follows that

$$\begin{aligned} & |\mu_1(\lim_n \xi_n) + \Sigma_{i=2}^{\infty} \mu_i \xi_{i-1}| \\ \leq\ & |\mu_1| |\lim_n \xi_n| + |\Sigma_{i=2}^{\infty} \mu_i \xi_{i-1}| \\ \leq\ & \sup_{i \geq 1} |\xi_i| (\Sigma_{n=1}^{\infty} |\mu_n|) = \|\xi\|_\infty \|\mu\|_1 < \infty, \end{aligned}$$

and hence that (c.1) defines a continuous linear functional on c with

(c.3) $$\|f\| \le \|\mu\|_1 .$$

To prove the converse, we first notice that every $\zeta = [\zeta_i]_{i \ge 1} \in c$ can be uniquely represented in the form

(c.4) $$\zeta = e(\lim_n \zeta_n) + \Sigma_{i=1}^{\infty}(\zeta_i - \lim_n \zeta_n)e^{(i)}$$

where $[\zeta_i - \lim_n \zeta_n]_{i \ge 1} \in c_0$. Now let $f \in c'$. Then (c.4) implies that

(c.5) $$\begin{aligned} <\zeta, f> &= (\lim_n \zeta_n)f(e) + \Sigma_{i=1}^{\infty} (\zeta_i - \lim_n \zeta_n) f(e^{(i)}) \\ &= (\lim_n \zeta_n)(f(e) - \Sigma_{i=1}^{\infty} f(e^{(i)})) + \Sigma_{i=1}^{\infty} f(e^{(i)})\zeta_i. \end{aligned}$$

Hence we assume naturally that

(c.6) $$\mu_1 = f(e) - \Sigma_{i=1}^{\infty} f(e^{(i)}) \text{ and } \mu_{i+1} = f(e^{(i)}) \quad \text{(for all } i \ge 1).$$

Then (c.5) becomes

$$<\zeta, f> = \mu_1(\lim_n \zeta_n) + \Sigma_{i=1}^{\infty} \zeta_i\mu_{i+1} \quad \text{(for all } \zeta = [\zeta_n] \in c).$$

We complete the proof by showing (see(c.3)) that

(c.7) $$\mu = [\mu_n] \in \ell^1 \text{ and } \|\mu\|_1 \le \|f\|.$$

Indeed, let us define

$$x_n = \begin{cases} \dfrac{|\mu_{n+1}|}{\mu_{n+1}} & \text{if } \mu_{n+1} \ne 0 \\ 1 & \text{if } \mu_{n+1} = 0 \end{cases} \quad \text{(for all } n \ge 1)$$

and

$$\epsilon_0 = \begin{cases} \dfrac{|\mu_1|}{\mu_1} & \text{if } \mu_1 \neq 0 \\ 1 & \text{if } \mu_1 = 0 \end{cases}$$

Then the sequence, defined by

$$x^{(n)} = (x_1, \cdots, x_n, \epsilon_0, \epsilon_0, \cdots),$$

belongs to c and satisfies

$$\|x^{(n)}\|_\infty = 1 \text{ and } \lim_n x_n = \epsilon_0 \text{ (where } x_{n+i} = \epsilon_0) \text{ (for all } i \geq 1).$$

By (c.5) and (c.6), we have

$$<x^{(n)}, f> = \epsilon_0\mu_1 + \Sigma_{i=1}^{n} \mu_{i+1} x_i + \Sigma_{i=n+1}^{\infty} \mu_{i+1} x_i$$

$$= \epsilon_0\mu_1 + \Sigma_{i=1}^{n} |\mu_{i+1}| + \Sigma_{i=n+1}^{\infty} \mu_{i+1}\epsilon_0$$

hence

$$\|f\| \geq |<x^{(n)}, f>|$$

$$\geq \left|\epsilon_0\mu_1 + \Sigma_{i=1}^{n} |\mu_{i+1}|\right| - |\Sigma_{i=n+1}^{\infty} \mu_{i+1}\epsilon_0|$$

$$\geq \Sigma_{i=1}^{n+1} |\mu_i| - \Sigma_{i=n+1}^{\infty} |\mu_{i+1}| \quad \text{(since } |\epsilon_0| = 1).$$

Letting $n \longrightarrow \infty$, we obtain $\Sigma_{i=1}^{\infty} |\mu_i| \leq \|f\|$, which proves our assertion (c.7).

(d) For $1 < p < \infty$, $(\ell^p)' \equiv \ell^q$ (where $\frac{1}{p} + \frac{1}{q} = 1$); more precisely, every $f \in (\ell^p)'$ can be represented in the form

(d.1) $<[\zeta_i], f> = \Sigma_{i=1}^{\infty} \zeta_i \eta_i$ (for all $\zeta = [\zeta_i] \in \ell^p$),

where $\eta = [\eta_i] \in \ell^q$ and

(d.2) $$\|f\| = \|\eta\|_q = (\Sigma_{i=1}^{\infty} |\eta_i|^q)^{\frac{1}{q}} .$$

Proof. The argument is very similar to that used in the proof of Examples (a) and (b), hence will be omitted.

For the Banach duals of $L^p[a,b]$, we mention the following results, whose proofs can be found in the book of Royden [1968, P.117–118, P.121–123].

(e) For $1 \leq p < \infty$, $(L^p[a,b])' \equiv L^q[a,b]$, (where $\frac{1}{p} + \frac{1}{q} = 1$); more precisely, every $f \in (L^p[a,b])'$ can be represented in the form

(e.1) $$\langle x, f\rangle = \int_a^b x(t)y(t)\, dm(t) \quad (\text{for all } x(\cdot) \in L^p[a,b]),$$

where $y(\cdot) \in L^q[a,b]$ is such that

(e.2) $$\|f\| = \|y\|_q = \begin{cases} (\int_a^b |y(t)|^q dm(t)\ )^{\frac{1}{q}} & \text{if } p \neq 1 \\ \operatorname{ess\,sup}_{[a,b]} |y(\cdot)| & \text{if } p = 1 . \end{cases}$$

The proof of the following important result, due to Riesz, can be found in the book of Rudin [1966, p.40 and p.130].

(f) Let Ω be a compact Hausdorff space and let $M(\Omega)$ be the set of all complex regular Borel measures (called complex Radon measures) on Ω . Then $M(\Omega)$ becomes a Banach space under the norm

(f.1) $$\|\mu\| = |\mu|(\Omega) = \sup \{ \Sigma_{i=1}^{n} |\mu(B_i)| : B_i \text{ are Borel sets with } B_i \cap B_j = \phi\ (i \neq j), \Omega = \bigcup_{i=1}^{n} B_i \}.$$

Moreover, $M(\Omega) \equiv (C(\Omega), \|\cdot\|_\infty)'$; namely, every $\Psi \in (C(\Omega), \|\cdot\|_\infty)'$ can be represented in the form

(f.2) $$\langle x,\Psi\rangle = \int_{\Omega} x(\cdot)\, d\mu_{\Psi} \quad \text{(for all } x(\cdot) \in C(\Omega)),$$

where $\mu_{\Psi} \in M(\Omega)$ is such that

(f.3) $$\|\Psi\| = \|\mu_{\Psi}\| = |\mu_{\Psi}|(\Omega).$$

By a similar argument given in the proof of (a),(b) and (d) of (3.9), one can verify the following more general result:

(3.g) <u>The spaces</u> $\ell^p(\Lambda)$ (II) : For a non–empty set Λ , we have

$$(\ell^1(\Lambda))' = \ell^\infty(\Lambda), \ (c_0(\Lambda))' \equiv \ell^1(\Lambda) \text{ and}$$

$$(\ell^p(\Lambda))' \equiv \ell^q(\Lambda), \text{ where } \frac{1}{p} + \frac{1}{q} = 1.$$

(3.h) <u>Some special properties of families in Banach space</u>: Let E be a Banach space, let Λ be a non–empty index set, and let $[x_i,\Lambda]$ be a family in E (i.e., $x_i \in E$ for all $i \in \Lambda$) with the index set Λ. We say that $[x_i,\Lambda]$ is:

(A) <u>weakly summable</u> if $[\langle x_i,u'\rangle,\Lambda] \in \ell^1(\Lambda)$ (for any $u' \in E'$);

(B) <u>summable</u> if the net $\{ \Sigma_{i\in\alpha} x_i, \alpha \in \mathcal{F}(\Lambda) \}$ converges in E; in this case, its limit (in E) is called the <u>unordered sum</u> of $[x_i,\Lambda]$, and denoted by $\Sigma_\Lambda x_i$ or $\Sigma_i x_i$;

(C) <u>absolutely summable</u> if $[\|x_i\|,\Lambda] \in \ell^1(\Lambda)$.

A family $[x_i',\Lambda]$ in E' is said to be weak* summable if

$$[\langle u,x_i'\rangle,\Lambda] \in \ell^1(\Lambda) \ \text{(for any } u \in E).$$

In particular, if $\Lambda = \mathbb{N}$ and if a sequence $\{x_n\}$ in E is weakly summable (resp. summable, absolutely summable), then the formal series $\sum_n x_n$ is said to be <u>weakly</u>

unconditionally convergent (resp. unconditionally convergent, absolutely convergent). If a sequence $\{x'_n\}$ in E' is weak* summable, then the formal series $\sum_n x'_n$ is said to be weak* unconditionally convergent.

(i) If $[x_i,\Lambda]$ is a weakly summable family in E and $[\lambda_i, \Lambda] \in c_0(\Lambda)$, then the family $[\lambda_i x_i, \Lambda]$ in E is summable ([see Pietsch [1972, p.26]).

(ii) Every absolutely summable family in E is summable (by the completeness of E (compared with (2.b) (iii)). However,as the famous theorem of Dvoretzky and Rogers says, the converse implication holds only for finite–dimensional Banach spaces (see Day [1973], Grothendieck [1956] or De Grande–De Kimpe [1977]).

(iii) Every summable family in a Banach space E contains at most countably many non–zero terms (compare with (2.e) (ii)) .

(iv) The completeness of E ensures that the formal series $\sum_n x_n$ (where $x_n \in E$) is unconditionally convergent if and only if it is bounded–multiplier convergent in the sense that for any $[\zeta_n] \in \ell^\infty$, $\lim_n \sum_{i=1}^n \zeta_i x_i$ exists in E (see Day [1973,p.78] or Lindenstrass /Tizafriri [1977, p.15]).

(vi) A sequence $\{x_n\}$ in E is absolutely summable if and only if $\sum_n \|x_n\| < \infty$. [By using (2.e) (i).]

(vii) If $[x_i,\Lambda]$ is a weakly summable family in E then the set $\{ \sum_{i\in\alpha} x_i : \alpha \in \mathcal{F}(\Lambda) \}$ is bounded; consequently,

$$\| [x_i,\Lambda] \|_\epsilon = \sup \{ \Sigma_\Lambda |<x_i,u'>| : u' \in U_{E'} \} < \infty.$$

Similarly, if $[x'_i, \Lambda]$ is a weak* summable family in E',then

$$\sup \{ \Sigma_\Lambda |<u, x_i'>| : u \in U_E \} < \infty.$$

(3.j) Operators in (or from) concrete Banach spaces (I): Let E and F be Banach spaces. If T is a linear map from E into either $\ell^1(\Lambda)$ or $c_0(\Lambda)$ or $\ell^\infty(\Lambda)$, then we denote the i–th coordinate of Tx by $(Tx)_i$, hence

$$(Tx)_i = <Tx, e^{(i)}> \quad \text{and} \quad Tx = [(Tx)_i,\Lambda] \quad (x \in E);$$

in particular, if $Tx \in \ell^1(\Lambda)$ (or $c_0(\Lambda)$), then

$$Tx = \Sigma_\Lambda (Tx)_i e^{(i)} \quad \text{in } \ell^1(\Lambda) \quad (\text{or } c_0(\Lambda)).$$

(i) A linear map $T : E \longrightarrow \ell^1(\Lambda)$ is continuous if and only if there exists a weak* summable family $[f_i,\Lambda]$ in E' such that

$$Tx = [<x,f_i>,\Lambda] = \Sigma_\Lambda <x,f_i> e^{(i)} \quad (\text{in } \ell^1(\Lambda)) \quad (\text{for all } x \in E).$$

In this case, we have $f_i = T'e^{(i)}$ $(i \in \Lambda)$ and

$$\|T\| = \sup \{ \Sigma_\Lambda |<x, f_i>| : x \in U_E \}.$$

(ii) A linear map $T : E \longrightarrow c_0(\Lambda)$ is continuous if and only if there exists a bounded, weak* null family $[g_i,\Lambda]$ in E' (i.e., for any $x \in E$, $[<x, g_i>, \Lambda] \in c_0(\Lambda)$) such that

$$Tx = [<x, g_i>,\Lambda] = \Sigma_\Lambda <x, g_i> e^{(i)} \quad (\text{in } c_0(\Lambda)) \quad (\text{for all } x \in E).$$

In this case, we have $g_i = T'e^{(i)}$ $(i \in \Lambda)$ and

$$\|T\| = \sup_{i\in\Lambda} \|g_i\|.$$ (Dasor [1976,p.106] for the case $\Lambda = \mathbb{N}$.)

(iii) A linear map $T : E \longrightarrow \ell^\infty(\Lambda)$ is continuous if and only if there exists a bounded family $[h_i, \Lambda]$ in E' such that

$$Tx = [<x, h_i>,\Lambda] \quad (\text{for all } x \in E).$$

In the case, we have $h_i = T'e^{(i)}$ $(i \in \Lambda)$ and

$$\|T\| = \sup_{i \in \Lambda} \|h_i\|. \qquad \text{(Pietsch [1980, p.337])}$$

(i)′ A linear map $S : \ell^1(\Lambda) \longrightarrow F$ is continuous if and only if there exists a bounded family $[y_i, \Lambda]$ in F such that

$$S([\zeta_i,\Lambda]) = \Sigma_\Lambda \zeta_i y_i \quad \text{(for all } [\zeta_i, \Lambda] \in \ell^1(\Lambda)).$$

In this case, we have $y_i = Se^{(i)}$ $(i \in \Lambda)$ and

$$\|S\| = \sup_{i \in \Lambda} \|y_i\|. \quad \text{(Köthe[1969, p.280],Pietsch[1980,p.34].)}$$

(ii)′ A linear map $S : c_0(\Lambda) \longrightarrow F$ is continuous if and only if there exists a weakly summable family $[y_i, \Lambda]$ in F such that

$$S([\lambda_i, \Lambda]) = \Sigma_\Lambda \lambda_i y_i \qquad \text{(for all } [\lambda_i, \Lambda] \in c_0(\Lambda)).$$

In this case, we have $y_i = Se^{(i)}$ $(i \in \Lambda)$ and

$$\|S\| = \sup \{ \Sigma_\Lambda |\langle y_i, g\rangle| : g \in U_{F'} \} \; (= \|[y_i,\Lambda]\|_\epsilon).$$

(Grothendieck[1956], Dazor [1976,pp.106–107] for the case $\Lambda = \mathbb{N}$.)

(iii)′ For any unconditionally convergent series $\Sigma_n y_n$ in F, the map, defined by

$$S([\mu_n]) = \Sigma_{i=1}^{\infty} \mu_n y_n \quad \text{(for all } [\mu_n] \in \ell^\infty),$$

is a continuous linear map from ℓ^∞ into F with $y_n = Se^{(n)}$ $(n \geq 1)$ such that

$$\|S\| \leq \sup \{ \Sigma_{n=1}^{\infty} |\langle y_n,g\rangle| : g \in U_{F'} \}.$$

As a consequence of (3.j) and Hahn–Banach's extension theorem (see (3.4)), one can verify that $\ell^\infty(\Lambda)$ (resp. $\ell^1(\Lambda)$) has the metric extension (resp. metric lifting) property, as defined in §6 (see(6.f)).

(3.k) Linear functionals (II): Let E be a real normed space, let $f, g \in E'$ be such that $\|f\| = \|g\| = 1$ and $\epsilon > 0$. If

$$|g(x)| \leq \frac{\epsilon}{2}\|x\| \quad \text{(for all } x \in f^{-1}(0)),$$

then one can use (3.3) to conclude that

$$\text{either } \|f - g\| \leq \epsilon \quad \text{or} \quad \|f + g\| \leq \epsilon.$$

[The preceding result is useful for proving Bishop–Phelps' supporting theorem, see Bishop/Phelps [1963].]

(3.m) Riesz's representation theorem for continuous linear functionals on Hilbert spaces. Let H be a Hilbert space with the topological dual H'. For any $\psi \in H'$, there exists a unique $y \in H$ such that

$$\psi(x) = [x,y] \quad \text{(for all } x \in H).$$

Moreover, the map $\phi : y \longmapsto \psi$ is a conjugate–linear isometry from H onto H' in the following sense:

$$\phi(\mu_1 y_1 + \mu_2 y_2) = \bar{\mu}_1\phi(y_1) + \bar{\mu}_2\phi(y_2) \quad \text{and} \quad \|y\| = \|\phi(y)\|.$$

Exercises

3–1. Prove (3.a).

3–2. Let E and F be normed spaces and $T \in L(E,F)$.

(a) Show that T is uniformly continuous on E.

(b) Denote by $\tilde{E}$ the completion of E. Verify that there exists a unique $\tilde{T} \in L(\tilde{E},\tilde{F})$ such that $\tilde{T}x = Tx$ (for all $x \in E$); moreover, $\|\tilde{T}\| = \|T\|$.

(c) Show that if $\Sigma_n x_n$ is a convergent series in E then $\Sigma_n Tx_n$ converges in F.

(d) Let M be a closed vector subspace of E and $Q_M : E \to E/_M$ the quotient map. Show that $\|Q_M\| = 1$.

3–3. Prove (3.b).

3–4. Prove (3.c).

3–5. Prove (3.d).

3–6. Prove (3.e).

3–7. Prove (3.f).

3–8. Prove (3.g).

3–9. Prove (3.h).

3–10. Prove (3.j).

3–11. Prove (3.k).

3–12. Prove (3.l).

3–13. Let E and F be normed spaces and $E \neq \{0\}$. Show that if L(E,F) is complete then so does F.

3–14. Let E, F be normed spaces, let T, $T_n \in L(E,F)$ and u, $x_n \in E$. Show that if $\lim_n \|T_n - T\| = 0$ and $\lim_n \|x_n - u\| = 0$ then $\lim_n \|T_n x_n - Tu\| = 0$.

3–15. Let E be a normed space and $\tilde{E}$ its completion. Show that $(\tilde{E})' \equiv E'$.

3–16. Let E be a normed space. Show that a linear functional f on E is continuous if and only if Ker f is closed in E.

3–17. (The dual result of (3.2)) Let E_0, E and F be B–spaces, let $T_0 \in L(E_0,F)$ and $T \in L(E,F)$ be operators. If T is one–to–one and $T_0(U_{E_0}) \subset T(U_E)$ (hence Im $T_0 \subset$ Im T), then there exists an operator $R \in L(E_0,E)$ such that $TR = T_0$.

4

The Uniform Boundedness Theorem and Banach's Open Mapping Theorem

First we recall the following important result.

Baire's category theorem. Let (X, d) be a complete metric space and $X = \bigcup_{n=1}^{\infty} M_n$ Then at least one $\overline{M}_k$ contains a non-empty open set. Consequently, every complete metric space is a second category space.

(4.1) **Theorem.** (The uniform boundedness principle). Let E be a Banach space, let F be a normed space and

$$\mathscr{R} \subset L(E, F).$$

If $\mathscr{R}$ is bounded pointwisely on E in the sense that for any $x \in E$ there is an $\lambda_x \geq 0$ such that

$$(1) \qquad \|Tx\| \leq \lambda_x \quad (\text{for all } T \in \mathscr{R}),$$

then $\mathscr{R}$ is bounded with respect to the operator norm (called uniformly bounded), that is, there is an $\lambda > 0$ such that

$$\sup \{ \|T\| : T \in \mathscr{R} \} \leq \lambda .$$

Proof. For any natural number $n \geq 1$, let

$$C_n = \{ x \in E : \sup_{T \in \mathscr{R}} \|Tx\| \leq n \}.$$

Then each C_n is closed in E (by the continuity) and $E = \bigcup_{n=1}^{\infty} C_n$ (by (1)), hence by Baire's category theorem, there is some $k \geq 1$ such that C_k contains a non-empty open set G. Consequently, there is some open ball

$$N(z, \epsilon) = \{ z \in E : \|x - z\| < \epsilon \} \text{ with } N(z, \epsilon) \subset C_k,$$

that is,

$$\|Tx\| \leq k \quad (\text{for all } T \in \mathscr{R} \text{ and } x \in N(z, \epsilon)).$$

As $N(z, \epsilon) = z + N(0, \epsilon)$, for any $u \in N(0,\epsilon)$, we have

$$\begin{aligned} \|Tu\| &\leq \| T(u + z) - Tz \| \\ &\leq \| T(u + z) \| + \|Tz\| \\ &\leq 2k \qquad\qquad (\text{for all } T \in \mathscr{R}), \end{aligned}$$

hence

$$\sup_{T \in \mathscr{R}} \|T\| \leq \frac{2k}{\epsilon}.$$

(4.2) **Corollary.** *Let* E *be a normed space, let* $B \subseteq E$ *and* $D \subset E'$.

(a) B *is bounded if and only if for any* $u' \in E'$,

$$\sup_{b \in B} |\langle b,u' \rangle| < \infty.$$

(b) *Assume that* E *is complete. Then* D *is bounded in the Banach dual* E' *if and only if for any* $x \in E$,

$$\sup_{d' \in D} |\langle x, d' \rangle| < \infty.$$

Proof. (a) E' is a Banach space and $\{ K_E b : b \in B \}$ is a family of continuous linear functionals on E', hence the result follows from (4.1).

(b) Follows from (4.1).

(4.3) **Theorem** (Banach–Steinhaus′ theorem). *Let* E *be a Banach space, let* F *be a normed space and suppose that* $\{T_n\}$ *is a sequence in* $L(E, F)$ *such that for any* $x \in E$, *there is some* $z_x \in F$ *for which* $z_x = \lim_n T_n x$. *Then the formula*

(1) $$Tx = z_x \ (= \lim_n T_n x) \quad (\underline{\text{for any}}\ x \in E),$$

<u>defines a unique</u> $T \in L(E, F)$ <u>such that</u>

(2) $$\|T\| \leq \liminf \|T_n\|.$$

Proof. Clearly formula (1) defines a unique linear map $T : E \longrightarrow F$. As $\lim_n \|T_n x\| = \|z_x\|$, it follows that $\{T_n\}$ is pointwise bounded on E, and hence from (4.1) that there is an $\lambda > 0$ such that $\sup_n \|T_n\| \leq \lambda$, so that

$$\|Tx\| = \lim_n \|T_n x\| \leq \liminf_n \|T_n\| \, \|x\|$$

$$\leq \lambda \|x\| \quad (\text{for all } x \in E).$$

Thus, $T \in L(E, F)$ and (2) holds.

<u>Remark</u>. In general, to ensure the continuity of the limit of a sequence of operators, we need uniform convergence. The preceding result shows that in the case of operators on a Banach space, pointwise convergence already implies the continuity of the limit.

The second application of Baire's category theorem is the so–called open mapping theorem. To do this, we require the following terminology.

(4.4) **Definition**. Let E and F be normed spaces over $\mathbf{K}$ and $T : E \longrightarrow F$ a linear map. We say that T is:

(a) <u>Open</u> if $T(U_E)$ (or $T(O_E)$) is a 0–neighbourhood in F for the norm–topology;

(b) <u>almost open</u> if $\overline{T(U_E)}$ (or $\overline{T(O_E)}$) is a 0–neighbourbood in F for the norm–topology, i.e., there is an $r > 0$ such that

$$r\,U_F \subset \overline{T(U_E)} \ \ (\text{or } r\,U_F \subset \overline{T(O_E)}) .$$

(4.5) **Lemma.** Let E and F be normed spaces over K and let $T : E \longrightarrow F$ be linear.

(a) If T(E) is of the second category (in particular, T(E) is complete), then T is almost open.

(b) Suppose that T is almost open. If E is complete and T is continuous, then T is open (and a fortiori, onto) hence F must be complete.

Proof. (a) As $T(E) = \bigcup_{n\geq 1} nT(U_E)$ (since $E = \bigcup_{n\geq 1} nU_E$) and T(E) is of the second category, it follows that one of the sets $\overline{nT(U_E)}$ has an interior point, and hence that $\overline{T(U_E)}$ also has an interior point, y say; consequently, $0 = y + (-y)$ is an interior point of $\overline{T(U_E)} + \overline{T(U_E)}$ (since $\overline{T(U_E)}$ is circled). Note that

$$\overline{T(U_E)} + \overline{T(U_E)} \subset \overline{T(U_E) + T(U_E)} = \overline{2T(U_E)} = 2\overline{T(U_E)} \ ;$$

it then follows that 0 is an interior point of $2\overline{T(U_E)}$, and hence that 0 must be an interior point of $\overline{T(U_E)}$; in other words, there exists an $r > 0$ such that

$$rU_F = \{ y \in F : \|y\| \leq r \} \subseteq \overline{T(U_E)} ,$$

that is, T is almost open.

(b) There exists an $r > 0$ such that $rU_F \subset \overline{T(U_E)}$ (by the almost openness of T); we claim that

$$\frac{r}{3}U_F \subset T(U_E),$$

so that the openness of T follows. To this end, we first notice that

(2) $$\frac{r}{2^n} U_F \subset \frac{1}{2^n} \overline{T(U_E)} = \overline{T\left(\frac{1}{2^n} U_E\right)} \qquad \text{(for all } n \geq 1).$$

Now for any $y \in rU_F$, there exists an $x_1 \in U_E$ such that

$$\|y - Tx_1\| < \frac{r}{2}$$

(since $rU_F \subset \overline{T(U_E)}$). As $y - Tx_1 \in \frac{r}{2} U_F$, it follows from (2) that there exists an $x_2 \in \frac{1}{2} U_E$ such that

$$\|y - Tx_1 - Tx_2\| < \frac{r}{2^2}.$$

As $y - Tx_1 - Tx_2 \in \frac{r}{2^2} U_F$, it follows from (2) (in this case, take $n = 2$) that there exists an $x_3 \in \frac{1}{2^2} U_F$ such that

$$\| y - Tx_1 - Tx_2 - Tx_3 \| < \frac{r}{2^3}.$$

Continue this process, we obtain a sequence $\{x_n\}$ in E such that

(3) $$x_n \in \frac{1}{2^{n-1}} U_F \quad \text{and} \quad \|y - (\Sigma_{i=1}^{n} Tx_i)\| < \frac{r}{2^n}.$$

As $\Sigma_{n=1}^{\infty} \|x_n\| \leq \Sigma_{n=1}^{\infty} \frac{1}{2^{n-1}} < \infty$, it follows from the completeness of E that there exists an $x \in E$ such that

(4) $$x = \Sigma_{n=1}^{\infty} x_n \quad \text{and} \quad \|x\| \leq \Sigma_{n=1}^{\infty} \|x_n\| \leq 2 < 3.$$

Now the continuity of T ensures that

$$Tx = \|\cdot\|\text{-}\lim T(\Sigma_{i=1}^{n} x_i) = \|\cdot\|\text{-}\lim \Sigma_{i=1}^{n} Tx_i,$$

it then follows from (3) that

$$y = \|\cdot\|\text{-}\lim_n \Sigma_{i=1}^{n} Tx_i = Tx ,$$

and hence from $\|x\| < 3$ that $y \in 3T(U_E)$; that is, $\frac{y}{3} \in T(U_E)$. This proves(1).

Finally, the bijection $\check{T} : E/\text{Ker } T \longrightarrow F$ (associated with T) is continuous and open, hence it is a topological isomorphism. Since Ker T is closed, it follows from the completeness of E that E/Ker T is complete, and hence from $F \cong E/\text{Ker } T$ that F is complete.

(4.6) **Theorem** (Banach's open mapping theorem). Let E and F be Banach spaces over K and $T \in L(E, F)$. If T is either almost open or onto, then T is open.

Proof. If T is almost open then T is open (by (4.5) (b)). Suppose now that T is onto. Then $F = T(E)$, hence the completeness of F ensures that T is almost open (by (4.5) (a)), so that T is open .

Let E and F be normed spaces over K and $T \in L^*(E, F)$. If the graph G(T) of T , defined by

$$G(T) = \{ (x,Tx) \in E \times F : x \in E \},$$

is closed in the product space $E \times F$, then we say that T has a closed graph. It is clear that G(T) is closed if and only if for any sequence $\{x_n\}$ in E, if $x_n \longrightarrow x$ in E and $Tx_n \longrightarrow y$ in F then $y = Tx$. Hence, every continuous linear map $T \in L(E,F)$ has a closed graph; but not converse. However, we have the following:

(4.7) **Theorem** (Banach's closed graph theorem). Let E and F be Banach spaces over K and $T : E \longrightarrow F$ a linear map. If T has a closed graph , then T is continuous.

<u>**Proof.**</u> G(T) is a closed vector subspace of the Banach space $E \times F$ equipped with the norm

$$\|(x,y)\|_2 = (\|x\|^2 + \|y\|^2)^{\frac{1}{2}} \quad ((x, y) \in E \times F),$$

hence $(G(T),\|\cdot\|_2)$ is a Banach space. On G(T) , let us define

$$\pi_1(x, Tx)= x \text{ and } \pi_2(x, Tx) = Tx \quad (\text{for all } x \in E).$$

Then $\pi_1 \in L(G(T),E)$ and $\pi_2 \in L(G(T), F)$. Moreover, π_1 is a bijection with

(1) $$T = \pi_2 \circ \pi_1^{-1}.$$

By Banach's open mapping theorem, π_1 is open, hence π_1^{-1} is continuous, and thus T is continuous (by (1)).

<u>**Remark.**</u> The uniform boundedness principle can be verified in terms of closed graph theorem.

Combining Banach's open mapping theorem and closed graph theorem, we obtain the following result.

(4.8)**Theorem.** <u>Let</u> E <u>and</u> F <u>be Banach spaces and let</u> $T : E \longrightarrow F$ <u>be linear</u>. <u>Suppose that</u> G(T) <u>is closed and that</u> T <u>is either almost open or onto</u>. <u>Then</u> T <u>is continuous and open</u>.

(4.9) <u>**Definition.**</u> Let E be a normed space. An operator $P \in L(E,E)$ is called a <u>projection</u> if

$$P^2 = P.$$

A vector subspace M of E is called a complemented subspace of E if there exists a projection $P \in L(E, E)$ such that

$$M = P(E).$$

Remark. Every complemented subspace M of E must be closed.

(4.10) **Lemma.** *Let E be a normed space, let M, N be closed vector subspaces of E such that*

(4.10.1) $$E = M + N \text{ and } M \cap N = \{ 0 \},$$

and let $\pi_M \in L^*(E,M)$ *and* $\pi_N \in L^*(E,N)$ *be such that*

$$\pi_M^2 = \pi_M \text{ and } \pi_N^2 = \pi_N .$$

Then π_M *is continuous* (that is, M is a complemented subspace of E) *if and only if the canonical map* $\Psi : M \times N \longrightarrow E$, *defined by*

$$\Psi(x_1,x_2)= x_1 + x_2 \qquad (\text{for all } x_1 \in M, x_2 \in N),$$

is a topological isomorphism from $M \times N$ *onto* E.(Notices that (4.10.1) ensures that Ψ is always bijective.)

Proof. We first observe that $I_E = \pi_M + \pi_N$, hence π_M is continuous if and only if π_N is continuous. On the other hand, the inverse of (the bijective map) Ψ is the product map $\pi_M \times \pi_N$, where $\pi_M \times \pi_N : E \longrightarrow M \times N$ is defined by

$$(\pi_M \times \pi_N)x = (\pi_M x, \pi_N x) \qquad (\text{for all } x \in E).$$

It then follows that $\Psi^{-1} = \pi_M \times \pi_N$ is continuous if and only if π_M is continuous; thus the result follows (since Ψ is always continuous).

(4.11) **Corollary.** Let E be a Banach space, and let M, N be closed vector subspaces of E such that

$$E = M + N \text{ and } M \cap N = \{0\}.$$

Then M is a complemented subspace. In particular, every finite–dimensional or finite–codimensional closed subspace of a Banach space is complemented.

Proof. Let $\pi_N \in L^*(E,N)$ be such that $\pi_N^2 = \pi_N$ and let $\hat{\pi}_N : E/M \longrightarrow N$ be the injection associated with π_N (since $M = \text{Ker}\ \pi_N$). Then $(\hat{\pi}_N)^{-1} = Q_M \circ J_N : N \longrightarrow E/M$ is always continuous and surjective. As N and E/M are complete, we conclude from Banach's open mapping theorem that $(\hat{\pi}_N)^{-1}$ is open, and hence that $\hat{\pi}_N$ is continuous; consequently, $\pi_N = \hat{\pi}_N \circ Q_M$ is continuous,thus the result follows from (4.10) .

(4.12) **Lemma.** Let E be a normed space. A closed vector subspace M of E is complemented if and only if there exists an $Q \in L(E,M)$ such that

$$I_M = QJ_M,$$

where $J_M : M \longrightarrow E$ is the canonical embedding. (In this case, Q must be onto.)

Proof. Necessity. There exists a projection $P \in L(E, E)$ such that $M = P(E)$. Regard P as an operator from E onto M, which is denoted by Q. Then we have $I_M = QJ_M$,

Sufficiency. The map, defined by

$$P = J_M Q ,$$

is a projection from E into E since

$$P^2 = (J_M Q)(J_M Q) = J_M I_M Q = J_M Q = P .$$

It remains to show that

$$M = P(E).$$

To do this, we first notice that $P(E) = J_M(Q(E))$, hence it suffices to show that Q is onto.

In fact, let $u \in M$ and $y = J_M(u)$. Then

$$Qy = Q(J_M(u)) = I_M(u) = u ,$$

which obtains our assertion.

In order to give another characterization of complemented subspaces, we require the following notations: Let E and F be normed spaces. An operator $T \in L(E, F)$ is said to admit a factorization through a Banach space G, in symbols

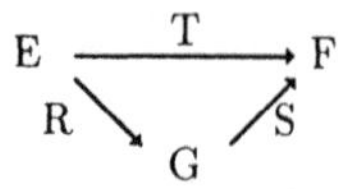

if there exists $R \in L(E, G)$ and $S \in L(G, F)$ such that $T = S \circ R$;in this case, we also say that T is G–factorable. In particular, if G is a vector subspace (resp. closed subspace) of c_0, then every G–factorable operator is also said to admit a factorization through a vector subspace (resp. a subspace) of c_0.

Notation: Let E and F be normed spaces. $E \prec F$ means that E is topologically isomorphic to some complemented subspace of F.

(4.13) **Theorem** (Stephani [1976]). Let G and E be Banach spaces over K Then $G \prec E$ if and only if the identity operator I_G on G is E–factorable.

Proof. Necessity. There exists a complemented subspace M of E such that $G \simeq M$ under some topological isomorphism T. By (4.12), there exists an onto operator $Q \in L(E,M)$ such that the following diagram commutes:

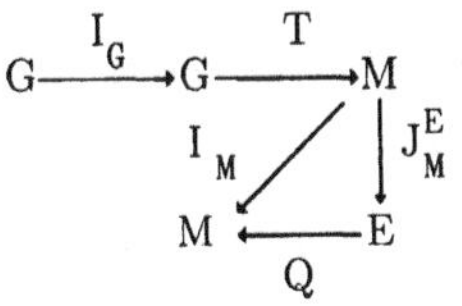

therefore, $R = J_M^E\, T \in L(G, E)$, $S = T^{-1}Q \in L(E, G)$ and

$$SR = T^{-1}QJ_M^E T = T^{-1}I_M T = I_G\,.$$

Sufficiency. There exist operators $R \in L(G,E)$ and $S \in L(E,G)$ such that $I_G = SR$. The map, defined by

$$P = RS,$$

is a projection from E into E since

$$P^2 = RSRS = RI_GS = P.$$

It remains to show that $G \simeq P(E)$ $(=R(S(E)))$. To this end, we claim that $S : E \longrightarrow G$ is onto and that $R : G \longrightarrow E$ is one–one with closed range.

In fact, let $x \in G$. Then $y = Rx \in E$ and

$$Sy = (SR)x = I_G(x) = x,$$

hence S is onto; consequently

$$P(E) = R(S(E)) = R(G), \tag{4.13.1}$$

thus the range of R is closed (since P is a projection) . On the other hand, let $Rx = 0$ (where $x \in G$). Then

$$x = I_G(x) = SR(x) = 0,$$

hence R is one–one. This proves our assertions.

Finally, (4.13.1) together with the injectivity of R shows that R is a bijective operator from the Banach space G onto the closed subspace $P(E) = R(G)$ of the Banach space E , hence R is a topological isomorphism from G onto P(E) (by Banach's open mapping theorem).

(4.a) <u>The inverse image of a complemented subspace under an isomorphism</u>: Let E and F be Banach spaces, let $T \in L(E, F)$, let M be a subspace of E and let

$$TJ_M^E : M \longrightarrow T(M) \text{ be a topological isomorphism.}$$

If T(M) is a complemented subspace of F , then so does M.

As an application of (3.j), one can verify the following two interesting results.

(4.b) <u>Factorable operators through concrete Banach spaces</u> (I): Let E and F be Banach spaces and $T \in L(E, F)$.

(i) T is $\ell^1(\Lambda)$–factorable if and only if T has a representation:

$$Tx = \Sigma_\Lambda \langle x, f_i \rangle y_i \quad (\text{for all } x \in E),$$

for some weak* summable family $[f_i, \Lambda]$ in E' and bounded family $[y_i, \Lambda]$ in F.

(ii) T is $c_0(\Lambda)$–factorable if and only if T has a representation

$$Tx = \Sigma_\Lambda <x, g_i>y_i \quad (\text{for all } x \in E),$$

for some bounded, weak* null family $[g_i, \Lambda]$ in E' and weakly summable family $[y_i, \Lambda]$ in F.

(iii) If there is a bounded sequence $\{h_n\}$ in E' and an unconditionally convergent series $\Sigma_n y_n$ in F such that

$$Tx = \Sigma_{n=1}^{\infty} <x, h_n>y_n \quad (\text{for all } x \in E),$$

then T is ℓ^∞–factorable.

<u>**Remark**</u>. It is amusing to take $F = E$ and $T = I_E$ in (i) through (iii).

(4.c) <u>Nuclear Operators</u> I (Grothendieck): Let E and F be Banach spaces and $T \in L(E,F)$. Then the following **statements** are equivalent.

(i) There exist a sequence $\{f_n\}$ in E' and a sequence $\{y_n\}$ in F with $\Sigma_{n=1}^{\infty} \|f_n\| \, \|y_n\| < \infty$ such that

(4.c.1) $$Tx = \Sigma_{n=1}^{\infty} <x, f_n>y_n \quad (\text{for all } x \in E).$$

(ii) There exists $[\lambda_n] \in \ell^1$ and bounded sequences $\{h_n\}$ and $\{z_n\}$ in E' and F respectively such that

(4.c.2) $$Tx = \Sigma_{n=1}^{\infty} \lambda_n <x, h_n>z_n \quad (\text{for all } x \in E).$$

(iii) The following diagram commutes :

$$\begin{array}{ccc} E & \xrightarrow{T} & F \\ R \downarrow & & \uparrow S \\ \ell^{\infty} & \xrightarrow[T_0]{} & \ell^{1} \end{array}$$

where $R \in L(E, \ell^{\infty})$, $S \in L(\ell^{1}, F)$ and $T_0 \in L(\ell^{\infty}, \ell^{1})$ is a diagonal operator of the form

$$T_0([\mu_n]) = [\sigma_n \mu_n] \qquad \text{(for all } [\mu_n] \in \ell^{\infty}) \tag{4.c.3}$$

for some $[\sigma_n] \in \ell^{1}$.

Operators satisfying one of these three equivalent statements are called <u>nuclear operators</u>, while (4.c.1) is called a <u>nuclear representation</u> of T.

Exercises

4–1. Let $\{\eta_n\}$ be a sequence of positive numbers such that

$$\Sigma_{n=1}^{\infty} |\zeta_n| \eta_n < \infty \text{ for all } [\zeta_n] \in c_o.$$

Show that $[\eta_n] \in \ell^{1}$. [Using the uniform boundedness theorem.]

4–2. Let E be a B–space, let F be a normed space and let $\{T_n\}$ be a sequence in L(E,F) which is pointwise convergent to zero [i.e. for any $x \in E$, we have $\lim_n T_n x = 0$ in F]. If K is a relatively compact subset of E, show that

$$\lim T_n x = 0 \text{ uniformly on } K \text{ (i.e., } \lim_n \sup_{x \in K} \|T_n x\| = 0).$$

[Using the uniform boundedness theorem.]

4–3. Prove (4.a).

4–4. Prove (4.b).

4–5. Prove (4.c).

4–6. Let E and F be Banach spaces and $T \in L(E,F)$. Show that if Im T is of the second category then Im T is closed in F.

4–7. Let $(E,\|\cdot\|)$ be a Banach space and $\{e_n\}$ a Schauder basis for E.

(a) For any $x = \Sigma_{j=1}^{\infty}\lambda_j e_j \in E$, let

$$p(x) = \sup\{ \|\Sigma_{j=1}^{n}\lambda_j e_j\| : n \geq 1\}.$$

Show that (E,p) is a Banach space, and hence deduce (from Banach's open mapping theorem) that there exists an $K \geq 1$ such that

$$\|x\| \leq p(x) \leq K\|x\| \quad \text{(for all } x \in E).$$

Moreover, the norm $p(\cdot)$ has the following property:

$$p(\Sigma_{j=1}^{m}\lambda_j e_j) \leq p(\Sigma_{j=1}^{n}\lambda_j e_j) \quad \text{whenever } m < n.$$

(b) For any $0 \neq u_o \in E$ and $T_o \in L(E,E')$, the linear functional f, defined by

$$f(x) = T_o(x)u_o \quad \text{(for all } x \in E),$$

is a continuous linear functional on $(E,\|\cdot\|)$.

(c) For any $n \geq 1$, the functional e_n', defined by

$$e_n'(\Sigma_{j=1}^{\infty}\lambda_j e_j) = \lambda_n,$$

is a continuous linear functional on $(E,\|\cdot\|)$. (We say that the sequence $\{e_n'\}$ is biorthogonal to $\{e_n\}$. [By part (a) and (b).]

4–8. Let E be a Banach space. A sequence $\{u_n\}$ in E is called a basis sequence if $\{u_n\}$ is a Schauder basis for the closed subspace $\overline{<\{u_n\}>}$ spanned by $\{u_n : n \geq 1\}$.

(a) (Nikolskii) A sequence $\{u_n\}$ in E is a basis sequence if and only if it satisfies the following two conditions.

(i) $u_n \neq 0$ (for all $n \geq 1$);

(ii) there exists a $\alpha \geq 1$ such that for any $[\lambda_n] \in \mathbb{K}^{\mathbb{N}}$,

$$\|\Sigma_{i=1}^{m}\lambda_i u_i\| \leq \alpha\|\Sigma_{i=1}^{n}\lambda_i u_i\| \qquad \text{(for all } m \leq n\text{)}.$$

(b) Suppose that M is a finite–dimensional vector subspace of E. Show that for any $\epsilon > 0$ there is an $x_\epsilon \in E$ with $\|x_\epsilon\| = 1$ such that

$$\|y\| \leq (1+\epsilon)\|y+\lambda x_\epsilon\| \qquad \text{(for all } y \in M \text{ and } \lambda \in \mathbb{K}\text{)}.$$

(c) Use parts (a) and (b) (or otherwise) to show that any infinite–dimensional Banach space has a basis sequence.

5

A Structure Theorem for Compact Sets in a Banach Space

Let E be a normed space and $A \subset E$. Recall that A is:

(a) bounded if there exists an $\lambda > 0$ such that $A \subset \lambda U_E$;

(b) totally bounded if for any $\epsilon > 0$, A is covered by a finite ϵ–net; namely, there is $\{ a_1, \cdots, a_n \} \subset A$ such that

$$A \subset \bigcup_{i=1}^{n} (a_i + \epsilon U_E).$$

(5.a) Bounded sets (I): Let E be a normed space and let $\mathcal{M}_{von}(E)$ be the family of all bounded subsets of E. Then $\mathcal{M}_{von}(E)$ has the following properties:

(VB1) $E = \cup \mathcal{M}_{von}(E)$.

(VB2) If $B \subset A$ and $A \in \mathcal{M}_{von}(E)$, then $B \in \mathcal{M}_{von}(E)$.

(VB3) $A_1 + \cdots + A_n \in \mathcal{M}_{von}(E)$ whenever $A_i \in \mathcal{M}_{von}(E)$ $(i=1,\cdots,n)$.

(VB4) $\lambda A \in \mathcal{M}_{von}(E)$ whenever $\lambda \in K$ and $A \in \mathcal{M}_{von}(E)$.

(VB5) $\Gamma A \in \mathcal{M}_{von}(E)$ whenever $A \in \mathcal{M}_{von}(E)$.

Moreover, the family

$$\mathcal{B}_{von}(E) = \{\Gamma A : A \in \mathcal{M}_{von}(E)\}$$

is a basis for $\mathcal{M}_{von}(E)$ in the sense that any element in $\mathcal{M}_{von}(E)$ is contained in some member in $\mathcal{B}_{von}(E)$. (Of course, $\overline{\Gamma A}$ is bounded if A is bounded.)

(5.b) Totally bounded sets and compact sets: Let E be a normed space, let $\mathcal{M}_p(E)$ (resp. $\mathcal{M}_c(E)$) the family of all totally bounded (resp. relatively compact) subsets of E.

(i) $\mathcal{M}_p(E)$ satisfies all properties (VB1) through (VB5) in (5.a) and

$$\overline{\Gamma A} \in \mathcal{M}_p(E) \qquad \text{(for all } A \in \mathcal{M}_p(E)).$$

(ii) $\mathcal{M}_c(E)$ satisfies (VB1) through (VB4) in (5.a), and

(VB5)* the circled hull of an $A \in \mathcal{M}_c(E)$ is relatively compact.

From (VB5)*, it is natural to ask whether the convex hull of a compact set is relatively compact, the answer is affirmative in Banach spaces as shown by the following :

(5.1) **Lemma.** Let E be a Banach space. The closed convex hull of a compact set B in E is compact.

Proof. Closed subsets of a Banach space E is complete, and a subset K of E is compact if and only if it is complete and totally bounded, thus it suffices to show that co B is totally bounded; but this is obvious since B is totally bounded.

Let E be a normed space and $\{u_n\}$ a null sequence in E (i.e., $\lim_n \|u_n\| = 0$). Then the set $\{u_n : n \geq 1\}$ is totally bounded (since the set $\{u_n : n \geq 1\} \cup \{0\}$ is compact), hence $\overline{\Gamma}(\{u_n : n \geq 1\})$ is totally bounded. Conversely, it is natural to ask whether every totally bounded set is contained in the closed disked hull of some null sequence. The answer is affirmative as the following important result shows:

(5.2) **Theorem** (Grothendieck). For any totally bounded set K in a normed space (E ,$\|\cdot\|$), there exists a null sequence $\{x_n\}$ in E (i.e., $\lim_n \|x_n\| = 0$) such that

$$K \subset \overline{\Gamma}(\{x_n : n \geq 1\}).$$

Proof The total boundedness of K ensures that there is a finite set $D_1 \subset K$ such that

(1) $$K \subset D_1 + \tfrac{1}{4}U_E.$$

The set, defined by

(2) $$K_2 = (K - D_1) \cap (\tfrac{1}{4} O_E),$$

is totally bounded, hence there is a finite set $D_2 \subset K_2$ such that

(3) $$K_2 \subset D_2 + \tfrac{1}{4^2} O_E.$$

The set, defined by

(4) $$K_3 = (K_2 - D_2) \cap (\tfrac{1}{4^2} O_E),$$

is totally bounded, there exists a finite set $D_3 \subset K_3$ such that

(5) $$K_3 \subset D_3 + \tfrac{1}{4^3} O_E.$$

Continue this process, we obtain a sequence $\{D_n\}$ of finite sets and a sequence $\{K_n\}$ of totally bounded sets (where $K_1 = K$) such that

(i) $$D_n \subset K_n \text{ with } K_n \subset D_n + \tfrac{1}{4^n} O_E;$$

(ii) $$K_{n+1} = (K_n - D_n) \cap (\tfrac{1}{4^n} O_E) \; (\subset \tfrac{1}{4^n} O_E).$$

Now for any $u \in K$, (1) and (2) show that there exists an $z_{i(1)} \in D_1$ such that

$$u - z_{i(1)} \in K_2,$$

it then follows from (3) that there exists an $z_{i(2)} \in D_2$ such that

$$u - z_{i(1)} - z_{i(2)} \in K_3.$$

Continue this process, we obtain

(5) $$u - \Sigma_{j=1}^{n} z_{i(j)} \in K_{n+1} \text{ with } z_{i(j)} \in D_j \; (j=1,..,n)$$

or, equivalently,

(6) $$u - \Sigma_{j=1}^{n} 2^{-j} y_{i(j)} \in K_{n+1} \text{ with } y_{i(j)} = 2^j z_{i(j)} \in 2^j D_j.$$

As $K_{n+1} \subset \frac{1}{4^n} O_E$ (see (ii)), it follows from (6) that

$$\lim_{n \to \infty} \|u - \Sigma_{j=1}^{n} 2^{-j} y_{i(j)}\| = 0, \text{ i.e.,}$$

(7) $$u = \Sigma_{n=1}^{\infty} 2^{-n} y_{i(n)} \text{ with } y_{i(n)} \in 2^n D_n$$

On the other hand, since $2^j D_j$ is finite, we may write

(8) $$\begin{aligned} 2D_1 &= \{x_1, \ldots, x_{k_1}\} \\ 2^2 D_2 &= \{x_{k_1+1}, \ldots, x_{k_2}\} \\ &\cdot \\ &\cdot \\ 2^n D_n &= \{x_{k_{n-1}+1}, \ldots, x_{k_n}\}. \end{aligned}$$

By (i) and (ii), we have

$$2^n D_n \subset 2^n K_n \subset 2^n \frac{1}{4^{n-1}} O_E = \frac{1}{2^{n-2}} O_E;$$

it then follows that the sequence $\{x_n\}$, defined by (8), is a null sequence in E such that each $u \in K$ is of the form (7); in other words,

$$u \in \overline{co}(\{0, y_{i(1)}, y_{i(2)}, \ldots\}) \subset \Gamma(\{x_n : n \geq 1\}).$$

This completes the proof.

The idea of the proof of (5.2) is essentially due to A.P. Robertson / W.J. Robertson; but it is interesting to compare the proof of (5.2) with Diestel [1984,p.4].

In order to give an important application of (5.2), we need the following notation: Let E be a normed space. For any absolutely convex bounded subset B of E, we write $E(B) = \langle B \rangle$ (the vector subspace of E spanned by B), that is

$$E(B) = \bigcup_n nB,$$

(since $B = \Gamma B$), and denote by γ_B the gauge of B defined on $E(B)$ (γ_B is finite on $E(B)$ since B is absorbing in $E(B)$). The boundedness of B ensures that there is some $\mu > 0$ such that

$$\|x\| \leq \mu \, \gamma_B(x) \qquad \text{(for all } x \in E(B)),$$

hence $\gamma_B(.)$ is a norm on $E(B)$ .The normed space $(E(B)), \gamma_B(\cdot))$ is always denoted by $E(B)$. The following gives a sufficiency for the completeness of $(E(B), \gamma_B)$.

(5.c) <u>Infracomplete sets</u>: Let E be a normed space and B an absolutely convex, bounded subset of E .If B is complete (and surely, compact),then $(E(B), \gamma_B))$ is complete. [An absolutely convex, bounded set B in E is said to be <u>infracomplete</u> if $(E(B), \gamma_B)$ is complete.]

As an application of (5.2), we have the following important result:

(5.3) **Theorem.** <u>Let</u> E <u>be a Banach space</u>. <u>For any compact subset</u> K <u>of</u> E, <u>there exists a compact, absolutely convex subset</u> B <u>of</u> E <u>such that</u> K <u>is compact in the Banach space</u> $(E(B), \gamma_B)$ (hence $K \subset \mu B$ for some $\mu > 0$).

Proof. By (5.2), there is a null sequence $\{x_n\}$ in E such that $K \subset \overline{\Gamma}(\{ x_n : n \geq 1 \})$. By (5.1), $\overline{\Gamma}(\{ x_n : n \geq 1 \})$ is compact (since $\{x_n : n \geq 1\} \cup \{0\}$ is compact). As $\lim_n \|x_n\| = 0$,one can find a sequence $\{\lambda_n\}$ of positive numbers such that

$$\lambda_n \longrightarrow \infty \text{ and } \lim_n \|\lambda_n x_n\| = 0.$$

[Take $\lambda_n = \max\{1, \|x_n\|^{-\frac{1}{2}}\}$ if $x_n \neq 0$ and $\lambda_n = n$ if $x_n = 0$.] Hence $B = \overline{\Gamma}(\{\lambda_n x_n : n \geq 1\})$ is compact in E [by (5.1) since $\lim_n \|\lambda_n x_n\| = 0$], thus $(E(B), \gamma_B)$ is a Banach space. As

$$\gamma_B(\lambda_n x_n) \leq 1 \text{ and } \lambda_n \longrightarrow \infty,$$

it follows that $\{x_n\}$ is a null sequence in the Banach space $(E(B), \gamma_B)$, and hence from the completeness of $(E(B), \gamma_B)$ that the γ_B–closed, absolutely convex hull of $\{x_n : n \geq 1\}$, denoted by C, is compact in $(E(B), \gamma_B)$ (see (5.1)). Therefore C is compact in $(E, \|\cdot\|)$ [since the canonical injection $J_B: (E(B), \gamma_B) \longrightarrow (E, \|\cdot\|)$ is continuous], thus $\overline{\Gamma}(\{x_n : n \geq 1\} \subset C$ [C is closed in E], consequently,

$$K \subset \overline{\Gamma}(\{x_n : n \geq 1\}) = C.$$

[On E(B), the $\|\cdot\|$–top $< \gamma_B(\cdot)$–topology, hence $C \subset \overline{\Gamma}(\{x_n : n \geq 1\})$]. We then conclude from the compactness of C in $(E(B), \gamma_B)$ that K is compact in the Banach space $(E(B), \gamma_B)$.

Let E and F be normed spaces. An operator $T \in L(E.F)$ is called a <u>compact operator</u> if $T(U_E)$ is relatively compact in F.

It is clear that the composite of two operators in which one of them is a compact operator must be compact. The converse is true as an immediate consequence of (5.3), as shown by the following:

(5.4) **<u>Corollary</u>** (Randtke [1972]). <u>Let</u> E <u>and</u> F <u>be Banach spaces and</u> $T \in L(E,F)$ <u>a compact operator</u>. <u>Then there exists a Banach space</u> G <u>and compact operators</u> $R \in L(E,G)$

<u>and</u> $S \in L(G,F)$ <u>such that</u> $T = SR$.

<u>**Proof.**</u> $\overline{T(U_E)}$ is compact in F, hence there exists an absolutely convex, compact subset D of F such that $\overline{T(U_E)} \subset D$ and $\overline{T(U_E)}$ is compact in the Banach space $(F(D),\gamma_D)$. Now let $G = (F(D),\gamma_D)$,

$$R(x) = Tx \qquad (\text{for all } x \in E)$$

and $S = J_D\colon (F(D),\gamma_D) \longrightarrow F$ the canonical embedding. Then R and S are compact operators (since $\overline{R(U_E)} = \overline{T(U_E)}$ and $S(D) = D$ is compact in F) with $T = SR$.

The following result is useful for giving a criterion for compact operators.

(5.d) <u>Compact sets in the Banach space</u> c_0: A subset K of c_0 is relatively compact if and only if there exists an $[\lambda_n] \in c_0$ with $\lambda_n > 0$ (for all $n \geq 1$) such that

$$K \subset \{ [\xi_n] \in c_0 : |\xi_n| \leq \lambda_n \text{ for all } n \geq 1\}.$$

<u>**Exercises**</u>

5–1. Prove (5.a).

5–2. Show that, in a normed space E, a subset K of E is compact if and only if it is totally bounded and complete.

5–3. Prove (5.b).

5–4. Prove (5.c).

5–5. Prove (5.d).

6

Topological Injections and Topological Surjections

We begin with the following:

(6.1) **Definition** (Pietsch [1980, p.26]). Let E and F be normed spaces over K and $T \in L(E,F)$. The injection modulus of T is given by

(6.1.1) $$j(T) = \sup \{ \tau \geq 0 : \tau\|x\| \leq \|Tx\| \quad (\text{for all } x \in E) \},$$

and the surjection modulus of T is given by

(6.1.2) $$q(T) = \sup \{ \tau \geq 0 : \tau U_F \subset T(U_E) \}.$$

In particular, if $T = O$ (the zero operator) then we put

$$j(O) = 0 \text{ and } q(O) = 0.$$

Remark. It is easily seen that

$$j(T) = \inf \{ \|Tx\| : \|x\| = 1 \} \quad (\text{for all } T \in L(E,F)).$$

It is also clear that

$$j(T) \leq \|T\| \text{ and } q(T) \leq \|T\| \quad (\text{for all } T \in L(E,F)).$$

and that $q(T) > 0$ if and only if T is open.

Most of the results of this section are taken from the excellent book written by Pietsch [1980].

(6.a) A characterization of injection modulus: Let E and F be normed spaces and $T \in L(E,F)$. Then the following statements are equivalent:

(i) $j(T) \geq \tau > 0$.

(ii) T is one–one and $\tau T^{-1}(U_F) \subset U_E$.

(iii) T is one–one and $(\tau U_F) \cap (TE) \subset TU_E$ (i.e. T is relatively open).

(iv) T has a bounded inverse $T^{-1}: T(E) \longrightarrow E$.

As an application of Banach's open mapping theorem, the surjection modulus can be calculated by the following:

(6.2) **Lemma.** Let E and F be Banach spaces and $T \in L(E,F)$. Then

$$q(T) = \sup\{ \tau \geq 0 : \tau U_F \subset \overline{T(U_E)} \}.$$

Proof. Suppose that

$$\overline{q(T)} = \sup\{ \tau \geq 0 : \tau U_F \subset \overline{T(U_E)} \}.$$

Then $q(T) \leq \overline{q(T)}$. To prove that $\overline{q(T)} \leq q(T)$, it suffices to show that for any $\beta > 0$ with $\beta U_F \subset \overline{T(U_E)}$, we have $\beta \leq q(T)$.

In fact, let β be such a number. Then $\overline{T(U_E)}$ is a 0–neighbourhood in F for the norm–topology, hence T is almost open, thus T open (by (4.5) (b)); consequently, $\delta T(U_E)$ is a 0–neighbourhood in F (for any $\delta > 0$). As

$$\beta U_F \subset \overline{T(U_E)} = \bigcap_{\mu>0} \{ T(U_E) + \mu U_F \} = \bigcap_{\delta>0} \{ T(U_E) + \delta T(U_E) \}$$

$$\subset (1 + \delta)T(U_E) \qquad \text{(for all } \delta > 0\text{)},$$

we conclude that $\frac{\beta}{1+\delta} \leq q(T)$, and hence that $\beta \leq q(T)$ (since δ was arbitrary), which obtains our requirement.

Remark. From the proof of the preceding result, we see that if E and F are only assumed to be normed spaces, then $T \in L(E,F)$ is almost open if and only if

$$\sup\{ \tau \geq 0 : \tau U_F \subset \overline{T(U_E)} \} > 0.$$

(6.3) **Defintion.** Let E and F be normed spaces over K and $T \in L(E,F)$. We say that T is a :

(a) *topological injection* (*or isomorphic embedding*), denoted by $T : E \rightarrowtail F$, if $j(T) > 0$;

(b) *metric injection* if $j(T) = 1 = \|T\|$;

(c) *topological surjection*, denoted by $T : E \twoheadrightarrow F$, if $q(T) > 0$;

(d) *metric surjection* if $q(T) = 1 = \|T\|$.

Remark. Suppose that $T \in L(E,F)$. Then T is a topological surjection if and only if it is open; T is an isomorphism if and only if it is a topological injection and topological surjection; and T is a metric isomorphism (i.e. isometry) if and only if it is a metric injection and a metric surjection.

(6.4) **Proposition.** *Let* E *and* F *be Banach spaces and* $T \in L(E,F)$. *Then the following statements are equivalent:*

(a) T *is a topological injection.*

(b) T *is one–one and has a closed range* T(E).

(c) T *admits a factorization*

$$\begin{array}{ccc} E & \xrightarrow{\quad T \quad} & F \\ & {}_{T_0}\searrow \quad \nearrow_{J_{T(E)}} & \\ & T(E) & \end{array}$$

where $T_0 : E \longrightarrow T(E)$ *is an isomorphism and* $J_{T(E)}$ *is the embedding map.*

In this case, we have

(6.4.1) $$j(T) = \|T_0^{-1}\|^{-1}.$$

Proof. (a) ⇒ (b): As $j(T) > 0$, for any α with $0 < \alpha < j(T)$, there exists an $r > 0$ such that

$$r > \alpha \quad \text{and} \quad r\|x\| \leq \|Tx\| \quad (\text{for all } x \in E). \tag{1}$$

It then follows that T is one–one [since $Tx = 0 \Rightarrow x = 0$].

To prove the closedness of $T(E)$, let $Tx_n \longrightarrow y$ in F. Then $\{x_n\}$ is a Cauchy sequence in E(by (1)), hence converges to some $x \in E$ (by the completeness of E), thus

$$\lim_n Tx_n = y = Tx \in T(E).$$

(b) ⇒ (c): By Banach's open mapping theorem, the statement (b) ensures that the map T_0, defined by

$$T_0x = Tx \qquad (\text{for all } x \in E),$$

is an isomorphism from E onto the Banach space $T(E)$. Thus the implication follows.

(c) ⇒ (a): Since T_0 has a bounded inverse $T_0^{-1}: T(E) \longrightarrow E$, it follows that

$$\|x\| = \|T_0^{-1}(Tx)\| \leq \|T_0^{-1}\| \; \|Tx\| \qquad (\text{for all } x \in E),$$

and hence that

$$j(T) \geq \|T_0^{-1}\|^{-1} > 0 . \tag{2}$$

Therefore T is a topological injection.

Finally, we assume that T is a topological injection. Then $j(T) \geq \|T_0^{-1}\|^{-1}$ (by (2)). In order to verify (6.4.1), it suffices to show that for any $r > 0$ with $r\|x\| \leq \|Tx\|$ $(x \in E)$, we have

$$r\|T_0^{-1}\| \leq 1.$$

Indeed, let $r > 0$ be such a number. Then

$$r\|T_0^{-1}\| \quad = \; r \sup\{ \; \|T_0^{-1}(Tx)\| : \|Tx\| \leq 1 \; \}$$

$$= \; \sup\{ \; r\|x\| : \|Tx\| \leq 1 \; \} \; \leq \; \sup\{ \; \|Tx\| : \|Tx\| \leq 1 \; \} \; \leq \; 1,$$

which obtains our assertion.

(6.5) **Proposition.** <u>Let</u> E <u>and</u> F <u>be Banach spaces and</u> $T \in L(E,F)$. <u>Then the following</u> <u>statements are equivalent</u>:

    (a) T <u>is a topological surjection.</u>

    (b) T <u>is onto.</u>

    (c) T <u>admits a factorization</u>

$$\begin{array}{ccc} E & \xrightarrow{\;\;T\;\;} & F \\ & Q\searrow \quad \nearrow T_0 & \\ & E/\mathrm{Ker}\ T & \end{array}$$

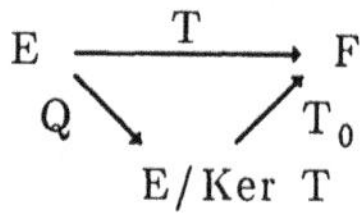

<u>where</u> $T_0 : E/\mathrm{Ker}\ T \longrightarrow F$ <u>is an isomorphism and</u> Q <u>is the quotient map.</u>
    <u>In this case, we have</u>

$$(6.5.1) \qquad q(T) = \; \|T_0^{-1}\|^{-1}.$$

Proof. (a) $\Rightarrow$ (b): Trivial (Open mappings must be onto).

(b) $\Rightarrow$ (c): Follows from Banach's open mapping theorem and the fact that T_0 is continuous and open (since T is onto and $E/\mathrm{Ker}\ T$ and F are Banach spaces).

(c) $\Rightarrow$ (a): As T_0 and Q are open, it follows from $T = T_0 Q$ that T is open.

Finally, to prove (6.5.1) , we first notice that if $\tau > 0$ is such that $\tau U_F \subset T(U_E)$, then

$$\|T_0^{-1}\| = \sup\{\ \|T_0^{-1}y\| : y \in U_F\ \}$$

$$\leq \sup\{\ \|T_0^{-1}(\tau^{-1}Tx)\| : x \in U_E\ \}$$

$$= \tau^{-1}\sup\{\ \|Qx\| : x \in U_E\ \} \leq \tau^{-1},$$

hence $\tau \leq \|T_0^{-1}\|^{-1}$, and consequently, $q(T) \leq \|T_0^{-1}\|^{-1}$. Conversely, let $r = \|T_0^{-1}\|^{-1}$. We claim that

$$rO_F \subset T(U_E),$$

so that $rU_F = \overline{rO_F} \subseteq \overline{T(U_E)}$; consequently, $r \leq q(T)$.

In fact, for any $y \in O_F$, there is an $x \in E$ with $ry = Tx$ [since T is onto]. As $T = T_0Q$ and T_0 is an isomorphism, it follows from $ry=Tx$ that $rT_0^{-1}y = T_0^{-1}Tx = Qx$, and hence that

$$\|Qx\| = r\|T_0^{-1}y\| \leq r\|T_0^{-1}\|\ \|y\| = \|y\| < 1.$$

Notice that $Q(O_E) = O_{E/\mathrm{Ker}\ T}$. There exists an $u \in O_E$ such that $Qx = Qu$; it then follows from $T=T_o \circ Q$ and $ry = Tx$ that

$$ry = Tx = (T_oQ)x = T_o(Qu) = Tu \in T(O_E) \subset T(U_E),$$

which obtains our assertion.

For criteria of topological injections (resp. topological surjections) between <u>normed spaces</u>, we quote the following two results:

(6.b) <u>Topological injections and subspaces</u> (see Junek [1983]): Let E and F be normed spaces and $T \in L(E, F)$. Then the following statements are equivalent

(i) T is a topological injection.

(ii) T is one–one and relatively open.

(iii) (Universal property of topological injections and subspaces) T is one–one,and for any normed space G and any $S \in L(G,F)$ with

$$S(G) \subset T(E)$$

there exists exactly one $S_0 \in L(G,E)$ such that $S = TS_0$ (i.e., S can be factorized through T from the left).

(6.c) Universal property of topological surjections and quotients (see Junek [1983]): Let E and F be normed spaces and $T \in L(E,F)$. Then T is a topological surjection (i.e., open) if and only if T is onto, and for any normed space F_0 and any $S \in L(E,F_0)$ with

$$\text{Ker } T \subset \text{Ker } S,$$

there exists exactly one $R_0 \in L(F,F_0)$ such that $S = R_0T$ (i.e. S can be factorized through T from the right).

We now describe the duality relations between topological injections and topological surjections as follows:

(6.6) **Theorem** (Pietsch [1980]). Let E and F be Banach spaces and $T \in L(E,F)$. Then

$$j(T) = q(T') \text{ and } j(T') = q(T).$$

In particular, T is a topological injection (resp. metric injection) if and only if its dual operator T' is a topological surjection (resp. metric surjection); T is a topological surjection (resp. metric surjection) if and only if T' is a topological injection (resp. metric injection). Consequently, T is an isomorphism (resp. metric isomorphism) if and only if T'

is an isomorphism (resp. metric isomorphism).

Proof. The equality $j(T) = q(T')$ clearly follows from

$$(1) \qquad \tau\|x\| \le \|Tx\| \quad (x \in E) \Leftrightarrow \tau U_{E'} \subset T'(U_{F'}).$$

Thus, we are going to show that (1) is true.

In fact, let $\tau > 0$ be such that $\tau\|x\| \le \|Tx\|$ $(x \in E)$. For any $u' \in U_{E'}$, the functional f, defined by

$$<Tx,f> = <x,u'> \quad (x \in E),$$

is a linear functional on $T(E)$ (since T is one–one). As $\tau\|x\| \le \|Tx\|$ $(x \in E)$, it follows that

$$|<Tx, f>| = |<x, u'>| \le \|x\| \, \|u'\| \le \|x\| \le \tau^{-1}\|Tx\|;$$

in other words, f is continuous with $\|f\|_{T(E)} \le \tau^{-1}$. By Hahn–Banach's extension theorem, f has an extension $y' \in F'$ with $\|y'\| = \|f\|_{T(E)} \le \tau^{-1}$. In particular,

$$<x, T'y'> = <Tx, y'> = <Tx, f> = <x, u'> \quad (x \in E)$$

thus $u' = T'y'$ and $\tau u' = T'(\tau y') \in T'(U_{F'})$, which shows that

$$\tau U_{E'} \subset T'(U_{F'}).$$

Conversely, let $\tau > 0$ be such that $\tau U_{E'} \subset T'(U_{F'})$. For any $x \in E$, we have

$$\begin{aligned} \tau\|x\| &= \tau \sup\{ |<x,u'>| : u' \in U_{E'} \} \\ &\le \tau \sup\{ |<x,\tau^{-1}T'v'>| : v' \in U_{F'} \} \\ &= \sup\{ |<Tx,v'>| : v' \in U_{F'} \} \\ &= \|Tx\|. \end{aligned}$$

Therefore (1) holds.

By (6.2), the equality $j(T') = q(T)$ follows from

$$\text{(2)} \qquad \tau U_F \subset \overline{T(U_E)} \quad \Leftrightarrow \quad \tau\|v'\| \leq \|T'v'\| \quad (v' \in F').$$

We are going to verify (2).

In fact, let $\tau > 0$ be such that $\tau U_F \subset \overline{T(U_E)}$ .For any $v' \in F'$, we have

$$\begin{aligned}\tau\|v'\| &= \tau \sup\{ |<y,v'>| : y \in U_F \} \\ &\leq \tau \sup\{ |\tau^{-1}<Tx, v'>| : x \in U_E \} \\ &= \sup\{ |<x,T'v'>| : x \in U_E \} = \|T'v'\| .\end{aligned}$$

Conversely, let $\tau > 0$ be such that

$$\text{(3)} \qquad \tau\|v'\| \leq \|T'v'\| . \quad (v' \in F').$$

To prove that $\tau U_F \subset \overline{T(U_E)}$, we assume on the contrary that there is an $y \in U_F$ such that

$$\tau y \notin \overline{T(U_E)}.$$

Then the strong separation theorem (see(8.3) in §8) ensures that there is an $g \in F'$ such that

$$\sup\{ |<Tx, g>| : x \in U_E \} \leq 1 < |<\tau y, g>|.$$

Therefore

$$\|T'g\| = \sup\{ |<x, T'g>| : x \in U_E \} \leq 1 < |<\tau y, g>| \leq \tau\|y\| \|g\| \leq \tau\|g\|,$$

which contradicts (3).

(6.d) The composition of topological injections: Let E, F and G be Banach spaces and given the following two operators:

$$E \xrightarrow{S} F \xrightarrow{T} G$$

Then TS is a topological injection if and only if

(i) $S : E \longrightarrow F$ is a topological injection, and

(ii) $TJ^{F}_{S(E)} : S(E) \longrightarrow G$ is a topological injection.

Dually, one can verify the following:

(6.e) The composition of topological surjections: Let E, F and G be Banach spaces and given the following two operators:

$$E \xrightarrow{S} F \xrightarrow{T} G$$

Then TS is a topological surjection if and only if

(i) $T: F \longrightarrow G$ is a topological surjection,and

(ii) $Q^{F}_{\mathrm{Ker}\ T} S : E \longrightarrow F/\mathrm{Ker}\ T$ is a topological surjection.

As other important applications of (3.j), we have the following:

(6.f) Extension property and lifting property: A Banach space F is said to have the extension property (resp. metric extension property) ,if for any Banach spaces E and E_0 , any topological injection (resp. metric injection), $J : E_0 \rightarrowtail E$ and any $S_0 \in L(E_0,F)$ there exists an extension $S \in L(E.F)$ such that

$$S_0 = SJ \text{ (resp. } S_0 = SJ \text{ and } \|S\| = \|S_0\|).$$

Dually, a Banach space E is said to have the lifting property (resp. metric lifting property) if for any Banach spaces F and F_0, any topological surjection (resp. metric surjection and $\epsilon > 0$) $Q : F \twoheadrightarrow F_0$ and any $T_0 \in L(E,F_0)$ there exists a lifting $T \in L(E,F)$ such that

$$T_0 = QT \text{ (resp. } T_0 = QT \text{ and } \|T\| \leq (1+\epsilon)\|T_0\|).$$

(i) For any index set Λ, the space $\ell^\infty(\Lambda)$ has the metric extension property; while the space $\ell^1(\Lambda)$ has the metric lifting property.

(ii) Let E be a Banach space. The map $J_E: E \rightarrowtail \ell^\infty(U_{E'})$, defined by

$$J_E x = [\langle x,f\rangle, f \in U_{E'}] \quad (x \in E),$$

is a metric injection from E into $\ell^\infty(U_{E'})$. (Consequently, any Banach space is metrically isomorphic to a subspace of some Banach space with the metric extension property.) The map $Q_E : \ell^1(U_E) \twoheadrightarrow E$, defined by

$$Q_E([\xi_x, x \in U_E]) = \Sigma_{U_E} \xi_x \cdot x \quad ([\xi_x, x \in U_E] \in \ell^1(U_E)),$$

is a metric surjection from $\ell^1(U_E)$ onto E. (Consequently, any Banach space is metrically isomorphic to a quotient space of some Banach space with the metric lifting property.)

(iii) $E \prec \ell^\infty(U_{E'})$ if and only if E has the extension property.

(iv) $E \prec \ell^1(\Lambda)$ (for some Λ) if and only if E has the lifting property.

Exercises

6–1. Prove (6.a).

6–2. Prove (6.b).

6–3. Prove (6.c).

6–4. Prove (6.d).

6–5. Prove (6.e).

6–6. Prove (6.f).

6–7. Denote by $\mathbb{A}$ a class of B–spaces containing all isomorphic copies of any $G \in \mathbb{A}$ (i.e., if E is a B–space such that $E \simeq G$ (isomorphic) for some $G \in \mathbb{A}$ then $E \in \mathbb{A}$).

Let E and F be B–spaces and $T \in L(E,F)$. Show that the following statements are equivalent.

(a) If M is a closed subspace of E such that $TJ_M^E : M \longrightarrow F$ is a topological injection, then $M \notin \mathbb{A}$.

(b) T has no bounded inverse on subspaces D of E with $D \in \mathbb{A}$.

(c) There exists no subspace D of E with $D \in \mathbb{A}$ such that $TJ_D^E : D \longrightarrow F$ is a topological injection.

(d) Let G be a B–space. If there exist topological injections $R_1 \in L(G,E)$ and $R_2 \in L(G,F)$ such that $R_2 = TR_1$ then $G \notin \mathbb{A}$.

Dually, the following statements are equivalent for any $T \in L(E,F)$:

(a$'$) If N is a closed subspace of F such that $Q_N^F T : E \longrightarrow F/N$ is open, then $F/N \notin \mathbb{A}$.

(c$'$) There exists no subspace N_o of F with $F/N_o \in \mathbb{A}$ such that $Q_{N_o}^F T : E \longrightarrow F/N_o$ is open.

(d$'$) Let G be a B–space. If there exist open operators $S_1 \in L(E,G)$ and $S_2 \in (F,G)$ such that $S_1 = S_2 T$, then $G \notin \mathbb{A}$.

6–8. Suppose that E, E_0, F and F_0 are B–spaces.

(a) Given two operators $S_0 \in L(E,F_0)$ and $S \in L(E,F)$. If $\|S_0 x\| \le \|Sx\|$ $(x \in E)$ and F_0 has the metric extension property, then there exists an operator $L \in L(F,F_0)$ such that $S_0 = LS$ and $\|L\| \le 1$.

(b) Dually given two operators $T_0 \in L(E_0,F)$ and $T \in L(E,F)$. If $T_0(U_{E_0}) \subset T(U_E)$ and E_0 has the metric lifting property, then for any $\epsilon > 0$, there exists an operator $R \in L(E_0,E)$ such that $T_0 = TR$ and $\|R\| \le 1 + \epsilon$.

(Hint : Use (3.2) and Ex. 3–17.)

7

Vector Topologies

A natural generalization of normed spaces is vector spaces equipped with a topology which is induced by a family of seminorms instead of one norm. Vector spaces with such a topology are special cases of a more general class of spaces, called topological vector spaces, as defined by the following:

(7.1) **Definition.** Let X be a vector space over $\mathbb{K}$. A topology $\mathcal{P}$ on X is called a vector topology if it is compatible with the algebraic operations of X in the following sense:

(VT1) the map $(x,y) \longrightarrow x + y : X \times X \longrightarrow X$ is continuous;

(VT2) the map $(\lambda,y) \longrightarrow \lambda y : \mathbb{K}\times X \longrightarrow X$ is continuous;

A vector space equipped with a vector topology is called a topological vector space (abbreviated TVS). Hereafter we shall generally denote a TVS by $(X,\mathcal{P})$ or $X_{\mathcal{P}}$ or simply by X if the vector topology on X does not require any special notation; and the term local base, denoted by $\mathcal{U}_X$, will always mean a local base at 0 for $\mathcal{P}$.

Two TVS X and Y over $\mathbb{K}$ are said to be topologically isomorphic, denoted by $X \cong Y$, if there exists a bijective linear map $T : X \longrightarrow Y$ which is a homeomorphism; T is referred to as a topological isomorphism from X onto Y.

For a TVS X, the whole topological structure of X is determined by a local base (hence a linear map $T : X \longrightarrow Y$ is continuous if and only if it is continuous at 0), as shown by the following simple, but important result which is similar to (2.a).

(7.a) Homeomorphism of translation: Let X be a TVS. For any $x_0 \in X$ and $0 \neq r_0 \in \mathbb{K}$, the translation,defined by

$$y \longrightarrow x_0 + r_0 y \quad (\text{for all } y \in X),$$

is a homeomorphism from X onto X.

Vector topologies can be characterized by means of a local base as shown by the following result.

(7.2) **Theorem.** In a TVS $X_{\mathscr{P}}$, there exists a local base $\mathscr{U}_X$ whose members have the following properties:

(NS1) every member in $\mathscr{U}_X$ is circled and absorbing;

(NS2) for any $V \in \mathscr{U}_X$ there is an $W \in \mathscr{U}_X$ such that $W + W \subset V$.

Conversely, if $\mathscr{U}$ is a filter base on X (i.e., any element in $\mathscr{U}$ is non–empty and for any $V_1, V_2 \in \mathscr{U}$ there is an $V \in \mathscr{U}$ with $V \subset V_1 \cap V_2$) which satisfies (NS1) and (NS2), then there exists a unique vector topology $\mathscr{T}$ on X such that $\mathscr{U}$ is a local base at 0 for $\mathscr{T}$.

Proof. The continuity of the map $(\lambda,x) \longrightarrow \lambda x$ at (0,0) ensures that the family of all circled 0–neighbourhoods in X is a local base at 0. [For any 0–neighbourhood W in X there is a 0–neighbourhood U and $\delta > 0$ such that $\lambda U \subset W$ (for all $|\lambda| \leq \delta$), hence the set $V = \cup\{ \lambda U : |\lambda| \leq \delta \}$ is a circled 0–neighbourhood with $V \subset W$.] For any fixed $x_0 \in X$, the continuity of $(\lambda, x_0) \longrightarrow \lambda x_0$ at $\lambda = 0$ ensures that every 0–neighbourhood (in particular, every $V \in \mathscr{U}_X$) is absorbing.

By (VT1),the continuity of the map $(x,y) \longrightarrow x + y$ at (0,0) ensures that $\mathscr{U}_X$ satisfies the condition (NS2).

Conversely, we first define a topology $\mathscr{T}$ on X by setting

$$\mathscr{T} = \{ \phi \neq G \subset X : x \in G \text{ implies } x + V \subset G \quad (\text{for some } V \in \mathscr{U}) \} \cup \{\phi\}.$$

Clearly $\mathscr{T}$ is the unique topology on X such that $\mathscr{U}$ is a local base at 0 for $\mathscr{T}$. We complete the proof by showing that $\mathscr{T}$ is a vector topology.

By (NS2), the continuity of the map $(x,y) \longrightarrow x + y$ at (x_0,y_0) follows from $(x_0 + W) + (y_0+W) \subset x_0 + y_0 + V$.

To prove the continuity of the map $(\lambda,y) \longrightarrow \lambda y$ at (λ_0,y_0), let $\lambda_0 y_0 + U$ be any $\lambda_0 y_0$–neighbourhood, where $U \in \mathscr{U}$. Suppose that n is a natural number with $|\lambda_0| \leq n$. Then (NS2) implies that there is an $V \in \mathscr{U}$ such that $V + \cdots + V$ (n+2 summands) is contained in U ; consequently, $nV + V + V \subset U$ (since $nV \subset V + \cdots + V$ (n summands)). As V is absorbing, there is a natural number $m \geq 1$ such that

(1) $$y_0 \in mV .$$

Suppose now that $\lambda \in K$ is such that $|\lambda - \lambda_0| \leq \frac{1}{m}$ and that $y \in y_0 + V$. We claim that

$$\lambda y \in \lambda_0 y_0 + U,$$

hence the assertion follows. Indeed, since V is circled, we have

(2) $$(\lambda - \lambda_0)y_0 \in (\lambda - \lambda_0)mV \subset V \quad (\text{by}(1)),$$

(3) $$(\lambda - \lambda_0)(y - y_0) \in (\lambda - \lambda_0)V \subset \frac{1}{m}V \subset V ,$$

(4) $$\lambda_0(y - y_0) \in \lambda_0 V \subset nV$$

(since $|\lambda_0| \leq n$); it then follows from (2),(3) and (4) that

$$\lambda y = \lambda_0 y_0 + (\lambda - \lambda_0)y_0 + (\lambda - \lambda_0)(y - y_0) + \lambda_0(y - y_0)$$

$$\in \lambda_0 y_0 + V + V + nV \subseteq \lambda_0 y_0 + U ;$$

this proves our assertion.

Remark. As a consequence of (7.2), we see that every 0–neighbourhood in $X_{\mathscr{P}}$ contains a closed 0–neighbourhood; hence there exists a local base $\mathscr{U}_X$ at 0 consisting of closed, circled and absorbing sets.

(7.3) **Definition.** A vector topology $\mathscr{P}$ on X is said to be locally convex if $\mathscr{P}$

admits a local base at 0 consisting of convex sets. A vector space equipped with a Hausdorff locally convex topology is called a locally convex space (abbreviated LCS).

Locally convex topologies can be characterized by means of local base as shown by the following result.

(7.4) **Theorem.** Let $\mathscr{P}$ be a locally convex topology on X. Then there exists a local base $\mathscr{U}_X$ whose members have the following properties:

(LC1) every member in $\mathscr{U}_X$ is absolutely convex and absorbing;

(LC2) if $V \in \mathscr{U}_X$ and $\lambda > 0$ then $\lambda V \in \mathscr{U}_X$.

Conversely, if $\mathscr{U}$ is a filter base on X which satisfies conditions (LC1) and (LC2), then there exists a unique locally convex topology $\mathscr{T}$ on X such that $\mathscr{U}$ is a local base at 0 for $\mathscr{T}$.

Proof. There exists, by (7.3), a local base $\mathscr{M}$ consisting of convex 0–neighbourhoods. If $V \in \mathscr{M}$, by (7.2) there is a circled 0–neighbourhood W such that $W \subset V$, hence $W \subset \text{co } W \subset V$, thus co W is a convex 0–neighbourhood which is obviously circled; consequently, $\mathscr{P}$ admits a local base $\mathscr{V}$ consisting of absolutely convex 0–neighbourhoods. Now the family, defined by

$$\mathscr{U}_X = \{ \lambda W : W \in \mathscr{V}, \lambda > 0 \},$$

has the required properties.

Conversely, the assertion follows from (7.2) and (7.3).

Locally convex topologies can be determined by a family of seminorms. To do this, we require the following terminology and result (see (7.b) below). A family $\mathbb{P}$ of seminorms on X is saturated if

$$\max_{1 \leq i \leq n} p_i \in \mathbb{P} \quad \text{whenever} \quad p_i \in \mathbb{P} \ (i = 1,2,\cdots n),$$

where $(\max_{1 \le i \le n} p_i)\, x = \max_{1 \le i \le n} p_i(x)$ (for all $x \in X$).

(7.b) <u>The continuity of seminorms</u>: Let $X_{\mathscr{P}}$ be a TVS, let V be an absolutely convex, absorbing subset of X and p_V the gauge of V. Then:

(i) $\text{Int } V \subset \{ x \in X : p_V(x) < 1 \} \subset V \subset \{ x \in X : p_V(x) \le 1 \} \subset \overline{V}$.

(ii) V is a $\mathscr{P}$-neighbourhood of 0 if and only if p_V is continuous; in this case,

$$\text{Int } V = \{ x \in X : p_V(x) < 1 \} \text{ and } \overline{V} = \{ x \in X : p_V(x) \le 1 \},$$

hence p_V is the gauge of $\text{Int } V$ as well as of $\overline{V}$ (where $\text{Int } V$ is the interior of V).

(7.5) **Theorem.** <u>A locally convex topology</u> $\mathscr{P}$ <u>on</u> X <u>can always be defined by a saturated family of seminorms</u>, e.g. <u>by the family of all</u> $\mathscr{P}$<u>-continuous seminorms on</u> X. <u>Conversely</u>, <u>if</u> $\mathbb{P}$ <u>is a family of seminorms on</u> X, <u>then there exists a coarsest locally convex topology</u> $\mathscr{T}$ <u>on</u> X <u>such that every element in</u> $\mathbb{P}$ <u>is</u> $\mathscr{T}$<u>-continuous</u>, <u>and a local base of closed</u> 0–<u>neighbourhoods is formed by the sets</u>

$$\{ x \in X : \max_{1 \le i \le n} p_i(x) \le \epsilon \} \quad (\epsilon > 0 \text{ and } p_i \in \mathbb{P}).$$

<u>Furthermore</u>, $\mathscr{T}$ <u>is Hausdorff if and only if for any</u> $0 \ne x \in X$ <u>there is a</u> $p \in \mathbb{P}$ <u>such that</u> $p(x) \ne 0$.

(2.2) holds for a TVS as shown by the following:

(7.c) <u>Simple properties of TVS</u>: Let $X_{\mathscr{P}}$ be a TVS, let $\mathscr{U}_X$ be a local base and $A, B \subset X$. Then:

(i) $\overline{A} = \cap \{ A + V : V \in \mathscr{U}_X \}$.
(ii) $A + G$ is open whenever G is open, hence

$$A + \text{Int } B \subset \text{Int } (A + B).$$

(iii) Let $X_{\mathscr{P}}$ be Hausdorff, let $K \subset X$ be compact, and let $B \subset X$ be closed. If $K \cap B = \phi$ then there is an $V \in \mathscr{U}_X$ such that

$$(K + V) \cap (B + V) = \phi .$$

Consequently, if $C \subset X$ is compact and B is closed then $C + B$ is closed in X.

(7.6) **Proposition.** Let $X_{\mathscr{P}}$ be a TVS and $B \subset X$.

(a) If B is open then so do ΓB and co B.

(b) Suppose that B is convex (resp. disked), and that $\text{Int } B \neq \phi$. Then

(7.6.1) $$\lambda \overline{B} + (1 - \lambda) \text{ Int } B \subset \text{Int } B \text{ whenever } \lambda \in [0,1).$$

$$(\text{resp. } \mu \overline{B} + \xi \text{Int } B \subset \text{Int } B \text{ whenever } |\mu| + |\xi| \leq 1 \text{ and } |\mu| \neq 1).$$

Cosequently, $\overline{B}$ and Int B are convex (resp. disked)

(7.6.2) $$\overline{B} = \overline{\text{Int } B} \quad \text{and} \quad \text{Int } B = \text{Int } \overline{B}.$$

<u>Proof.</u> (a) An element in ΓB is of the form $\Sigma_{i=1}^{n} \mu_i x_i$, where $x_i \in B$ and $\Sigma_{i=1}^{n} |\mu_i| = \mu \leq 1$ with $\mu_i \neq 0$. The openness of B ensures that there is a circled 0–neighbourhood V in X such that $x_i + V \subset B$ $(i = 1,\cdots,n)$. We claim that

$$\Sigma_{i=1}^{n} \mu_i x_i + \mu V \subset \Gamma B ,$$

hence the openness of ΓB follows. Indeed, for any $y \in V$, since $x_i + \frac{|\mu_i|}{\mu_i} y \in B$, it follows that

$$\Sigma_{i=1}^{n} \mu_i x_i + \mu y = \Sigma_{i=1}^{n} \mu_i \left(x_i + \frac{|\mu_i|}{\mu_i} y\right) \in \Gamma B,$$

which proves our assertion.

The proof of the openness of $\text{co } B$ is similar to that of ΓB.

(b) For any fixed $\lambda \in [0,1)$, since $\lambda \overline{B} + (1 - \lambda) \text{ Int } B$ is open, it suffices to show that

$$\lambda \overline{B} + (1 - \lambda) \text{ Int } B \subseteq B.$$

Indeed, for any $x \in \text{Int } B$, $(1 - \lambda)(\text{Int } B - x)$ is an open 0–neighbourhood, it follows (see (7.c) (i)) that

$$\begin{aligned} \lambda \overline{B} &= \overline{\lambda B} \subseteq \lambda B + (1 - \lambda)(\text{Int } B - x) \\ &\subset \lambda B + (1 - \lambda)B - (1 - \lambda)x \subseteq B - (1 - \lambda)x \end{aligned}$$

(on account of the convexity of B), and hence that $\lambda \overline{B} + (1 - \lambda) \text{ Int } B \subseteq B$ (since $x \in \text{Int } B$ was arbitrary).

By (7.6.1), it is easily seen that $\text{Int } B$ is convex; while the convexity of $\overline{B}$ follows from the continuity of sums and multiplications by scalars.

To prove (7.6.2), we first observe that

$$\overline{\text{Int } B} \subset \overline{B} \quad \text{and} \quad \text{Int } B \subset \text{Int } \overline{B}.$$

In (7.6.1), letting $\lambda \longrightarrow 1$,we obtain $\overline{B} \subset \overline{\text{Int } B}$, hence $\overline{B} = \overline{\text{Int } B}$. On the other hand, to verify that $\text{Int } \overline{B} \subset \text{Int } B$, it suffices to show that

$$0 \in \operatorname{Int} \overline{B} \nRightarrow 0 \in \operatorname{Int} B.$$

Indeed, there is a circled 0–neighbourhood V such that $V \subsetneq \overline{B}$, hence $0 \in \overline{\operatorname{Int} B}$ (on account of $\overline{B} = \overline{\operatorname{Int} B}$ and $0 \in V \subset \overline{B}$); consequently, $V \cap \operatorname{Int} B \neq \phi$. Now we take $y \in V \cap \operatorname{Int} B$, then $-y \in V \subset \overline{B}$, and apply (7.6.1) to conclude that

$$0 = \tfrac{1}{2}(-y) + \tfrac{1}{2}y \in \tfrac{1}{2}\overline{B} + \tfrac{1}{2}\operatorname{Int} B \subset \operatorname{Int} B.$$

Remark. If B is closed, then ΓB and $\operatorname{co} B$ need not be closed. For instance, if B is the closed set in $\mathbb{R}^2$ which consists of the points $(-1,0)$, $(1,0)$ and the y–axis, then the real absolutely convex hull of B is not closed.

(7.7) **Definition.** Let $X_{\mathscr{P}}$ be a TVS and A, B two subsets of X. We say that A *absorbs* B (or B is *absorbed* by A) if there exists an $\lambda > 0$ such that

$$B \subset \mu A \quad (\text{for all } |\mu| \geq \lambda).$$

A subset D of X is said to be *bounded* if it is absorbed by any 0–neighbourhood in X.

In a normed space $(E, \|\cdot\|)$, the boundedness of a subset B means that $B \subset \mu U_E$ for some $\mu > 0$, hence the definition of boundedness is a generalization of the notion of bounded sets in a normed space; moreover, (5.a) holds for TVS as shown by the following:

(7.8) **Proposition.** *Let* $X_{\mathscr{P}}$ *be a TVS, and let* $\mathfrak{M}_{von}(\mathscr{P})$ *be the family of all bounded subsets of* X. *Then*:

(VB1) $X = \cup\, \mathfrak{M}_{von}(\mathscr{P})$.

(VB2) *If* $B \subset A$ *and* $A \in \mathfrak{M}_{von}(\mathscr{P})$, *then* $B \in \mathfrak{M}_{von}(\mathscr{P})$.

(VB3) $A_1 + \cdots + A_n \in \mathfrak{M}_{von}(\mathscr{P})$ whenever $A_i \in \mathcal{M}_{von}(\mathscr{P})$ $(i=1,2,\cdots,n)$.

(VB4) $\lambda A \in \mathfrak{M}_{von}(\mathscr{P})$ whenever $A \in \mathfrak{M}_{von}(\mathscr{P})$ and $\lambda \in K$.

(VB5) The circled hull of any $A \in \mathfrak{M}_{von}(\mathscr{P})$ belongs to $\mathfrak{M}_{von}(\mathscr{P})$.

If $X_{\mathscr{P}}$ is assumed to be a LCS, then (VB5) can be replaced by

$(VB5)^*$ $\Gamma A \in \mathfrak{M}_{von}(\mathscr{P})$ whenever $A \in \mathfrak{M}_{von}(\mathscr{P})$.

The family $\mathfrak{M}_{von}(\mathscr{P})$ is called the von Neumann bornology.

The proof of this result is trivial, hence will be omitted.

Remark. It is clear that $\overline{A} \in \mathfrak{M}_{von}(\mathscr{P})$ whenever $A \in \mathfrak{M}_{von}(\mathscr{P})$.

Let $X_{\mathscr{P}}$ be a TVS. By a fundamental system of bounded sets is meant a family $\mathscr{B}$ of $\mathscr{P}$-bounded sets in X such that every bounded set is contained in a suitable member of $\mathscr{B}$. Clearly, the family of all ($\mathscr{P}$-closed) circled bounded sets forms a fundamental system of bounded sets. If, in addition, $X_{\mathscr{P}}$ is a LCS, then the family of all ($\mathscr{P}$-closed) absolutely convex bounded sets forms a fundamental system of bounded sets.

(7.9) **Proposition.** Let $X_{\mathscr{P}}$ be a TVS and $B \subset X$. Then the following statements are equivalent:

(a) B is bounded.

(b) For any sequence $\{x_n\}$ in B and any sequence $\{\lambda_n\}$ in K with $\lambda_n \longrightarrow 0$, one has $0 = \mathscr{P}\text{-}\lim_n \lambda_n x_n$.

(c) Every sequence in B is bounded.

Proof. (a) $\Rightarrow$ (b): Let V be a circled 0–neighbourhood in X. There is an $\alpha > 0$ such that $x_n \in \alpha V$ (for all $n \geq 1$), hence $\lambda_n x_n \in \lambda_n \alpha V$ $(n \geq 1)$. As $\lambda_n \longrightarrow 0$, it follows that $\alpha\lambda_n \longrightarrow 0$, and hence that there exists a natural number $n_0 \geq 1$ such that $|\alpha\lambda_n| \leq 1$ (for all $n \geq n_0$); consequently,

$$\lambda_n x_n \in \lambda_n \alpha V \subset V \quad (\text{for all } n \geq n_0)$$

since V is circled.

(b) ⇏ (c): Suppose that there exists a sequence $\{y_n\}$ in B which is not bounded. Then there is a circled 0–neighbourhood V such that

$$\{ y_n : n \geq 1 \} \not\subset m^2V \quad \text{for all } m \geq 1.$$

For each $m \geq 1$, let $z_m \in \{ y_n : n \geq 1 \}$ be such that $z_m \notin m^2V$. Then $\{z_m\}$ is a sequence in B and $\frac{1}{m} \longrightarrow 0$ for which the sequence $\{ \frac{1}{m} z_m \}$ does not converge to 0.

(c) ⇏ (a): Suppose that B is not bounded. Then there exists a circled 0–neighbourhood U such that

$$B \not\subset n^2U \quad (\text{for all } n \geq 1).$$

For each $n \geq 1$, let $x_n \in B \backslash n^2U$. Then $\{x_n\}$ is a sequence in B which is not bounded.

In a normed space $(E, \|\cdot\|)$, the unit ball O_E is bounded and also a convex 0–neighbourhood for the $\|\cdot\|_E$–top. The converse is a criterion for a TVS to be normable in the sense that a TVS $X_{\mathscr{P}}$ is <u>normable</u> if its topology $\mathscr{P}$ is defined by some norm on X, i.e., if there exists a norm $\|\cdot\|$ on X such that the balls

$$U_r = \{ x \in X : \|x\| \leq r \} \quad (r > 0)$$

form a local base at 0 for $\mathscr{P}$.

(7.10) **Proposition.** <u>A TVS $X_{\mathscr{P}}$ is normable if and only if it is Hausdorff and has a bounded convex $\mathscr{P}$–neighbourhood of</u> 0.

<u>**Proof.**</u> The necessity is obvious. To prove the sufficiency, let U be a bounded, convex $\mathscr{P}$–neighbourhood of 0. Then there is a circled 0–neighbourhood V in X such that $V \subset U$, hence $W = \text{co } V$ is a bounded, absolutely convex 0–neighbourhood such that

$W \subset U$. As $\mathscr{P}$ is Hausdorff, it follows from the boundedness of W that the gauge p_W of W is a norm on X. [If $p_W(x) = 0$, then $nx \in \operatorname{Ker} p_W \subset W$ $(n \geq 1)$, hence $\frac{1}{n}(nx) = x \longrightarrow 0$ by the boundedness of W.] Clearly $\{ n^{-1}W : n \geq 1 \}$ is a local base for $\mathscr{P}$; in other words, $\mathscr{P}$ is defined by the norm p_W.

Remark. The preceding result also tells us that a TVS $X_{\mathscr{P}}$ is normable if and only if it is a LCS and has a bounded 0–neighbourhood.

Let $X_{\mathscr{P}}$ be a TVS. A net $\{ x_d, d \in D \}$ in X is called a Cauchy net if for any 0–neighbourhood V, there is an d_0 such that

$$x_\lambda - x_\mu \in V \quad (\text{for all } \lambda,\mu \geq d_0).$$

A filter base $\mathfrak{M}$ on X is called a Cauchy filter base if for any 0–neighbourhood V there exists an $N \in \mathfrak{M}$ such that

$$N - N \subset V.$$

(7.11) **Definition.** A Hausdorff TVS $X_{\mathscr{P}}$ is said to be:

(a) complete if every Cauchy net in X is convergent;

(b) sequentially complete if every Cauchy sequence in X converges;

(c) quasi–complete if every bounded, Cauchy net in X converges.

(7.d) The completion: Let $(X,\mathscr{P})$ be a Hausdorff TVS. There is a complete Hausdorff TVS $(\tilde{X},\tilde{\mathscr{P}})$ containing X as a dense subspace; $\tilde{X}$ is uniquely up to a topological isomorphism; moreover, if $\mathscr{U}_X$ is a local base at 0 for $\mathscr{P}$, then $\{ \tilde{V} : V \in \mathscr{U}_X \}$ is a local base for $\tilde{\mathscr{P}}$, where $\tilde{V}$ is the $\tilde{\mathscr{P}}$-closure of V in $\tilde{X}$. The space $(\tilde{X},\tilde{\mathscr{P}})$ is called the completion of $(X,\mathscr{P})$.

(7.e) Totally bounded sets and compact sets in TVS: Let $X_{\mathscr{P}}$ be a TVS and $B \subset X$. B is said to be totally bounded if for any $\mathscr{P}$-neighbourhood V of 0 there is a finite subset $\{b_1, \cdots, b_n\}$ of B such that

$$B \subset \bigcup_{i=1}^{n} (b_i + V).$$

A subset B of a Hausdroff TVS $X_{\mathscr{P}}$ is precompact if the closure of B in the completion $\tilde{X}$ of X is compact. [Clearly, this is equivalent to say that B is regarded as a relatively compact subset of the completion $(\tilde{X}, \tilde{\mathscr{P}})$ of $(X, \mathscr{P})$.] Suppose now that $X_{\mathscr{P}}$ is a Hausdorff TVS. Then the following assertions are true:

(i) A subset B of X is precompact $\Leftrightarrow$ B is totally bounded. $\Leftrightarrow$ Every ultrafilter on B is a Cauchy filter.

(ii) A subset B of X is compact if and only if it is totally bounded and complete.

(iii) The circled hull of a totally bounded (resp.compact) set is totally bounded (resp.compact).

(iv) The convex hull (resp.absolutely convex hull)of a finite family of compact convex (resp. compact absolutely convex) sets in $X_{\mathscr{P}}$ is compact.

(7.12) **Definition.** A TVS $X_{\mathscr{P}}$ is said to be metrizable if its topology $\mathscr{P}$ is metrizable, namely there is a metric d on X such that for any $x \in X$ the sets

$$O_{\alpha}(x) = \{ u \in X : d(u,x) < \alpha \} \quad (\alpha > 0)$$

form a local base at x for $\mathscr{P}$.

Complete, metrizable LCS are called Fréchet spaces (abbreviated F–spaces).

(7.f) Metrizable TVS (I):(i) For a TVS $X_{\mathscr{P}}$, the following statements are equivalent:

(A) $X_{\mathscr{P}}$ is metrizable.

(B) $\mathscr{P}$ is Hausdorff and admits a countable local base $\{V_n\}$ of circled sets with

$$V_{n+1} + V_{n+1} \subset V_n \quad \text{(for all } n \geq 1).$$

(C) $\mathscr{P}$ is generated by some F–norm q in the following sense: a mapping $q : X \longrightarrow [0,+\infty)$ satisfies:

(F1) $q(x) = 0$ if and only if $x = 0$.

(F2) $q(x + y) \leq q(x) + q(y)$ (for all $x, y \in X$).

(F3) $q(\lambda x) \leq q(x)$ (for all $x \in X$ and $|\lambda| \leq 1$).

(F4) $q(\lambda_j x) \longrightarrow 0$ as $\lambda_j \longrightarrow 0$.

(ii) A LCS $X_{\mathscr{P}}$ is metrizable if and only if $\mathscr{P}$ is defined by a sequence $\{p_n : n \geq 1\}$ of seminorms which satisfies the following properties:

(1) $p_1 \leq p_2 \leq \cdots$ ($\{p_n : n \geq 1\}$ is increasing).

(2) If $p_n(x) = 0$ (for all $n \geq 1$), then $x = 0$.

In this case, $\mathscr{P}$ is determined by an F–norm $|\cdot|$, which is given by

$$|x| = \Sigma_{n=1}^{\infty} \frac{1}{2^n} \frac{p_n(x)}{1+p_n(x)} \quad \text{(for all } x \in X).$$

The result (2b)(iii), which is a characterization of completeness of normed spaces, can be generalized to the case of metrizable spaces, as shown by the following result.

(7.13) **Theorem.** <u>Let</u> $X_{\mathscr{P}}$ <u>be a metrizable TVS and let</u> q <u>be an</u> F–<u>norm on</u> X <u>determined by</u> $\mathscr{P}$. <u>Then</u> $X_{\mathscr{P}}$ <u>is complete if and only if a formal series</u> Σx_n <u>in</u> X <u>which is absolutely convergent</u> (i.e., $\Sigma_n q(x_n) < \infty$), <u>must be convergent</u> (i.e., $u = \mathscr{P}\text{-}\lim_n \Sigma_{j=1}^{n} x_j$); <u>in this case, we have</u>

$$q(u) \leq \Sigma_n q(x_n).$$

<u>**Proof**</u>. <u>**Necessity**</u>: The sequence $\{s_n\}$ in X, defined by $s_n = \Sigma_{j=1}^{n} x_j$, is a Cauchy

sequence (since $q(s_{n+k} - s_n) \leq \Sigma_{i=n+1}^{n+k} q(x_i)$), hence converges to a $u \in X$. Clearly the F–norm q is $\mathscr{P}$-continuous,hence

$$q(u) = \lim_n q(s_n) \leq \lim_n \Sigma_{i=1}^{n} q(x_i) = \Sigma_i q(x_i) .$$

<u>Sufficiency</u>. Let $\{y_n\}$ be a Cauchy sequence in X. Using the Cauchy property,one can construct by induction a subsequence $\{y_{n_i}\}$ of $\{y_n\}$ such that

$$q(y_{n_{i+1}} - y_{n_i}) \leq \frac{1}{2^i} \quad \text{(for all } i \geq 1),$$

hence $\Sigma_{i=1}^{\infty} q(y_{n_{i+1}} - y_{n_i}) < \infty$. By the assumption, there is a unique $z \in X$ such that

$$z = \mathscr{P}\text{-}\lim_{r\to\infty} \Sigma_{i=1}^{r} (y_{n_{i+1}} - y_{n_i}) = \mathscr{P}\text{-}\lim_{r\to\infty} (y_{n_{r+1}} - y_{n_1}),$$

(since $\Sigma_{i=1}^{r} (y_{n_{i+1}} - y_{n_i}) = y_{n_{r+1}} - y_{n_1}$); consequently,

$$\mathscr{P}\text{-}\lim_k y_{n_k} = z + y_{n_1} = u ;$$

in other words, the subsequence $\{y_{n_k}\}$ of the Cauchy sequence $\{y_n\}$ is convergent to u, hence $u = \mathscr{P}\text{-}\lim_n y_n$.

Banach's open mapping theorem and closed graph theorem can be generalized to the case of complete metrizable TVS as mentioned in the following:

(7.g) <u>Complete metrizable TVS:</u> Let X and Y be TVS. A linear map $T : X \longrightarrow Y$ is said to be <u>almost open</u> if the closure $\overline{T(V)}$, in Y, is a 0–neighbourhood in Y whenever V is a 0–neighbourhood in X.

(i) (Banach's open mapping theorem) Let X and Y be complete metrizable TVS and let $T : X \longrightarrow Y$ be continuous and linear. If T is either almost open or onto, then T is open.

(ii) (Banach's closed graph theorm) Let X and Y be complete metrizable TVS and let $T : X \longrightarrow Y$ be linear. If the graph of T is closed, then T is continuous.

(iii) (The uniform boundedness theorem) Let X and Y be F–spaces and $\mathfrak{R}$ a family of continuous linear maps from X into Y. If $\mathfrak{R}$ is bounded pointwisely in the sense that for any $x \in X$, the set $\{T(x) : T \in \mathfrak{R}\}$ is bounded in Y, then for any continuous seminorm q on Y , there exists a continuous seminorm p on X and $\lambda > 0$ such that

$$\sup_{T\in\mathfrak{R}} q(Tx) \le \lambda p(x) \quad (\text{for all } x \in X).$$

For any two TVS X and Y, the set consisting of all continuous linear maps (called operators) from X into Y, denoted by $L(X,Y)$, is a vector subspace of $L^*(X,Y)$. In particular, if $Y = K$, then we write X' for $L(X,K)$. X' is called the topological dual of X.

(3.a) and the induced mapping theorem can be generalized to a more general case as shown by the following:

(7.h) Continuity of linear maps: Let X and Y be LCS.

(i) The following statements are equivalent for any $T \in L^*(X,Y)$:

(A) T is continuous at 0.

(B) T is continuous on X.

(C) For any 0–neighbourhood W in Y there is some 0–neighbourhood V in X such that $T(V) \subseteq W$.

(D) For any continuous seminorm q on Y there exists some continuous seminorm p on X such that $q(Tx) \le p(x)$ (for all $x \in X$).

(ii) (Induced mapping theorem) Let Z be another LCS. Given

$$T \in L(X,Y) \quad \text{and} \quad S \in L(X,Z).$$

Suppose that S is surjective and that for any continuous seminorm q on Y there is some continuous seminorm r on Z such that

$$q(Tx) \le r(Sx) \quad (\text{for all } x \in X).$$

Then there exists a unique $R \in L(Z,Y)$ such that $T = RS$.

We conclude this section with a fundamental method for constructing new TVS from given ones.

Let $\{(X_\alpha, \mathcal{P}_\alpha) : \alpha \in \Lambda\}$ be a family of TVS. Recall that the product topology $\prod_\alpha \mathcal{P}_\alpha$ on $\prod_\alpha X_\alpha$ is the coarsest topology for which all projections $\pi_\beta : \prod_\alpha X_\alpha \longrightarrow X_\beta$ are continuous. For each $\alpha \in \Lambda$, let $\mathcal{U}_\alpha$ be a local base for $\mathcal{P}_\alpha$. Then a local base at $o \in \prod_\alpha X_\alpha$ for $\prod_\alpha \mathcal{P}_\alpha$ is given by all sets of the form:

$$\prod_\alpha V_\alpha \text{ with } V_\alpha \in \mathcal{U}_\alpha \text{ and } V_\alpha = X_\alpha \quad \text{except for a finite number of indices.}$$

Clearly, each $\prod_\alpha V_\alpha$ is circled and absorbing; it is easy to check that the condition (NS2) of (7.2) holds ,hence $\prod_\alpha \mathcal{P}_\alpha$ is a vector topology, and thus $(\prod_\alpha X_\alpha, \prod_\alpha \mathcal{P}_\alpha)$ is a TVS. Moreover, it is easily seen that if each $\mathcal{P}_\alpha$ is locally convex, then so is $\prod_\alpha \mathcal{P}_\alpha$. In particular, if

$$X_\alpha = X \text{ and } \mathcal{P}_\alpha = \mathcal{P} \text{ for all } \alpha \in \Lambda,$$

then we write $\mathcal{P}^\Lambda$ for $\prod_\alpha \mathcal{P}_\alpha$ on the product space X^Λ.

Let $(X, \mathcal{P})$ be a TVS and M a vector subspace of X. Then the relative topology on M, denoted by $\mathcal{P}_M$, is clearly a vector topology. If $\mathcal{P}$ is metrizable (or locally convex), then so is $\mathcal{P}_M$. Vector subspaces of X, equipped with the relative topologies, are referred to as <u>subspaces</u> of X.

Let $Q_M : X \longrightarrow X/M$ be the quotient map. It is known that the quotient topology on X/M, denoted by $\hat{\mathcal{P}}$ (more precisely $\mathcal{P}/M$), is the finest topology for which

Q_M is continuous; in other words, a set B in X/M is $\hat{\mathscr{P}}$-open if and only if $Q_M^{-1}(B)$ is $\mathscr{P}$-open. Moreover, we have the following result:

(7.14) **Proposition.** Let $(X,\mathscr{P})$ be a TVS and M a vector subspace of X.

(a) The quotient topology $\hat{\mathscr{P}}$ on X/M is a vector topology.

(b) The quotient map Q_M is open.

(c) $\hat{\mathscr{P}}$ is Hausdorff if and only if M is closed.

(d) If $\mathscr{P}$ is locally convex then so does $\hat{\mathscr{P}}$.

Proof. For any $\mathscr{P}$-open set A in X, as

$$Q_M^{-1}(Q_M(A)) = A + M,$$

it follows that (see (7.c) (ii)) $Q_M(A)$ is $\hat{\mathscr{P}}$-open, and hence that Q_M is an open map. Accordingly, the family

$$Q_M(\mathscr{U}_X) = \{ Q_M(V) : V \in \mathscr{U}_X \}$$

is a local base at $\hat{0} = 0 + M$ for $\hat{\mathscr{P}}$. Since Q_M is surjective, the images, under Q_M, of circled, absorbing sets are circled and absorbing. It is easily seen that the condition (NS2) of (7.2) holds, hence $\hat{\mathscr{P}}$ is a vector topology. Therefore (a) and (b) hold.

(c) The necessity is obvious since $M = Q_M^{-1}(\hat{0})$ and $\{\hat{0}\}$ is $\hat{\mathscr{P}}$-closed. To prove the sufficiency, let $\hat{0} \neq \hat{x} \in X/M$. Then $\hat{x} = Q_M(x)$ for some $x \in X$. Since $U = X \setminus M$ is an open neighbourhood of x, it follows that $Q_M(U)$ is an $\hat{\mathscr{P}}$-open set for which $\hat{0} \notin Q_M(U)$ and $\hat{x} \in Q_M(U)$. Therefore $\hat{\mathscr{P}}$ is Hausdorff.

(d) Since the image, under Q_M, of any absolutely convex subset of X is absolutely convex, part (d) follows.

(7.j) Metrizable TVS (II): Let $(X,\mathscr{P})$ be a complete metrizable TVS and M a closed vector subspaces of X. Then the quotient space $(X/M, \hat{\mathscr{P}})$ is a complete metrizable TVS.

Note: In (7.j), the metrizable condition is essential; namely, there is a complete Hausdorff TVS X with a closed vector subspace M such that the quotient space X/M is not complete (see Köthe [1969, §31.6]).

(7.15) **Examples.** (a) Let X be a vector space, and $\mathfrak{U}$ the family of all absolutely convex, absorbing subsets of X. Then $\mathfrak{U}$ is a filter base on X which satisfies (LC1) and (LC2) of (7.4), hence $\mathfrak{U}$ defines a locally convex topology $\mathscr{P}_{\text{fin}}$ on X, which is the finest one; thus $\mathscr{P}_{\text{fin}}$ is called the finest locally convex topology. It is also clear that $\mathscr{P}_{\text{fin}}$ is determined by the family of all seminorms on X.

(b) Let X be a vector space and let X' be a vector subspace of X^*. For any $f \in X'$, the functional, defined by

$$p_f(x) = |f(x)| \quad (\text{for all } x \in X),$$

is a seminorm on X. The locally convex topology on X, defined by the family $\{p_f : f \in X'\}$ of seminorms, is called the weak topology, and denoted by $\sigma(X,X')$.

Dually, for any $x \in X$, we define a seminorm q_x on X' by

$$q_x(f) = |f(x)| \quad (\text{for all } f \in X').$$

The locally convex topology on X' determined by $\{q_x : x \in X\}$ is called the weak topology (or weak*–topology) on X' and denoted by $\sigma(X',X)$.

(c) The space $C(\Omega)$ and $H(\Omega)$: Let Ω be a non–empty open subset of $\mathbb{C}$ (or $\mathbb{R}^n$). We denote by $C(\Omega)$ the vector space of all complex–valued continuous functions on Ω, and by $H(\Omega)$ the vector subspace of $C(\Omega)$ that consists of holomorphic functions on Ω. It is well–known that there exists a sequence $\{K_n\}$ of compact sets with

$$\Omega = \bigcup_{m=1}^{\infty} K_m \text{ and } K_m \subset \text{Int } K_{m+1} \quad (\text{for all } m \geq 1)$$

such that each compact subset of Ω is contained in some K_m. For any $m \geq 1$, the functional p_m, defined by

$$p_m(x) = \sup \{|x(t)| : t \in K_m\} \quad (\text{for all } x \in C(\Omega)),$$

is a seminorm on $C(\Omega)$. The metrizable locally convex topology on $C(\Omega)$, determined by $\{p_m : m \geq 1\}$, is called the <u>compact–open topology</u>. One can show that $C(\Omega)$ is complete [since the convergence for the compact–open topology is the convergence uniformly on compact sets K_m, and each $(C(K_m),p_m)$ is a Banach space].

Since sequences of holomorphic functions that converge uniformly on compact sets have holomorphic limits, it follows that $H(\Omega)$ is close in $C(\Omega)$, and hence that $H(\Omega)$ is complete. In complex function theory, a relatively compact subset of $C(\Omega)$ is usually called a <u>normal family</u>. Montel's classical theorem says that any subset of $H(\Omega)$ which is uniformly bounded on each compact subset of Ω, is a normal family. It is easily seen from (7.7) that a subset $H(\Omega)$, which is uniformly bounded on each compact subset of Ω, is bounded for the compact–open topology.Therefore, Montel's theorem can be restated by using the language of TVS that any bounded subset of $H(\Omega)$ is relatively compact.

(d) (Köthe sequence spaces). A collection P of sequences $a = [a_n]$ of positive numbers is called a <u>Köthe set</u> if it satisfies the following two conditions:

(i) for any $a = [a_n], b = [b_n] \in P$ there is an $c = [c_n] \in P$

such that $a_n,b_n \leq c_n$ $(\forall\, n \geq 1)$;

(ii) for any $k \geq 1$, there is some $a = [a_n] \in P$ such that

$a_k > 0$.

For any Köthe set P, the set defined by

$$\lambda(P) = \{\alpha = [\alpha_n] \in \mathbb{C}^{\mathbb{N}} : p_a(\alpha) = \sum_{n=1}^{\infty} a_n|\alpha_n| < \infty, \forall\, a = [a_n] \in P\},$$

is clearly a vector subspace of $\mathbb{C}^{\mathbb{N}}$ which is called the <u>Köthe sequence space</u> (or the <u>Köthe</u>

sequence space determined by P). For any $a = [a_n] \in P$,

$$p_a(\alpha) = \Sigma_{n=1}^{\infty} a_n |\alpha_n| \quad \text{for all } \alpha = [\alpha_n] \in \lambda(P),$$

is clearly a seminorm on $\lambda(P)$. The family $\{p_a : a \in P\}$ determines a Hausdorff locally convex topology on $\lambda(P)$ which is denoted by $\mathscr{T}_P$ and called the natural topology determined by P. One can show that $(\lambda(P), \mathscr{T}_P)$ is complete.

The set P, defined by

$$P = \{ [n^k] : k = 1,2,\cdots \},$$

is obviously a (countable) Köthe set, the natural topology on $\lambda(P)$ is determined by a countable family $\{p_k : k \geq 1\}$ of seminorms, where

$$p_k(\alpha) = \Sigma_{n=1}^{\infty} n^k |\alpha_n| \quad (\text{for all } \alpha = [\alpha_n] \in \lambda(P)).$$

$(\lambda(P), \mathscr{T}_P)$ is called the Fréchet space of rapidly decreasing sequences, and is denoted by s.

(7.k) Let E be a normed space with its Banach dual E', and let $\{x_n\}$ be a sequence in E. If $x_n \longrightarrow 0$ for $\sigma(E,E')$ and the set $\{x_n : n \geq 1\}$ is relatively compact in E (or sequentially compact), then $\lim_n \|x_n\| = 0$.

Exercises

7–1. Prove (7.a).

7–2. Prove (7.b).

7–3. Prove (7.c).

7–4. Prove (7.d).

7–5. Let X be a TVS and V a circled subset of X. Show that $\overline{V}$ is circled, and that if $0 \in \text{Int } V$ then Int V is circled.

7–6. Let X be a vector space, and let $\mathscr{P}$ and $\mathscr{T}$ be Hausdorff vector topologies on X. Suppose that $\mathscr{P} \leq \mathscr{T}$ and that $\mathscr{T}$ has a local base at 0 consisting of $\mathscr{P}$–complete sets in X. Show that $(X,\mathscr{T})$ is complete.

7–7. Prove (7.e).

7–8. Let $(X,\mathscr{P})$ be a metrizable LCS whose topology $\mathscr{P}$ is defined by an increasing sequence $\{p_n : n \geq 1\}$ of seminorms, and suppose that

$$|x| = \Sigma_{n=1}^{\infty} 2^{-n} \frac{p_n(x)}{1+p_n(x)} \quad \text{(for any } x \in X\text{)}.$$

Show that $|\cdot|$ is a F–norm (for definition, see (7.f)(c)) defined the topology $\mathscr{P}$.

7–9. Prove (7.g).

7–10. Prove (7.h).

7–11. Prove (7.j).

7–12. Prove (7.k).

7–13. (Schauder bases in TVS). Let $(X,\mathscr{P})$ be a LCS (or more general, a Hausdorff TVS). A sequence $\{e_n\}$ in X is called a <u>Schauder basis</u> for $(X,\mathscr{P})$ if for any $x \in X$ there exists a unique sequence $\{\zeta_n\}$ in K such that

$$x = \Sigma_{j=1}^{\infty} \zeta_j e_j \quad \text{(for the topology } \mathscr{P}\text{)},$$

and each of the maps $u_n' : X \longrightarrow K$, defined by

$$u_n'(x) = \zeta_n,$$

is a continuous linear functional on X. Usually, we use (e_n, u_n') to denote a Schauder basis for $(X,\mathscr{P})$.

(a) For any Köthe space $(\lambda(P),\mathscr{T}_P)$, let $\delta_i^{(j)}$ be the Kronecker delta. Then the sequence $\{\delta^{(n)}\}$, defined by

$$\delta^{(n)} = [\delta_i^{(n)}]_{i\geq 1} \text{ (for all } n\geq 1\text{)},$$

is a Schauder basis for $(\lambda(P), \mathcal{T}_P)$.

(b) Suppose that (e_n, u'_n) is a Schauder basis for $(X, \mathcal{P})$, and let us define

$$\lambda_X = \{[u'_n(x)] : x \in X\} \text{ and } P_{(S)} = \{a^V = [p_V(e_n)] : V \in \mathcal{U}_X\}.$$

Show that:

(i) The map $S : \lambda_X \longrightarrow X$, defined by

(13.a) $$S([u'_n(x)]) = x \quad (\text{for all } x \in X),$$

is an algebraic isomorphism.

(ii) $P_{(S)}$ is a Köthe set.

(c) A Schauder basis (e_n, u'_n) for $(X, \mathcal{P})$ is called an <u>absolute basis</u> (or <u>continuous ℓ^1–basis</u>) if for any $V \in \mathcal{U}_X$ there is an $W \in \mathcal{U}_X$ such that

$$\Sigma_{n=1}^{\infty} |u'_n(x)| p_V(e_n) \le p_W(x) \quad (\text{for all } x \in X).$$

(i) (e_n, u'_n) is an absolute basis for $(X, \mathcal{P})$

$\Leftrightarrow$ for any $V \in \mathcal{U}_X$, the map

$$\psi_V : x \longrightarrow [u'_n(x)\, p_V(e_n)] : X \longrightarrow \ell^1 \text{ is continuous.}$$

$\Leftrightarrow$ The map

$$R : x \longrightarrow [u'_n(x)] : (X, \mathcal{P}) \longrightarrow (\lambda(P_{(S)}), \mathcal{T}_{P_{(S)}}) \text{ is continuous.}$$

(ii) Suppose now that (e_n, u'_n) is an absolute basis for $(X, \mathcal{P})$. If $(X, \mathcal{P})$ is sequentially complete, show that $\lambda_X = \lambda(P_{(S)})$ and that

$$(X, \mathcal{P}) \cong (\lambda(P_{(S)}), \mathcal{T}_{P_{(S)}}) \quad \text{under S defined by (13.a).}$$

8

Separation Theorems and Krein-Milman's Theorem

We start with the following interesting result:

(8.1) **Lemma.** Let X be a TVS and $0 \neq f \in X^*$. Then $f(G)$ is open in K whenever G is open in X.

Proof. One can assume that G is non–empty and that $0 \neq x_0 \in X$ is such that $f(x_0) = 1$. For any $a \in G$, it is required to show that $f(a)$ is an interior point of $f(G)$. Indeed, since $G - a$ is an open 0–neighbourhood, it absorbs x_0; namely, there exists an $\alpha > 0$ such that

(1) $$\lambda x_0 \in G - a \quad \text{whenever } \lambda \in K \text{ with } |\lambda| \leq \alpha.$$

Now for any $\beta \in K$ with $|\beta - f(a)| \leq \alpha$, we have $(\beta - f(a))x_0 \in G - a$ (by(1)),hence

$$\beta - f(a) = f((\beta - f(a))x_0) \in f(G) - f(a)$$

which implies $\beta \in f(G)$; thus $f(a)$ is an interior point of $f(G)$.

The following two separation theorems, that have become standard tools in functional analysis, are actually geometric forms of the Hahn–Banach extension theorem.

(8.2) **Theorem.**(The first separation theorem). Let X be a TVS and A, B two non–empty convex sets in X. If

$$\text{Int } A \neq \phi \text{ and } B \cap \text{Int } A = \phi,$$

then there exists a real $0 \neq f \in X'$ which separates A and B in the sense that

(1) $$\sup f(A) \; [= \sup_{a \in A} f(a)] \leq \inf f(B) \; [= \inf_{b \in B} f(b)].$$

Furthermore, if in addition, A and B are disjoint open sets, then A and B

are strictly separated by f in the sense that there is an $\lambda \in \mathbb{R}$ such that

$$f(a) < \lambda < f(b)(\text{or } f(b) < \lambda < f(a)) \quad (\text{for all } a \in A \text{ and } b \in B).$$

Proof. By (7.6) (b), Int A is convex and open, hence $-B + \text{Int } A$ is a non–empty open convex set which does not contain 0 [since $B \cap \text{Int } A = \phi$]. Let $e \in -B + \text{Int } A$. Then $-e - B + \text{Int } A$ is an open, convex 0–neighbourhood, hence its gauge, denoted by p, is a continuous sublinear functional on X such that

$$(2) \qquad -e - B + \text{Int } A = \{ x \in X : p(x) < 1 \}.$$

$\{ \lambda e : \lambda \in \mathbb{R} \}$ is a real vector subspace of X, and the functional g, defined by

$$g(\lambda e) = -\lambda \quad (\text{for all } \lambda \in \mathbb{R}),$$

is a real linear functional on $\{ \lambda e : \lambda \in \mathbb{R} \}$ such that

$$(3) \qquad g(\lambda e) \le p(\lambda e) \quad (\text{for all } \lambda \in \mathbb{R}).$$

[Indeed, we first notice from (2) that $p(-e) \ge 1$ since $-e \notin -e - B + \text{Int } A$ (on account of $0 \notin -B + \text{Int } A$); now for any $\lambda \in \mathbb{R}$, if $\lambda > 0$ then

$$\lambda p(-e) = p(-\lambda e) \ge \lambda = -g(\lambda e) = g(-\lambda e),$$

if $\lambda < 0$ then $-\lambda > 0$ and

$$g(\lambda e) = -\lambda \le -\lambda p(-e) = p(\lambda e).]$$

It then follows from Hahn–Banach's extension theorem that there is a real $0 \ne f \in X^*$ such that

$$f(\lambda e) = g(\lambda e) \quad (\text{for all } \lambda \in \mathbb{R}) \text{ and } f(x) \le p(x) \quad (x \in X),$$

hence f is continuous on X [by the continuity of p]. For any $b \in B$ and $a \in \text{Int } A$, we have

$$f(a-b) = f(-e+a-b) + f(e) \leq p(-e+a-b) + g(e) \lneq 1 + g(e) = 0,$$

thus (1) holds by the continuity of f.

Finally, we assume that A and B are <u>disjoint open</u> convex subsets of X. Then (8.1) shows that f(A) and f(B) are non–empty <u>open convex</u> subsets of $\mathbb{R}$ with $f(A) \cap f(B) = \phi$, and surely disjoint open intervals in $\mathbb{R}$; therefore there is an $\lambda \in \mathbb{R}$ such that

$$f(a) < \lambda < f(b) \quad (\text{or } f(b) < \lambda < f(a)) \quad (\text{for all } a \in A \text{ and } b \in B);$$

namely, f separates strictly A and B.

From the first separation theorem, we see that for any two disjoint <u>convex</u> sets A and B, if they can be separated by some <u>convex</u> 0–neighbourhood V, that is $(A + V) \cap (B + V) = \phi$, then we can apply the first separation theorem to get a non–zero $f \in X'$ strictly separating A and B. Therefore, it is important for us to find some sufficient condition ensuring that this can be done; (7.c) (iii) has already given such conditions.

(8.3) **Theorem** (The strong (or second) separation theorem). <u>Let</u> X <u>be a LCS</u>, <u>and let</u> K, B <u>be two non–empty disjoint convex subsets of</u> X. <u>If</u> K <u>is compact and</u> B <u>is closed</u> ,<u>then there exists a real</u> $0 \neq f \in X'$ <u>and</u> $\lambda \in \mathbb{R}$ <u>such that</u> f <u>strongly separates</u> K <u>and</u> B <u>in the sense that</u>

(8.3.1) $$\max f(K) < \lambda \leq \inf f(B) \qquad (\text{or } \sup f(B) \leq \lambda < \min f(K)).$$

<u>**Proof**</u>. It is known [see(7.c)(iii) and (7.4)] that there is an open convex 0–neighbourhood V in X such that $(K + V) \cap (B + V) = \phi$. As $K + V$ and $B + V$ are open and convex, it follows from (8.2) that there is a real $0 \neq f \in X'$ and $\lambda \in \mathbb{R}$ such that

$$\text{either } f(K + V) < \lambda < f(B + V) \text{ or } f(B + V) < \lambda < f(K+V),$$

hence we have either

$$f(K) < \lambda < f(B + V) \quad \text{or} \quad f(B + V) < \lambda < f(K).$$

It then follows from the compactness of K that (8.3.1) holds.

Separation theorems have a lot of important applications, we mention some as follows:

(8.4) **Corollary.** Let X be a LCS. For any $0 \neq x_0 \in X$, there is a $0 \neq f \in X'$ such that $f(x_0) \neq 0$.

Proof. Follows immediately from (8.3) since $\{ 0 \}$ and $\{ x_0 \}$ are non–empty, disjoint compact convex sets.

Let $X_{\mathscr{P}}$ be a LCS with the topological dual X'. The preceding result shows that $\{ 0 \} \neq X' \subset X^*$, hence one can define the weak topology $\sigma(X,X')$ on X (see (7.15)(b)). It is clear that $\sigma(X,X') < \mathscr{P}$, hence every $\sigma(X,X')$–closed subset of X must be $\mathscr{P}$–closed. The converse is true for convex sets as shown by the following:

(8.5) **Corollary.** Let $X_{\mathscr{P}}$ be a LCS. Any $\mathscr{P}$–closed convex subset B of X is $\sigma(X,X')$–closed; consequently, the $\mathscr{P}$–closure and the $\sigma(X,X')$–closure of any convex subset of X are identical.

Proof. For any $x_0 \notin B$, the strong separation theorem ensures that there exists a real $0 \neq f \in X'$ and $\lambda \in \mathbb{R}$ such that

$$f(x_0) < \lambda \leq \inf f(B),$$

so that x_0 is an interior point of $X \backslash B$ for $\sigma(X,X')$, thus B is $\sigma(X,X')$–closed.

(8.a) <u>Consequences of separation theorems :</u> (i) Let E be a normed space and $B \subset E$. If

$$\|f\| = \sup_{b \in B} |f(b)| \quad \text{(for all } f \in E'\text{)},$$

then $U_E = \overline{\Gamma B}$.

(ii) Let X be a LCS and M a closed vector subspace of X. $x_0 \notin M$ if and only if there exists a $0 \neq f \in X'$ such that

$$f(x_0) \neq 0 \quad \text{and} \quad M \subset f^{-1}(0).$$

Moreover, we have $M = (M^{\perp})^{\top}$, where $M^{\perp}$ is the annihilator of M, defined by

$$M^{\perp} = \{ f \in X' : M \subset f^{-1}(0) \},$$

and $(M^{\perp})^{\top}$ (in X) is the annihilator of $M^{\perp}$.

As another important application of separation theorems, we are going to verify Krein–Milman's theorem. To do this, we need the following terminology:

Let X be a vector space and $\phi \neq K \subset X$. A point $a \in K$ is called an <u>extreme point</u> of K if the following holds: If $x, y \in K$ are such that

$$a = \lambda x + (1-\lambda)y \quad \text{for some } \lambda \in (0,1),$$

then $a = x = y$. Denoted by ∂K the set consisting of all extreme points of K. More general, a non–empty subset B of K is called an <u>extremal subset</u> of K if the following holds: If $x, y \in K$ are such that

$$\lambda x + (1-\lambda)y \in B \quad \text{(for some } \lambda \in (0,1)\text{)},$$

then $x,y \in B$.

(8.b) <u>A characterization of extreme points</u>: Let X be a vector space, let K be a non–empty convex subset of X and $u_o \in K$. Then the following statements are equivalent:

(i) u_o is an extreme point of K.

(ii) If $x,y \in K$ are such that $u_o = \frac{1}{2}(x+y)$, then $u_o = x = y$.

(iii) Let $x,y \in K$ be such that $x \neq y$, let $\lambda \in [0,1]$ and $u_o = \lambda x + (1-\lambda)y$. Then we have either $\lambda = 0$ or $\lambda = 1$.

(iv) $K \setminus \{x_o\}$ is convex.

Although the notion of extreme points is purely algebraic , we have the following interesting result, which is useful to give examples of extreme points of closed unit balls in some concrete Banach spaces.

(8.6) Proposition. <u>Let</u> X <u>be a TVS and</u> K <u>a convex subset of</u> X. <u>Then</u> $(\text{Int}K) \cap (\partial K) = \phi$ [i.e., any interior point of K (when it exists) is not an extreme point of K.]

<u>Proof</u>. If $\text{Int}K = \phi$, the result is trivial. Suppose now that $\text{Int}K \neq \phi$ and that $x \in \text{Int}K$. Then there exists a 0–neighbourhood V in X such that $x+V \subset K$. As the map $\mu \to \mu x : \mathbb{R} \to X$ is continuous at $\mu = 1$, for this x–neighbourhood $x + V$, there is an $r > 0$ such that

$$\mu x \in x + V \text{ whenever } |\mu - 1| \leq r \ (\mu \in \mathbb{R});$$

in particular, we have

$$(1+r)x \in x + V \subset K \quad \text{and} \quad (1-r)x \in x + V \subset K.$$

Consequently, we have

$$x = \tfrac{1}{2}(1+r)x + \tfrac{1}{2}(1-r)x,$$

which implies that x is not an extreme point of K.

As a consequence of (8.6), we see that any <u>open</u> convex subset of a TVS does not have any extreme point.

(8.7) <u>Examples</u>. (a) In the Banach space $\ell_n^\infty(\mathbb{R}) = (\mathbb{R}^n, \|\cdot\|_\infty)$, let $U = \{[\zeta_i] \in \mathbb{R}^n : \|[\zeta_i]\|_\infty = \max_{1 \leq i \leq n} |\zeta_i| \leq 1\}$; one has

$$\partial U = \{[\zeta_i] \in U : |\zeta_i| = 1 \text{ for } i=1,..,n\}.$$

<u>Proof</u>. By (8.6), if $[\zeta_i] \in \partial U$ then $\|[\zeta_i]\|_\infty = 1$, moreover we claim that $|\zeta_i| = 1$ $(i=1,..,n)$. Suppose not, there is some $k \in \{1,..,n\}$ such that $|\zeta_k| < 1$. Then one can choose $r > 0$ such that $|\zeta_k|+r < 1$. Now consider the following two vectors

$$y = (\zeta_1,..,\zeta_{k-1},\zeta_k+r,\zeta_{k+1},..\zeta_n) \quad z = (\zeta_1,..,\zeta_{k-1},\zeta_k-r,\zeta_{k+1},..\zeta_n)$$

Then $y, z \in U$ are such that

$$y \neq z \text{ and } [\zeta_i] = \tfrac{1}{2}(y+z)$$

which contradicts the fact that $[\zeta_i] \in \partial U$.

Conversely, let $[\zeta_i] \in U$ be such that $|\zeta_i| = 1$ $(i=1,..,n)$ and let $[\lambda_i], [\mu_i] \in U$ be such that

$$[\zeta_i] = \tfrac{1}{2}[\lambda_i] + \tfrac{1}{2}[\mu_i].$$

Then for any $i \in \{1,..,n\}$,

$$\tfrac{1}{2}\lambda_i + \tfrac{1}{2}\mu_i = 1 \text{ or } -1.$$

As $1, -1$ are extreme points of $[-1,1]$, it follows that

$$\lambda_i = \mu_i = 1 \text{ or } -1,$$

and hence that $[\zeta_i] = [\lambda_i] = [\mu_i]$; in other words, $[\zeta_i] \in \partial U$.

(b) In the Banach space l^1, we have

$$\partial U_{l^1} = \{\lambda e^{(n)} : \lambda \in \mathbb{C}, |\lambda| = 1 \text{ and } n = 1...\}$$

Proof. Suppose that $|\lambda| = 1$ and $\lambda e^{(n)} = \frac{1}{2}(y+z)$, where $y=[y_i]$, $z=[z_i] \in U_{l^1}$. Then

$$\lambda = \tfrac{1}{2}y_n + \tfrac{1}{2}z_n \text{ and } 0 = \tfrac{1}{2}y_i + \tfrac{1}{2}z_i \ (i \neq n). \tag{1}$$

As λ (with $|\lambda| = 1$) is an extreme point of U_K, it follows that $\lambda = y_n = z_n$, and hence from

$$1 = |\lambda| \le \Sigma_{k=1}^{\infty} |y_k| \le 1 \text{ and } 1 = |\lambda| \le \Sigma_{k=1}^{\infty} |z_k| \le 1$$

that all $y_i = z_i = 0$ (for $i \neq n$); consequently,

$$\lambda e^{(n)} = y = z;$$

thus $\lambda e^{(n)} \in \partial U_{l^1}$.

Conversely, if $[\zeta_i] \in \partial U_{l^1}$, then (8.6) shows that $\|[\zeta_i]\|_1 = 1$; moreover, we claim that there is some n such that

$$\zeta_i = 0 \quad \text{for all } i \neq n$$

(hence $[\zeta_i] = \zeta_n e^{(n)}$ with $|\zeta_n| = 1$ (since $\|[\zeta_i]\|_1 = 1$)).

In fact, suppose, on the contrary, that there are j and k with $j < k$ such that $\zeta_j \neq 0$ and $\zeta_k \neq 0$. Then

$$0 < \Sigma_{i=1}^{j} |\zeta_i| = \mu < 1 \quad \text{and} \quad \Sigma_{i>j}^{\infty} |\zeta_i| = 1 - \mu$$

(since $\|[\zeta_i]\|_1 = \Sigma_{i=1}^{\infty}|\zeta_i| = 1$). Let

$$y = \frac{1}{\mu}(\zeta_1 \ldots, \zeta_j, 0, 0 \ldots) \quad \text{and} \quad z = \frac{1}{1-\mu}(0, ..0, \zeta_{j+1}, \ldots).$$

Then $y, z \in U_{\ell}$ are such that

$$y \neq z \quad \text{and} \quad [\zeta_i] = \mu y + (1-\mu)z,$$

where $\mu \in (0,1)$; this contradicts the assumption that $[\zeta_i] \in \partial U_{\ell}$.

(c) In the Banach space c_o, we have

$$\partial U_{c_o} = \phi.$$

<u>Proof</u>. It suffices to show that any $0 \neq [\zeta_i] \in U_{c_o}$ is not an extreme point of U_{c_o}. Indeed, since $\lim_j \zeta_j = 0$, it follows from $\|[\zeta_i]\|_{\infty} \neq 0$ that there is some $k \in \mathbb{N}$ such that

$$|\zeta_k| < \|[\zeta_i]\|_{\infty} \leq 1,$$

and hence that one can choose $r > 0$ such that

$$0 < |\zeta_k| + r < 1.$$

Now consider the vectors $y = [y_i]$ and $z = [z_i]$ with

$$y_i = \begin{cases} \zeta_i & \text{if } i \neq k \\ \zeta_k + r & \text{if } i = k \end{cases} \qquad z_i = \begin{cases} \zeta_i & \text{if } i \neq k \\ \zeta_k - r & \text{if } i = k \end{cases}.$$

then $y,z \in U_{c_o}$ are such that

$$y \neq z \quad \text{and} \quad [\zeta_i] = \tfrac{1}{2}y + \tfrac{1}{2}z,$$

hence $[\zeta_i]$ is not an extreme point of U_{c_o}.

(d) In the Banach space $L^1(\mathbb{R})$, one has

$$\partial U_{L^1(\mathbb{R})} = \phi.$$

Proof. By (8.6), it suffices to consider any $u \in U_{L^1(\mathbb{R})}$ with $\|u\|_1 = 1$ not an extreme point of $U_{L^1(\mathbb{R})}$. Indeed, one can choose $r \in \mathbb{R}$ such that

$$\int_{-\infty}^{r} |u(t)|\,dm(t) = \tfrac{1}{2}.$$

Now let

$$y = 2u\aleph_{(-\infty,r)} \quad \text{and} \quad z = 2u\aleph_{[r,+\infty)},$$

where $\aleph_{(-\infty,r)}$ and $\aleph_{[r,+\infty)}$ are respectively the characteristic functions of $(-\infty,r)$ and $[r,+\infty)$. Then $\|y\|_1 = \|z\|_1$ are such that

$$y \neq z \quad \text{and} \quad u = \tfrac{1}{2}y + \tfrac{1}{2}z,$$

hence u is not an extreme point of $U_{L^1(\mathbb{R})}$.

(8.c) Extreme points of closed unit balls in the concrete B–spaces : (i) For $1 < p < \infty$, one has

$$\partial U_{L^p(\mathbb{R})} = \{f \in L^p(\mathbb{R}) : \|f\|_p = 1\} \quad \text{and}$$

$$\partial U_{\ell^p} = \{[\zeta_n] \in \ell^p : \|[\zeta_n]\|_p = 1\}.$$

(ii) $\partial U_{L^\infty[0,1]} = \{f \in L^\infty[0,1] : |f(t)| = 1 \text{ almost all } t \in [0,1]\}$. In ℓ^∞, let $A = \{\frac{e^{(n)}}{n} : n \geq 1\} \cup \{0\}$ and $K = \overline{co}A$.. Then $\partial K = A$.

(iii) Let Ω be a compact Hausdorff space, and let $M(\Omega)$ be the Banach space of all complex regular Borel measures [see (3.9)(f)]. Then

$$\partial U_{M(\Omega)} = \{\lambda\delta_t : \lambda \in \mathbb{C},\ |\lambda| = 1 \text{ and } t \in \Omega\},$$

where $\delta_t \in M(\Omega)$ is the Dirac measure on Ω, i.e.,

$$\delta_t(B) = \begin{cases} 1 & \text{if } t \in B \\ 0 & \text{if } t \notin B \end{cases}$$

for any Borel set $B \subset \Omega$.

(iv) $\partial U_{C_{\mathbb{R}}([0,1])} = \{x \in C_{\mathbb{R}}[0,1] : x(t) \equiv 1 \text{ or } -1 \ (t \in [0,1])\}$; and more generally one can show that

$$\partial U_{C(\Omega)} = \{x \in C(\Omega): |x(t)| = 1 \ (\forall\, t \in \Omega)\},$$

where Ω is a compact Hausdorff space.

(8.8) Lemma. Let X be a LCS and K a non–empty compact convex subset of X. For any real $0 \neq f \in X'$, the set, defined by

$$K_f = \{ x \in K : f(x) = \inf f(K) \},$$

is a non–empty, convex, extremal subset of K (called a face in K).

Proof. The convexity of K and the linearity of f imply that K_f is convex, which is non–empty (by the compactness of K). On the other hand, the definition of infimum and the linearity of f imply that K_f is extremal.

(8.9) **Theorem (Krein–Milman).** Let X be a LCS and K a non–empty, compact convex subset of X. Then

(a) $C_0 \cap \partial K \neq \phi$ for any non–empty closed, convex, extremal subset C_0 of K (called a closed face in K).

(b) $K = \overline{co}(\partial K)$.

Consequently, any non–empty compact convex set in a LCS always has extreme points .

Proof. (a) Let $\mathscr{C} = \{ C \subset K : C$ closed face in K with $C \subseteq C_0 \}$. Then $\mathscr{C}$ is an inductive ordered set under downward inclusion. [If $\mathscr{M}$ is a chain in $\mathscr{C}$ and if $C_{\alpha_i} \in \mathscr{M}$

$(i = 1,2,\cdots,n)$, then $\bigcap_{i=1}^{n} C_{\alpha_i}$ is one of $C_{\alpha_1},\cdots,C_{\alpha_n}$, that is $\mathscr{M}$ has the finite intersetion property, thus $\cap \mathscr{M} \neq \phi$ by the compactness of K; consequently $\cap \mathscr{M} \in \mathscr{C}$ is such that $\cap \mathscr{M} \leq C_\alpha$ (for all $C_\alpha \in \mathscr{M}$).] By Zorn's lemma, $\mathscr{C}$ contains a minimal element C. (Remember that C is a closed face in K with $C \subseteq C_0$). Now we show that

$$C = \{e\} \quad \text{for some } e \in K,$$

thus $e \in \partial K$ and $e \in \partial K \cap C_0$ (on account of $e \in C \cap \partial K \subset C_0 \cap \partial K$).

In fact, if C contains elements x and y with $x \neq y$, then (8.3) shows that there is a real $0 \neq f \in X'$ such that $f(x) \neq f(y)$. By (8.6), the set

$$C_f = \{ z \in C : f(z) = \inf (C) \}$$

is a closed face in C, (hence in K) with $C_f \subset C_0$. As $f(x) \neq f(y)$ it follows that $C_f \subsetneq C$, which contradicts the minimality of C.

(b) Let $D = \overline{co}\, \partial K$. Then $D \subset K$.If $D \neq K$, one can take $x_0 \in K\backslash D$. By (8.3), there is a real $0 \neq h \in X'$ such that

(1) $$h(x_0) < \min h(D).$$

On the other hand, (8.8) shows that the set, defined by

$$K_h = \{ z \in K : h(z) = \inf h(K) \},$$

is a closed face in K, thus there is an $e \in (\partial K) \cap K_h$ [by part (a)]; consequently,

(2) $$h(e) = \inf h(K) \leq h(x_0) < \min h(D)$$

[by(1)]. Notice that $e \in \partial K \subset D$, so that

$$\min h(D) \leq h(e),$$

which contradicts (2).

Remark. The condition that X be a locally convex space cannot be dropped from the above theorem; in fact, Roberts[1977] (see Kalton/Peck [1980]) gives an example of a compact convex subset of $L^p[0,1]$, $0 < p < 1$, with no extreme points.

Example (e). Even in a finite–dimensional Banach space, the set of extreme points of a compact, convex set K need not be closed. For instance, in $\mathbb{R}^3$, let

$$D = \{(x,y,z) \in \mathbb{R}^3 : z = 0,\ (x-1)^2 + y^2 = 1\}$$

and

$$K = \mathrm{co}(D \cup \{(0,0,1),\ (0,0,-1)\}),$$

then K is compact convex and $(0,0,0) \in \overline{\partial K} \setminus (\partial K)$, hence ∂K is not closed.

As a consequence of Krein – Milman's theorem, we obtain:

(8.d) Let X, Y be LCS, let $T \in L(X,Y)$ and K a non–empty compact convex subset of X. Then $\partial(T(K)) \subset T(\partial K)$.

A partial converse of Krein – Milman's theorem is the following:

(8.e) Let X be a LCS and $\phi \neq A \subset X$.

(i) (Milman) If $\overline{\mathrm{co}}\,A$ is compact, then $\partial(\overline{\mathrm{co}}\,A) \subset \overline{A}$.

(ii) Let $K \subset X$ be compact convex with $A \subset K$. Then $K = \overline{\mathrm{co}}\,A$ if and only if $\partial K \subset \overline{A}$; consequently, $\overline{\partial K}$ is the minimal closed subset of K whose closed convex hull is K.

(8.f) Extrema at extreme points: Let X be a real LCS and K a non–empty compact convex subset of X.

(i) For any $0 \neq f \in X'$ there exists an $e \in \partial K$ such that

$$f(e) = \max f(K).$$

Moreover, $\overline{\partial K}$ is the smallest closed subset of K on which every $0 \neq f \in X'$ attains its maximum over K.

(ii) (Bauer) For any upper semicontinuous convex function $g : K \longrightarrow \mathbb{R}$, there is a $u \in \partial K$ such that

$$g(u) = \max g(K).$$

For any lower semicontinuous concave function $h : K \longrightarrow \mathbb{R}$, there is a $w \in \partial K$ such that

$$h(w) = \min h(K).$$

Exercises

8–1. Prove (8.a).

8–2. Let X be a LCS, and let K and B be two non–empty disjoint convex subsets of X. If K is totally bounded and closed, and B is complete, show that there exist a real $0 \neq f \in X'$ and $\lambda \in \mathbb{R}$ such that f strongly separates K and B (i.e., $\sup f(K) < \lambda \leq \inf f(B)$) (compared with (8.3)).

8–3. (Support theorem). Let X be a TVS and $C \subset X$. A point $x_o \in C$ is called a supporting point if there is a non–zero real continuous linear functional f on X such that $f(x_o) = \sup f(C)$. (f is also called a supporting functional of C.)

Suppose now that K is a closed convex subset of X such that Int $K \neq \phi$. Show that every boundary point of K is a supporting point of K, and that interior points of K are not supporting points.

8–4. (Approximation theorem for continuous affine functions). Let X be a LCS and K a non–empty convex subset of X. A function $f : K \longrightarrow \mathbb{R}$ is said to be affine (resp. convex) if

$$f(\lambda x+(1-\lambda)y) \underset{(\leq)}{=} \lambda f(x)+(1-\lambda)f(y) \quad (x,y \in K,\ \lambda\in[0,1]) \quad .$$

A function $g : K \longrightarrow \mathbb{R}$ is <u>concave</u> if $-g$ is convex.

Suppose now that K is a compact convex subset of X. Show that for any continuous affine function f on K and $\epsilon > 0$, there exist $h \in X'$ and $\lambda \in \mathbb{R}$ such that

$$\|f-(h+\lambda)|_K\|_\infty = \sup\{|f(x)-h(x)-\lambda| : x \in K\} \leq \epsilon.$$

[Hint: The graphs G(f) and G(f+ϵ) are disjoint compact convex sets in the product space $X \times \mathbb{R}$, and then apply the strong separation theorem.]

8–5. (Separation of convex and concave functions). Let X be a LCS, let K be a compact convex set in X, let f (resp. g) be an upper (resp. lower) semi–continuous, real–valued function on K and suppose that $f < g$ (i.e. $f(x) < g(x)$ for all $x \in K$).

(a) Write

$$L(f) = \{(x,\mu) \in K\times\mathbb{R} : f(x) \geq \mu\} \quad \text{and} \quad U(g) = \{(x,\lambda) \in K\times\mathbb{R} : g(x) \leq \lambda\}.$$

Show that there exists an open convex 0–neighbourhood W in the product space $X\times\mathbb{R}$ such that

$$(L(f) + W) \cap (U(g) + W) = \phi.$$

(b) Suppose, in addition, that f is concave and g is convex. Then use part (a) and the first separation theorem (or otherwise) to show that there exists a continuous affine function φ on K such that $f < \varphi < g$.

8–6. Prove (8.b).

8–7. Prove (8.c).

8–8. Prove (8.d).

8–9. Let X be a real LCS and K a non–empty compact convex subset of X.

(a) If $\phi \neq A \subset K$, show that $\overline{\text{co } A} = K$ if and only if

$$\inf f(A) = \min f(K) \text{ (for any } f \in X'\text{)}.$$

(b) For any lower semi–continuous concave (resp. convex) function h on K, there exists an $u_o \in \partial K$ such that

$$h(u_o) = \inf h(K) \text{ (resp. } h(u_o) = \sup f(K)\text{)}.$$

8–10 (Hardy–Littlewood–Polya's theorem) An n×n–matrix $(p_{ij})_{n\times n}$ is called a <u>doubly stochastic matrix</u> if it satisfies the following conditions:

(*) $$p_{ij}\geq 0 \ (i,j = 1,\cdots,n),\ \Sigma_{j=1}^{n} p_{ij}= 1 \ (i = 1,\cdots,n) \ \text{and} \ \Sigma_{i=1}^{n} p_{ij}= 1 \ (j = 1,\cdots,n).$$

Suppose that 2n real numbers a_i, b_i $(1\leq i\leq n)$ are given and satisfy

$(*)_1$ $$\Sigma_{i=1}^{n} a_i \leq \Sigma_{i=1}^{k} b_i \ (1 \leq k \leq n-1) \ \text{and} \ \Sigma_{i=1}^{n} a_i = \Sigma_{i=1}^{n} b_i \ .$$

Then show that:

(a) For any n positive numbers $c_1,\cdots,c_n$, let π be a permutation of $\{1,\cdots,n\}$ with

$$c_{\pi(1)} \geq c_{\pi(2)} \geq \cdots\cdot \geq c_{\pi(n)} \geq 0.$$

Then the n×n–matrix (p^*_{ij}), defined by

$$p^*_{ij} = \begin{cases} 1 & \text{if } i=\pi(j) \\ 0 & \text{if } i\neq \pi(j) \end{cases},$$

is a doubly stochastic matrix such that

$$\Sigma_{i=1}^{n} c_i a_i \leq \Sigma_{i=1}^{n}\Sigma_{j=1}^{n} c_i p^*_{ij} b_j \ .$$

(b) For any n real numbers $r_1,\cdots,r_n$ (not necessarily positive), using part (a) (or otherwise), the condition (*) and $\Sigma_{i=1}^{n} a_i = \Sigma_{i=1}^{n} b_i$ to deduce that there is a doubly stochastic n×n–matrix (p_{ij}) such that

$$\Sigma_{i=1}^{n} c_i a_i \leq \Sigma_{i=1}^{n}\Sigma_{j=1}^{n} c_i p_{ij} b_j$$

(c) Employ the strong separation theorem and part (b) to show that there is a doubly stochastic n×n–matrix (q_{ij}) such that

$$a_i = \Sigma_{j=1}^{n} q_{ij} b_j \quad (i=1,\cdots,n).$$

[Hint: Let P be the set of all doubly stochastic n×n–matrices. let $a = \begin{pmatrix} a_1 \\ \vdots \\ a_n \end{pmatrix}$ and

$b = \begin{pmatrix} b_1 \\ \vdots \\ b_n \end{pmatrix}$. Then $K = \{(p_{ij})\, b : (p_{ij}) \in P\}$ is a compact convex set in $\mathbb{R}^n$. Use the strong separation theorem and part (b) to show that $a \in K$.]

9

Projective Topologies and Inductive Topologies

This section is devoted to give the fundamental methods for constructing new TVS from given ones. These methods are standard in Functional Analysis, and also enable us to understand the topological structure of a vector space.

(9.1) **Definition.** Let X be a vector space, let $\{(Y_i, \mathscr{T}_i) : i \in \Lambda\}$ be a family of TVS and $T_i \in L^*(X, Y_i)$ (for all $i \in \Lambda$). The coarsest topology $\mathscr{P}_{pro}$ on X for which all T_i are continuous is called the projective (or initial, kernel) topology with respect to (w.r.t. for short) $\{ (Y_i, \mathscr{T}_i, T_i) : i \in \Lambda \}$.

Remark : Denote by $\mathscr{F}(\Lambda)$ the family of all non-empty finite subsets of Λ. If $\mathscr{U}_i$ is a local base at 0 for $\mathscr{T}_i$, then a local base at 0 for $\mathscr{P}_{pro}$ is given by

(9.1.1)
$$\mathscr{U}_{pro} = \{ \bigcap_{i \in \alpha} T_i^{-1}(V_i) : \alpha \in \mathscr{F}(\Lambda), V_i \in \mathscr{U}_i \},$$

hence $\mathscr{P}_{pro}$ is always a vector topology. Moreover, the following statements hold:

(i) If all $\mathscr{T}_i$ $(i \in \Lambda)$ are Hausdorff, then $\mathscr{P}_{pro}$ is Hausdorff if and only if $\bigcap_{i \in \Lambda} \operatorname{Ker} T_i = \{ 0 \}$.

(ii) If all $\mathscr{T}_i$ are locally convex, then so is $\mathscr{P}_{pro}$; in this case,if each $\mathscr{T}_i$ is determined by a family $\mathbb{P}_i$ of seminorms, then $\mathscr{P}_{pro}$ is defined by the following family of seminorms

(9.1.2)
$$\mathbb{P}_{pro} = \{ \max_{i \in \alpha} p_i \circ T_i : \alpha \in \mathscr{F}(\Lambda), p_i \in \mathbb{P}_i \}$$

(iii) If Λ is countable and each $\mathscr{T}_i$ $(i \in \Lambda)$ is metrizable then $\mathscr{P}_{pro}$ is semimetrizable.

(iv) $\mathscr{P}_{pro}$ has the following important property: if G is any topological space and $\varphi : G \longrightarrow X$ is a mapping, then φ is continuous (for $\mathscr{P}_{pro}$) if and only if

$T_i \circ \varphi : G \longrightarrow (Y_i, \mathcal{T}_i)$ is continuous (for all $i \in \Lambda$).

(9.2) **Examples.** (a) Subspaces: Let $(X, \mathcal{P})$ be a TVS, let M be a vector subspace of X and $J_M : M \longrightarrow X$ the canonical embedding. Then the relative topology $\mathcal{P}_M$ on M (induced by $\mathcal{P}$) is the projective topology w.r.t. $(X, \mathcal{P}, J_M)$. Moreover, if $\mathcal{P}$ is Hausdorff, then so is $\mathcal{P}_M$ since $\text{Ker } J_M = \{0\}$.

(b) Least upper bound topologies : Let X be a vector space, let $\{ \mathcal{T}_i : i \in \Lambda \}$ be a family of vector topologies on X and let $I_i : X \longrightarrow X$ be the identity map $(i \in \Lambda)$. Then the least upper bound topology $\sup_{i \in \Lambda} \mathcal{T}_i$ on X of $\{ \mathcal{T}_i : i \in \Lambda \}$ is the projective topology w.r.t $\{ (X, \mathcal{T}_i, I_i) : i \in \Lambda \}$. As $\bigcap_{i \in \Lambda} \text{Ker } I_i = \{0\}$, it follows that if each $\mathcal{T}_i$ is Hausdorff then so is $\sup_{i \in \Lambda} \mathcal{T}_i$.

(c) Weak topologies: Let X be a LCS with the topological dual X'. Then the weak topology $\sigma(X, X')$ on X is the projective topology w.r.t $\{(K, |\cdot|, f) : f \in X'\}$ [where $|\lambda|$ is the absolute value of $\lambda \in K$].

(d) Product spaces: Let $\{(Y_i, \mathcal{T}_i) : i \in \Lambda\}$ be a family of TVS, let $X = \prod_{i \in \Lambda} Y_i$ and $\pi_i : X \longrightarrow Y_i$ the i–th projection $(i \in \Lambda)$. Then the product topology $\prod_{i \in \Lambda} \mathcal{T}_i$ on X is the projective topology w.r.t. $\{(Y_i, \mathcal{T}_i, \pi_i) : i \in \Lambda\}$. As $\bigcap_{i \in \Lambda} \text{Ker } \pi_i = \{0\}$, it follows that $\prod_{i \in \Lambda} \mathcal{T}_i$ is Hausdorff if and only if each $\mathcal{T}_i$ is Hausdorff.

Projective topologies can be identified with the relative topology induced by the product topologies as shown by the following:

(9.3) **Propsition.** Let X be a vector space, let $\{ (Y_i, \mathcal{T}_i, \pi_i) : i \in \Lambda \}$ be a family of Hausdorff TVS and let $T_i \in L^*(X, Y_i)$ $(i \in \Lambda)$ be such that $\bigcap_{i \in \Lambda} T_i^{-1}(\{0\}) = \{0\}$. Suppose that $\mathcal{P}_{pro}$ is the projective topology on X w.r.t. $\{(Y_i, \mathcal{T}_i, T_i) : i \in \Lambda\}$ and that $T : X \longrightarrow \prod_{i \in \Lambda} Y_i$ is defined by

$$Tx = [T_i(x), i \in \Lambda] \quad (\underline{\text{for all}}\ \ x \in X\ )$$

<u>Then</u> $(X, \mathscr{P}_{pro})$ <u>and</u> $(T(X), (\prod_{i\in\Lambda} \mathscr{T}_i)|_{T(X)})$ <u>are topologically isomorphic under</u> T.

<u>Proof.</u> T is clearly linear and one–one [since $\bigcap_{i\in\Lambda} \operatorname{Ker} T_i = \{0\}$], and also satisfies

$$\pi_i \circ T = T_i \ \text{ and } \ T_i \circ T^{-1} = \pi_i|_{T(X)} \ \text{(the restriction on } T(X))(i\in\Lambda),$$

the resullt then follows from Remark (iv) and the fact that $\prod_{i\in\Lambda} \mathscr{T}_i$ is the projective topology.

(9.4) **<u>Definition.</u>** Let Γ be a directed set, let $\{ (Y_i, \mathscr{T}_i) : i \in \Gamma \}$ be a family of TVS, and for any $i, j \in \Gamma$ with $i \leq j$, let $T_{ij} \in L((Y_j, \mathscr{T}_j), (Y_i, \mathscr{T}_i))$. Then the vector subspace of $\prod_i Y_i$, defined by

(9.4.1) $$\varprojlim T_{ij} Y_j = \{ x \in \prod_i Y_i : \pi_i(x) = T_{ij}(\pi_j(x)), i \leq j \}$$

$$= \bigcap \{\operatorname{Ker}(\pi_i - T_{ij}\, \pi_j) : i \leq j\,,\, i, j \in \Gamma\}\,,$$

is called the <u>projective spectrum</u> (of $\{ Y_i : i \in \Gamma \}$ under T_{ij}), while the space $\varprojlim T_{ij} Y_j$, equipped with the relative topology induced by $\Pi_i \mathscr{T}_i$, is called the <u>topological projective limit</u> of $\{ (Y_i, \mathscr{T}_i) : i \in \Gamma \}$ <u>under the continuous maps</u> T_{ij}, and denoted by $\varprojlim T_{ij}\,(Y_j, \mathscr{T}_j)$.

<u>Remark.</u> It can happen that $\varprojlim T_{ij} Y_j = \{0\}$; but it can be shown that if $\Gamma = \mathbb{N}$ and $T_{ij} Y_j = Y_i$ (where $i \leq j$), then $\pi_i(\varprojlim T_{ij} Y_j) = Y_i$ (see Köthe [1969, p.229]). Moreover, (9.4.1) shows that $\varprojlim T_{ij} Y_j$ is always closed in the product space $(\prod_i Y_i, \prod_i \mathscr{T}_i)$.

(9.a) Let $\{(Y_i, \mathscr{T}_i) : i \in \Gamma\}$ be a family of TVS for which Γ is directed, and for any $i,j \in \Gamma$ with $i \leq j$, let $T_{ij} \in L(Y_j, Y_i)$.

(i) For any $i \in \Gamma$, let π_{iR} be the restriction of π_i on $\varprojlim T_{ij}Y_j$. Then the topology on the topological projective limit $\varprojlim T_{ij}(Y_j, \mathcal{T}_j)$ is the projective topology $\varprojlim T_{ij}Y_j$ w.r.t. $\{(Y_i, \mathcal{T}_i, \pi_{iR}) : i \in \Gamma\}$.

(ii) If each $\mathcal{T}_i$ is locally convex then so is $\varprojlim T_{ij}(Y_j, \mathcal{T}_j)$.

(iii) If each $\mathcal{T}_i$ is complete and Hausdorff, then so is $\varprojlim T_{ij}(Y_j, \mathcal{T}_j)$.

(9.2) **Example.** (e) The product space is an example of topological projective limits: Let $\{ (Y_i, \mathcal{T}_i) : i \in \Lambda \}$ be a family of TVS with Λ as an index set, and let $\mathcal{F}(\Lambda)$ be the family of all non–empty finite subsets of Λ which is directed under the set inclusion. For any $\alpha \in \mathcal{F}(\Lambda)$, let

$$Y_\alpha = \prod_{i \in \alpha} Y_i, \qquad \mathcal{T}_\alpha = \prod_{i \in \alpha} \mathcal{T}_i$$

and let $\pi_\alpha : \prod_{i \in \Lambda} Y_i \longrightarrow Y_\alpha$ be the projection, defined by

$$\pi_\alpha(x) = [\pi_i(x), i \in \alpha] \quad (\text{for all } x \in \prod_{i \in \Lambda} Y_i).$$

If $\alpha, \beta \in \mathcal{F}(\Lambda)$ are such that $\alpha \leq \beta$ (i.e., $\alpha \subseteq \beta$), we write

$$\pi_{\alpha\beta} : Y_\beta \longrightarrow Y_\alpha \quad \text{the projection.}$$

Then

$$\pi_\alpha = \pi_{\alpha\beta} \circ \pi_\beta \quad \text{whenever } \alpha \leq \beta,$$

thus $(\prod_{j \in \Lambda} Y_j, \prod_{j \in \Lambda} \mathcal{T}_j) = \varprojlim \pi_{\alpha\beta}(Y_\beta, \mathcal{T}_\beta)$(of $\{(Y_\alpha, \mathcal{T}_\alpha) : \alpha \in \mathcal{F}(\Lambda)\}$ under the continuous maps $\pi_{\alpha\beta}$).

It is clear that the topological projective limit of Banach spaces is, in general, not normable, but locally convex spaces can be regarded as the "projective closure" of Banach spaces as shown by the following:

(9.b) Every LCS is a topological projective limit of Banach spaces : Let $(X,\mathscr{P})$ be a LCS and let $\{p_i : i \in \Gamma\}$ be a family of continuous seminorms on X which determines $\mathscr{P}$ and satisfies $p_i \leq p_j$ (whenever $i \leq j$). For any $i \in \Gamma$ let

$$X_i = X/p_i^{-1}(0)\,, \quad \|\cdot\|_i = \hat{p}_i(\cdot) \quad \text{(the quotient norm on } X_i\text{)},$$

$\tilde{X}_i$(or($\tilde{X}_i,\tilde{\|\cdot\|}_i$)) the completion of $(X_i,\|\cdot\|_i)$ and $Q_i : X \longrightarrow \tilde{X}_i$ the quotient map. For $i \leq j$, let $Q_{ij} : \tilde{X}_j \longrightarrow \tilde{X}_i$ be the canonical map. Then

$$(1) \qquad Q_i = Q_{ij} \circ Q_j \quad \text{whenever } i \leq j,$$

and $\mathscr{P}$ is the projective topology w.r.t. $\{ (X_i,\|\cdot\|_i,Q_i) : i \in \Gamma \}$. Moreover, (1) shows that the map $T : X \longrightarrow \prod_{i\in\Lambda} X_i$, defined by

$$Tx = [Q_i(x), i \in \Gamma] \quad (x \in X),$$

is a topological isomorphism from $(X,\mathscr{P})$ onto a subspace of the topological projective limit $\underleftarrow{\lim}\, Q_{ij}(\tilde{X}_j,\tilde{\|\cdot\|}_j)$.

Next, let us turn to the second general method of topologizing a vector space which is, in some sense, dual to that of taking projective topologies.

(9.5) **Definition.** Let Y be a vector space, let $\{ (X_i,\mathscr{P}_i) : i \in \Lambda \}$ be a family of LCS and $S_i \in L^*(X_i,Y)$ (for all $i \in \Lambda$). The finest locally convex topology $\mathscr{T}_{ind}$ on Y for which all S_i are continuous is called the inductive topology with respect to $\{ (X_i,\mathscr{P}_i,S_i) : i \in \Lambda \}$.

Remark. (i) It should be noted that $\mathscr{T}_{ind}$ is well–defined. [The set $\mathfrak{M}$ consisting of all locally convex topologies on Y for which all S_i are continuous is non–empty ($\mathfrak{M}$ contains the indiscrete topology), hence $\sup \mathfrak{M}$ is the inductive topology w.r.t. $\{(X_i, \mathscr{P}_i, S_i) : i \in \Lambda\}$. (see (9.2) (b)).]

Remark (ii). If $\mathscr{U}_i$ is a local base at 0 for $\mathscr{P}_i$, then a local base at 0 for $\mathscr{T}_{ind}$ is given by

(9.5.1) $$\mathscr{U}_{ind} = \{U \subseteq Y : U = \Gamma U \text{ and absorbing with } S_i^{-1}(U) \in \mathscr{U}_i \ (i \in \Lambda)\};$$

if, in addition, $Y = \langle \bigcup_{i \in \Lambda} S_i(X_i) \rangle$ (denoted by $Y = \sum_i S_i(X_i)$), then a local base at 0 for $\mathscr{T}_{ind}$ is given by

$$\mathscr{U}_{ind} = \{ U \subset Y : U = \Gamma U \text{ with } S_i^{-1}(U) \in \mathscr{U}_i \quad (i \in \Lambda) \}.$$

Moreover, the following statements hold:

(a) $\mathscr{T}_{ind}$ is, in general,not Hausdorff even all $\mathscr{P}_i$ are Hausdorff.

(b) $\mathscr{T}_{ind}$ has the following important property [compared with Remark (iv) of (9.1)] : Let Z be a LCS and $T : (Y, \mathscr{T}_{ind}) \longrightarrow Z$ be linear. Then T is continuous (for $\mathscr{T}_{ind}$) if and only if all $T \circ S_i : (X_i, \mathscr{P}_i) \longrightarrow Z$ are continuous ($i \in \Lambda$).

(9.6) **Examples:** (a) Quotient spaces: Let $(X, \mathscr{P})$ be a LCS, let M be a vector subspace of X and $Q_M : X \longrightarrow X/M$ the quotient map. Then the quotient topology $\hat{\mathscr{P}}$ on X/M is the inductive topology on X/M w.r.t. $(X, \mathscr{P}, Q_M)$.

(b) Greatest lower bound topologies: Let Y be a vector space, let $\{ \mathscr{P}_i : i \in \Lambda \}$be a family of all locally convex topologies on Y and $I_i : Y \longrightarrow Y$ the identity map. Then the inductive topology $\mathscr{T}_{ind}$ on Y w.r.t $\{ (Y, \mathscr{P}_i, I_i) : i \in \Lambda \}$ is the greatest lower bound of $\{ \mathscr{P}_i : i \in \Lambda \}$.

(c) The finite–dimensional topology: Let Y be a vector space and $\{ M_i : i \in \Lambda \}$ a family of all finite–dimensional vector subspace of Y. On each M_i, we consider the unique Hausdroff locally convex topology $\mathcal{P}_i$ and the canonical embedding $J_i : M_i \longrightarrow Y$. Then the inductive topology $\mathcal{T}_{ind}$ on Y w.r.t $\{ (M_i, \mathcal{P}_i, J_i) : i \in \Lambda \}$ is the finest locally convex topology.

(d) Locally convex direct sum spaces: Let $\{(X_i, \mathcal{P}_i) : i \in \Lambda\}$ be a family of LCS and $J_i : X_i \longrightarrow \underset{j\in\Lambda}{\oplus} X_j$ the i–th canonical injection. Then the inductive topology on $\underset{j}{\oplus} X_j$ w.r.t $\{ (X_i, \mathcal{P}_i, J_i) : i \in \Lambda \}$ is called the locally convex direct sum topology, and denoted by $\underset{j}{\oplus} \mathcal{P}_j$. The space $(\underset{j}{\oplus} X_j, \underset{j}{\oplus} \mathcal{P}_j)$ (or denoted by $\underset{j}{\oplus} (X_j, \mathcal{P}_j)$) is called the locally convex direct sum space. As $\underset{j\in\Lambda}{\oplus} X_i = \langle \underset{j\in\Lambda}{\cup} J_j(X_j) \rangle$, it follows that $\{\Gamma(\underset{i\in\Lambda}{\cup} J_i(V_i) : V_i \in \mathcal{U}_{X_i}, i \in \Lambda\}$ is a local base for $\underset{i}{\oplus} \mathcal{P}_i$.

The following result, dual to (9.3), shows that inductive topologies can be identified with a quotient topology of some locally convex direct sum topology.

(9.7) **Proposition.** Let Y be a vector space, let $\{ (X_i, \mathcal{P}_i) : i \in \Lambda \}$ be a family of LCS, let $S_i \in L^*(X_i, Y)$ and

(9.7.1) $$Y = \Sigma S_i(X_i) \quad (= \langle \underset{i\in\Lambda}{\cup} S_i(X_i) \rangle \text{ the linear hull}).$$

Suppose that $\mathcal{T}_{ind}$ is the inductive topology on Y w.r.t. $\{ (X_i, \mathcal{P}_i, S_i) : i \in \Lambda \}$, that $S : \underset{i\in\Lambda}{\oplus} X_i \longrightarrow Y$ is defined by

(9.7.2) $$S(x) = \Sigma_i S_i(\pi_i(x)) \quad (\text{whenever } x = \Sigma_i J_i(\pi_i(x)) \in \underset{j}{\oplus} X_j),$$

and that $\check{S} : \underset{j}{\oplus} X_j/\text{Ker } S \longrightarrow Y$ is the injection (in fact bijection by (9.7.1)) associated with S. Then $(\underset{i}{\oplus} X_i/\text{Ker } S, \underset{i}{\oplus} \mathcal{P}_i/\text{Ker } S)$ and $(Y, \mathcal{T}_{ind})$ are topologically isomorphic under $\check{S}$.

Proof. (9.7.2) implies that

(1) $$S \circ J_i = S_i \quad (\text{for all } i \in \Lambda)$$

(where $J_i : X_i \longrightarrow \oplus_j X_j$ is the i–th canonical injection), and the definition of $\check{S}$ implies that

$$(\check{S})^{-1} \circ S_i = Q_S \circ J_i : X_i \longrightarrow \oplus_j X_j/\mathrm{Ker}\, S \quad (i \in \Lambda)$$

are continuous (where $Q_S : \oplus_i X_i \longrightarrow \oplus_j X_j/\mathrm{Ker}\, S$ is the quotient map). The result then follows from Remark (ii) (b) of (9.5) and (9.6) (d).

As a consequence of the preceding result, we obtain the following:

(9.c) <u>Inductive topologies</u>: Let $\{(X_i, \mathcal{P}_i) : i \in \Lambda\}$ be a family of LCS, let M be a vector subspace of $\oplus_i X_i$, let $Q_M : \oplus_i X_i \longrightarrow \oplus_i X_i/\ M$ be the quotient map and $Q_{M,i} = Q_M \circ J_i \ (i \in \Lambda)$. Then

(i) $\oplus_j X_j/M = \Sigma_i Q_{M,i}(X_i)$ (= linear hull of $\cup_{i \in \Lambda} Q_{M,i}(X_i)$).

(ii) The quotient topology $\oplus_i \mathcal{P}_i/M$ is the inductive topology on $\Sigma_i Q_{M,i}(X_i)$ w.r.t $\{(X_i, \mathcal{P}_i, Q_{M,i}) : i \in \Lambda\}$.

(9.8) **Definition.** Let Γ be a directed set, let $\{(X_i, \mathcal{P}_i) : i \in \Gamma\}$ be a family of LCS, and for any $i, j \in \Gamma$ with $i \leq j$ let

$$S_{ji} : (X_i, \mathcal{P}_i) \longrightarrow (X_j, \mathcal{P}_j) \quad \text{be linear and continuous.}$$

Let us define a vector subspace H of $\oplus_j X_j$ by

(9.8.1) $$H = \text{linear hull of} \cup \{\, \mathrm{Im}(J_i - J_j \circ S_{ji}) : i \leq j\ , i,j \in \Gamma \,\}.$$

Then the quotient space $\oplus_j X_j \,/\, H$ is called the <u>inductive spectrum</u> (of $\{X_i : i \in \Gamma\}$ under S_{ji}), and denoted by $\varinjlim S_{ji} X_i$, that is

$$\varinjlim S_{ji}X_i = \underset{j}{\oplus} X_j \,/\, H .$$

Suppose that H is closed in $(\underset{i}{\oplus} X_i, \underset{i}{\oplus} \mathscr{P}_i)$. Then the (Hausdorff) quotient space $(\underset{i}{\oplus} X_i/H, \underset{i}{\oplus} \mathscr{P}_i/H)$ is called the <u>topological inductive limit of</u> $\{(X_i, \mathscr{P}_i) : i \in \Gamma\}$ <u>under the continuous maps</u> and denoted by $\varinjlim S_{ji}(X_i, \mathscr{P}_i)$; that is,

$$\varinjlim S_{ji}(X_i, \mathscr{P}_i) = (\underset{j}{\oplus} X_j \,/\, H, \underset{j}{\oplus} \mathscr{P}_j \,/\, H) ,$$

(9.d) <u>Topological inductive limits:</u> Let $\{(X_i, \mathscr{P}_i) : i \in \Gamma\}$, S_{ji} and H be as in (9.8). Denote by $Q_H : \underset{j}{\oplus} X_j \longrightarrow \underset{j}{\oplus} X_j \,/\, H$ the quotient map and $Q_{H,i} = Q_H \circ J_i$ $(i \in \Gamma)$. Then (9.c) (i) shows that

$$\varinjlim S_{ji}X_i = \sum_i Q_{H,i}\,(X_i).$$

Suppose that H is closed in $(\underset{j}{\oplus} X_j, \underset{j}{\oplus} \mathscr{P}_j)$. Then (9.c) (ii) shows that

$$\underset{j}{\oplus} \mathscr{P}_j \,/\, H = \text{the inductive topology on } \sum_i Q_{H,i}(X_i) \text{ w.r.t. } \{(X_i, \mathscr{P}_i, Q_{H,i}) : i \in \Gamma\}.$$

Moreover, it is worth noting that the completeness of $\varinjlim S_{ji}(X_i, \mathscr{P}_i)$ does not follow from the completeness of each $(X_i, \mathscr{P}_i)$.

(9.e) <u>Strict inductive limits:</u> The requirements for the construction of a topological inductive limit are often realized in the following special form: Let Y be a vector space and let $\{M_i : i \in \Gamma\}$ be a family of vector subspaces of Y which is directed under set inclusion and satisfies $Y = \bigcup_{i \in \Gamma} M_i$. Then Γ is directed under "$\leq$" defined by

$$i \leq j \text{ if and only if } M_i \subseteq M_j \quad (i,j \in \Lambda).$$

For any $i, j \in \Gamma$ with $i \leq j$, let

$$J_{ji} : M_i \longrightarrow M_j \text{ and } J_i : M_i \longrightarrow Y$$

be the canonical embedding. Then it is not hard to show that

(9.e.1) $$\varinjlim J_{ji}M_i \sim Y \quad \text{(algebraically isomorphic).}$$

Moreover, on each M_i, let a Hausdorff locally convex topology $\mathscr{P}_i$ be given for which

$$\mathscr{P}_j\big|_{M_i} \leq \mathscr{P}_i \quad \text{whenever } i \leq j.$$

Suppose that the inductive topology $\mathscr{T}_{ind}$ on Y with respect to $\{ (M_i, \mathscr{P}_i, J_i) : i \in \Gamma\}$ is Hausdorff. Then

(9.c.2) $$\varinjlim J_{ji}(M_i, \mathscr{P}_i) \simeq (Y, \mathscr{T}_{ind}) \quad \text{(topologically isomorphic)}$$

In these circumstances, $(Y, \mathscr{T}_{ind})$ is referred to as the <u>inductive inclusion limit of</u> $\{(M_i, J_i) : i \in \Gamma\}$. If, in addition.

$$\mathscr{P}_j\big|_{M_i} = \mathscr{P}_i \quad \text{for all } i \leq j \quad (i, j \in \Gamma),$$

then $(Y, \mathscr{T}_{ind})$ is called the <u>strict inductive inclusion limit of</u> $\{(M_i, \mathscr{P}_i) : i \in \Gamma\}$. In particular, the strict inductive inclusion limit of an <u>increasing sequence of Banach spaces</u> (resp. F–<u>spaces</u>) is called a (LB)–<u>space</u> (resp. (LF)–<u>space</u>).

(i) Every (LB)–space (or (LF)–space) is complete, but not metrizable.

(ii) (Dieudonné–Schwartz). Let $\{(X_n, \mathscr{P}_n) : n \geq 1\}$ be an increasing sequence of LCS such that $Y = \bigcup_{n \geq 1} X_n$, let

$$J_{n+1,n} : (X_n, \mathscr{P}_n) \longrightarrow (X_{n+1}, \mathscr{P}_{n+1}) \text{ and } J_n : X_n \longrightarrow Y$$

be the canonical embedding, and let $\mathscr{T}_{ind}$ be the inductive topology on Y w.r.t. $\{(X_n, \mathscr{P}_n, J_n) : n \geq 1\}$. Suppose that $\mathscr{P}_{n+1}\big|_{X_n} = \mathscr{P}_n$ $(n \geq 1)$. Then $\mathscr{T}_{ind}$ is Hausdorff and

$$\mathcal{T}_{\mathrm{ind}}|_{X_n} = \mathcal{P}_n \quad \text{(for all } n \geq 1),$$

hence $(Y, \mathcal{T}_{\mathrm{ind}})$ is the strict inductive inclusion limit of $\{(X_n, \mathcal{P}_n) : n \geq 1\}$. (For a proof, see Schaefer [1966,p.58].)

(9.6) **Example:** (e) The locally convex direct sum is an example of inductive limit

The proof is similar to that of (9.2) (e), hence will be omitted.

(9.f) The finest extension : Let $(X, \mathcal{P})$ be a LCS, let $\mathcal{C}$ be a family of $\mathcal{P}$–compact subsets of X, and for any $K \in \mathcal{C}$, let $J_K : K \longrightarrow X$ be the canonical injection and

$$\mathcal{P}_K = \mathcal{P}|_K \quad \text{(the relative topology on } K\text{)}.$$

The finest topology on X for which all maps $\{J_K : K \in \mathcal{C}\}$ are continuous is called the finest extension of $\mathcal{P}$ w.r.t. $\mathcal{C}$, and denoted by $i(\mathcal{P}, \mathcal{C})$. Clearly, $\mathcal{P} \leq i(\mathcal{P}, \mathcal{C})$ hence $i(\mathcal{P}, \mathcal{C})$ is Hausdorff. $i(\mathcal{P}, \mathcal{C})$–closed (resp. $i(\mathcal{P}, \mathcal{C})$–open) sets in X are said to be almost $\mathcal{P}$–closed (resp. almost $\mathcal{P}$–open). The $\mathcal{P}$–closedness of each $K \in \mathcal{C}$ ensures that a subset B of X is almost $\mathcal{P}$–closed if and only if

$$B \cap K \qquad \text{is } \mathcal{P}\text{–closed in } K \quad (\text{for any } K \in \mathcal{C}).$$

(i) (Collin [1955]). The topology $i(\mathcal{P}, \mathcal{C})$ is, in general, not a vector topology, it is only a semi–vector topology in the following sense: the maps

$$(x,y) \longrightarrow x - y : X \times X \longrightarrow X$$

and

$$(\lambda,y) \longrightarrow \lambda y : K\times X \longrightarrow X$$

are separately continuous for $i(\mathscr{P}, \mathscr{C})$ and the usual topology on K

(ii) Denote by

$$\mathscr{U}_a = \{ V = \Gamma V \subset X : V \text{ is a } i(\mathscr{P}, \mathscr{C})\text{–neighbourhood of } 0 \}.$$

Then $\mathscr{U}_a$ determines a finest locally convex topology on X which is coarser than $i(\mathscr{P}, \mathscr{C})$; this topology is called the locally convex topology on X associated with $i(\mathscr{P}, \mathscr{C})$, and denoted by $i_{lf}(\mathscr{P}, \mathscr{C})$. It is clear that

$$\mathscr{P} \le i_{lf}(\mathscr{P}, \mathscr{C}) \text{ and } i_{lf}(\mathscr{P}, \mathscr{C})\big|_K = i(\mathscr{P}, \mathscr{C})\big|_K \quad (K \in \mathscr{C}).$$

(iii) If $(X, \mathscr{P})$ is metrizable and if $\mathscr{C}$ is the family of all $\mathscr{P}$–compact subsets of X, then $\mathscr{P} = i(\mathscr{P}, \mathscr{C})$.

Exercises

9–1. Let X be a vector space, and let $\mathscr{P}_{pro}$ be the projective topology on X w.r.t. TVS $\{(Y_i, \mathscr{T}_i, T_i) : i \in \Lambda\}$.

(a) Suppose that all $\mathscr{T}_i$ are Hausdorff. Show that the following statements are equivalent:

(i) $\mathscr{P}_{pro}$ is Hausdorff.

(ii) For any $0 \neq x \in X$. there exist an $i \in \Lambda$ and a $\mathscr{T}_i$–neighourhood U_i of 0 in Y_i such that $T_i(x) \notin U_i$.

(iii) For any $0 \neq x \in X$ there is $i \in \Lambda$ such that $T_i(x) \neq 0$.

(b) If G is another topological space and $\varphi : G \longrightarrow X$ is a mapping, then φ is continuous (for $\mathscr{P}_{pro}$) if and only if all $T_i \circ \varphi : G \longrightarrow (Y_i, \mathscr{T}_i)$ are continuous (for all $i \in \Lambda$).

(c) An $B \subset (X, \mathscr{P}_{pro})$ is bounded (resp. totally bounded) if and only if each $T_i(B) \subset (Y_i, \mathscr{T}_i)$ is bounded (resp. totally bounded) (for all $i \in \Lambda$).

(d) A filter $\mathscr{F}$ on X is a $\mathscr{P}_{pro}$–Cauchy filter if and only if each $T_i(\mathscr{F})$ is a $\mathscr{T}_i$–Cauchy filter basis in $Y_i (i \in \Lambda)$.

9–2. Prove (9.a).

9–3. Prove (9.b).

9–4. Prove (9.c).

9–5. Prove (9.d).

9–6. Prove (9.e).

10

Normed Spaces Associated with a Locally Convex Space

This section is devoted to a study of two methods for constructing auxiliary normed spaces from a given LCS. These two methods were systematically employed by Grothendieck and will be extremely useful in what follows.

Let $(X,\mathscr{P})$ be a LCS. For any disked $\mathscr{P}$-neighbourhood V of O, let $N(V) = \operatorname{Ker} p_V$ (called the <u>null space</u> of p_V), let $X_V = X/N(V)$, let $Q_V : X \longrightarrow X_V$ be the quotient map, and let $x(V)$ (or $\hat{x}$) be the equivalence class $x + N(V)$ modulo $N(V)$. Then the quotient seminorm $\hat{p}_V$ of p_V, defined by

$$\hat{p}_V(x(V)) = \inf \{p_V(x + v) : v \in N(V)\},$$

is actually a norm on X_V such that $\hat{p}_V(x(V)) = p_V(x)$ $(x \in X)$ and

$$\{ x(V) \in X_V : \hat{p}_V(x(V)) < 1\} \subset Q_V(V) \subset \{x(V) \in X_V : \hat{p}_V(x(V)) \leq 1 \}.$$

For simplicity of notation, we write X_V for the normed space $(X_V,\hat{p}_V)$. If p is a continuous seminorm on X and $V = \{ x \in X : p(x) \leq 1 \}$, then we set

$$X_p = X_V\,,\ Q_p = Q_V\,,\ x(p) = x(V) \text{ and } \hat{p} = \hat{p}_V\,.$$

It should be noted that X_V is, in general, not complete, even if X is; its completion will be denoted by $\tilde{X}_V$.

(10.a) <u>The Banach dual space</u> of X_V : Let X be a LCS, let V be a disked 0–neighbourhood, let $X_V' = (X_V,\hat{p}_V)'$ and

$$V^0 = \{ f \in X' : \operatorname{Re} f(x) \leq 1 \ (x \in V)\}.$$

Suppose that $X'(V^0) = \bigcup_{n\geq 1} nV^0$ and that

$$\gamma_{V^0}(f) = \sup \{ |f(x)| : x \in V \} \quad (\text{for all } f \in X'(V^0)).$$

Then $(X'(V^0), \gamma_{V^0})$ is metrically isomorphic to the Banach dual space $(X_V', \|\cdot\|)$ of $(X_V, \hat{p}_V)$ (and surely complete) and

$$\gamma_{V^0}(f) = \inf \{ \lambda > 0 : f \in \lambda V^0 \} \quad (\text{for all } f \in X'(V^0)).$$

We shall see that the construction of the Banach space $(X'(V^0), \gamma_{V^0})$ is a special case of the following general method for constructing normed spaces from a given LCS.

Let $(X, \mathscr{P})$ be a LCS. For any disked $\mathscr{P}$-bounded subset B of X, let $X(B) = \langle B \rangle$ (the linear hull of B). Then $X(B) = \bigcup_n nB$, and B is absorbing in $X(B)$, hence the gauge γ_B of B, defined by

$$\gamma_B(x) = \inf \{ \lambda > 0 : x \in \lambda B \} \quad (\text{for all } x \in X(B)),$$

is finite on $X(B)$ and

$$\{ x \in X(B) : \gamma_B(x) < 1 \} \subset B \subset \{ x \in X(B) : \gamma_B(x) \leq 1 \};$$

also the boundedness of B ensures that $\mathscr{P}|_{X(B)}$ (the relative topology induced by $\mathscr{P}$) is coarser than the $\gamma_B(\cdot)$–topology, hence γ_B is actually a norm on $X(B)$, and the canonical embedding $J_B : (X(B), \gamma_B) \longrightarrow (X, \mathscr{P})$ is always continuous. For simplicity of notation, we write $X(B)$ for the normed space $(X(B), \gamma_B)$.

(10.b) <u>The Banach dual space of</u> $X(B)$: Let $X_{\mathscr{P}}$ be a LCS, let B be a disked $\mathscr{P}$-bounded subset of X, let $X(B)' = (X(B), \gamma_B)'$, let

$$B^0 = \{ f \in X' : |f(b)| \leq 1 \ (b \in B) \},$$

and p_{B^0} the gauge of B^0. Define J_B' on X' by setting

$$J_B'(f) = f \circ J_B \quad (\text{for any } f \in X').$$

Then the following holds:

(i) B^0 is an absolutely convex, absorbing subset of X'.

(ii) $\operatorname{Ker} J_B' = \operatorname{Ker} p_{B^0}$.

(iii) The injection $\widehat{J_B'}$ associated with J_B' is a metric injection from the normed space $(X'_{B^0}, \hat{p}_{B^0})$ into the Banach dual space $(X(B)', \|\cdot\|)$ of $X(B)$.

(iv) $\widehat{J_B'}(U_{X'_{B^0}})$ is $\sigma(X(B)', X(B))$–dense in $U_{X(B)'}$.

Let $(X, \mathscr{P})$ be a LCS. If B and C are absolutely convex $\mathscr{P}$-bounded subsets of X such that $B \subseteq C$, then

$$X(B) \subseteq X(C) \text{ and } \gamma_C \leq \gamma_B \text{ on } X(B),$$

thus the canonical embedding map from $X(B)$ into $X(C)$, denoted by $J_{C,B}$, is continuous for which

$$J_B = J_C J_{C,B}.$$

$J_{C,B}$ has a unique continuous extension from the Banach space $\tilde{X}(B)$ into the Banach space $\tilde{X}(C)$, which is again denoted by $J_{C,B}$ (also called the canonical embedding).

Dually, if W and V are disked $\mathscr{P}$-neighbourboods of 0 in $X_{\mathscr{P}}$ such that $W \subset V$, then

$$p_V \leq p_W \text{ and } N(W) \subseteq N(V),$$

hence the canonical map $Q_{V,W} : X_W \longrightarrow X_V$, defined by

$$Q_{V,W}(x(W)) = x(V) \quad (\text{for all } x \in X),$$

is a continuous linear map such that

$$Q_V = Q_{V,W} \circ Q_W .$$

$Q_{V,W}$ has a unique continuous extension from the Banach space $\tilde{X}_W$ into the Banach space $\tilde{X}_V$ which is again denoted by $Q_{V,W}$ (also called the canonical map)

Let $X_{\mathscr{P}}$ be a LCS. An absolutely convex $\mathscr{P}$-bounded subset B of X is called an infracomplete set (or a completant, Banach disk) if the normed space $(X(B), \gamma_B)$ is complete.

(10.1) **Lemma.** Let $(X, \mathscr{P})$ be a LCS and B a $\mathscr{P}$-bounded, disked subset of X. Then one of the following conditions ensures that B is infracomplete:

(a) B is compact.

(b) B is complete.

(c) B is sequentially compact (i.e. every sequence in B has a subsequence convergent to a point in B).

(d) B is sequentially complete.

Proof. If B satisfies one of the conditions (a), (b) and (c), then B must be sequentially complete; thus we have only to verify the result assuming that B be sequentially complete.

In fact, let $\{x_n, n \geq 1\}$ be a Cauchy sequence in the normed space $(X(B), \gamma_B)$. Then $\{x_n\}$ is γ_B-bounded and $\mathscr{P}$-Cauchy in X, thus there is an $\lambda > 0$ such that $x_n \in \lambda B$ (for all n). As λB is sequentially complete for $\mathscr{P}$, there is a $u \in \lambda B$ such that $u = \mathscr{P}\text{-}\lim_n x_n$. On the other hand, for any $\epsilon > 0$, since ϵB is a $\gamma_B(\cdot)$-neighbourhood of 0 in $X(B)$, there exists a $k \geq 1$ such that

$$x_n - x_m \in \epsilon B \quad (\text{for all } n,m \geq k).$$

It then follows from $u = \mathscr{P}\text{-}\lim\limits_n x_n$ that

$$u - x_m \in \epsilon B \quad (\text{for all } m \geq k),$$

and hence that $u = \gamma_B(\cdot) - \lim\limits_m x_m$ (since ϵB is a $\gamma_B(\cdot)$–neighbourhood of 0 in $X(B)$). This shows that $(X(B), \gamma_B)$ is complete.

It is clear that the image of a bounded set under an operator must be bounded; it is natural to ask whether this assertion is still true replacing the bounded sets by infracomplete sets . The answer is affirmative as shown by the following:

(10.2) **Proposition.** Let X and Y be LCS, let $T \in L(X,Y)$, let B be a bounded, disked subset of X and $C = T(B)$. Denote by γ_B(resp. γ_C) the gauge of B (resp. of C) defined on $X(B)$ (resp. on $Y(C)$), and by $\hat{\gamma}_B$ the quotient norm on $X(B)/\text{Ker } T_B$, where

(10.2.1) $$T_B = T \circ J^X_{X(B)} : X(B) \longrightarrow Y.$$

Then the injection $\hat{T}_B$ associated with T_B satisfies

(10.2.2) $$\gamma_C(\hat{T}_B(\hat{x})) = (\hat{\gamma}_B(\hat{x})) \quad (\text{for all } \hat{x} \in X(B)/\text{Ker } T_B).$$

Consequently, $(X(B)/\text{Ker } T_B, \hat{\gamma}_B)$ and $(Y(C), \gamma_C)$ are metrically isomorphic under $\hat{T}_B$, and the image of any infracomplete set in X under T is infracomplete.

Proof. Let $Q_B : X(B) \longrightarrow X(B)/\text{Ker } T_B$ be the quotient map. Then

$$Q_B(x) = \hat{x} = x(\text{Ker } T_B) \text{ and } T_B = \hat{T}_B Q_B .$$

Notice that

$$\gamma_C(\hat{T}_B(\hat{x})) = \gamma_C(T_B x) = \gamma_C(Tx) \quad (\text{for all } x \in X(B)),$$

and that

$$\hat{\gamma}_B(\hat{x}) = \inf \{ \gamma_B(x + u) : u \in \text{Ker } T_B \} \quad (x \in X(B)).$$

Now for any $u \in \text{Ker } T_B$ and $\lambda > 0$ with $x + u \in \lambda B$, we have $Tx = T(x+u) \in \lambda C$, hence $\gamma_C(Tx) \leq \lambda$. As $\lambda > 0$ and $u \in \text{Ker } T_B$ were arbitrary, we conclude that $\gamma_C(Tx) \leq \hat{\gamma}_B(\hat{x})$, and hence that

(1) $$\hat{\gamma}_C(\hat{T}_B(\hat{x})) \leq \hat{\gamma}_B(\hat{x}).$$

Conversely, for any $x \in X(B)$ and $\mu > 0$ with $Tx \in \mu C$, there exists a $z \in B$ and $w \in \text{Ker } T_B$ such that $x - \mu z = w$. As $x - w = \mu z \in \mu B$, it follows that $\gamma_B(x - w) \leq \mu$, and hence that $\gamma_B(x - w) \leq \gamma_C(Tx)$ (since $\mu > 0$ was arbitrary); consequently,

(2) $$\hat{\gamma}_B(\hat{x}) \leq \gamma_B(x - w) \leq \gamma_C(Tx) = \gamma_C(\hat{T}_B(\hat{x})).$$

Combining (1) and (2), we obtain our assertion (10.2.2).

It is clear that $\hat{T}_B : X(B)/\text{Ker } T_B \longrightarrow Y(C)$ is bijective (since $C = T(B)$), thus $\hat{T}_B$ is a metric isomorphism by (10.2.2).

Finally, if B is assumed to be infracomplete, then $(X(B),\gamma_B)$ is a Banach space, hence $(X(B)/\text{Ker } T_B,\hat{\gamma}_B)$ is a Banach space [by the continuity of T], thus $(Y(C),\gamma_C)$ is complete [since $(X(B)/\text{Ker } T_B,\hat{\gamma}_B)$ and $(Y(C),\gamma_C)$ are metrically isomorphic].

Exercises

10–1. Prove (10.a).

10–2. Prove (10.b).

10–3. Let X be a LCS and let A, B be two disks in X. Show that:

(a) $X(A+B) = X(A) + X(B)$ and

$$r_{A+B}(x) = \inf \{\max \{r_A(u), r_B(v)\} : x = u+v,\ u \in X(A),\ v \in X(B)\}$$

for any $x \in X(A+B)$.

(b) $X(A) \times X(B)$ is a seminormed space under the product seminorm of r_A and r_B, defined by

$$\|(u,v)\| = \max \{r_A(u), r_B(v)\} \text{ (for any } (u,v) \in X(A) \times X(B)).$$

(c) The map $\psi : X(A) \times X(B) \to X(A+B)$, defined by

$$\psi(u,v) = u+v \text{ (for all } (u,v) \in X(A) \times X(B)),$$

is a surjective linear map, hence its bijection $\check{\psi}$ associated with ψ is an isometry from the quotient seminormed space $(X(A) \times X(B)/_{\mathrm{Ker}\ \psi}, \widehat{\|\cdot\|})$ onto the seminormed space $(X(A+B), r_{A+B})$

(d) If A and B are assumed to be bounded, then r_{A+B} is a norm.

(e) If A and B are infracomplete (bounded) disks then A+B and μB ($\mu \in K$) are infracomplete.

10–4. Let X and Y be LCS and $T \in L(X,Y)$. Suppose that U is a disked 0–neighbourhood in Y and that $V = T^{-1}(U)$. Show that

(a) $p_U(Tx) = p_V(x)$ (for all $x \in X$);

(b) $x \in p_V^{-1}(0)$ if and only if $Tx \in p_U^{-1}(0)$ (consequently, T is compatible with the equivalence relations in X and Y induced by $p_V^{-1}(0)$ and $p_U^{-1}(0)$ respectively);

(c) the map $\check{T}$ obtained from T by passing to the quotient is an operator from the normed space X_V into the normed space Y_U with $\|\check{T}\| = 1$.

10–5. Let X and Y be LCS, let V be a disked 0–neighbourhood in X, let B be a bounded disk in Y. Let

$$P(V,B) = \{S \in L^*(X,Y) : S(V) \subset B\} \quad \text{and} \quad L^b(P(V,B)) = \bigcup_{n \geq 1} nP(V,B).$$

Show that:

(a) P(V,B) is a disk in L(X,Y).

(b) $p_V^{-1}(0) \subset \text{Ker } T$ (for any $T \in L^b(P(V,B))$).

(c) For any $T \in L^b(P(V,B))$, there exists a unique operator $T_{(B,V)}$ from the normed space X_V into the normed space $Y(B)$ such that

$$T = J_B \, T_{(B,V)} \, Q_V$$

(where $Q_V : X \to X_V$ is the quotient map and $J_B : Y(B) \to Y$ is the canonical embedding), hence the map $\psi : L(X_V, Y(B)) \to L^b(P(V,B))$, defined by

(*) $$\psi(S) = J_B S Q_V \text{ (for all } S \in L(X_V, Y(B)))$$

is an algebraic isomorphism.

(d) Denote by $r_{(V,B)}$ the gauge of $P(V,B)$ defined on $L^b(P(V,B))$ and by $B^\circ = \{g \in Y' : |g(b)| \leq 1 \text{ (for all } b \in B)\}$. If B is closed in Y, then the map ψ, defined by (*), is an isometry from the normed space $(L(X_V, Y(B)), \|\cdot\|)$ onto the normed space $(L^b(P(V,B)), r_{(V,B)})$.

11

Bounded Sets and Compact Sets in Metrizable Topological Vector Spaces

It is known from the previous section that for any bounded, absolutely convex set B in a LCS X , one can construct a normed space X(B) whose norm–topology on B is finer than the relative topology induced by the original locally convex topology. It is natural to ask that under what conditions these two topologies on B are identical. The following result gives some sufficient condition to answer this question.

(11.1) **Proposition.** Let $(X, \mathscr{P})$ be a metrizable LCS. For any bounded subset A of X, there exists a bounded, $\mathscr{P}$-closed, disked subset B of X and $\mu > 0$ with $A \subset \mu B$ such that

$$\mathscr{P}\big|_A = \gamma_B(\cdot)\text{–top}\big|_A ,$$

where $\gamma_B(\cdot)\text{–top}\big|_A$ is the relative topology on A induced by the norm–top on $(X(B), \gamma_B)$ and similarly for $\mathscr{P}\big|_A$.

Proof. Let $\{ V_n : n \geq 1 \}$ be a local base for $\mathscr{P}$ consisting of closed, absolutely convex sets which satisfy $V_{n+1} + V_{n+1} \subset V_n$ $(n \geq 1)$. For any $n \geq 1$, the boundedness of A ensures that there is an ζ_n such that $A \subset \zeta_n V_n$ so that $A \subset \bigcap_{n=1}^{\infty} \zeta_n V_n$. Now we choose a sequence $\{\eta_n\}$ of positive numbers such that $\lim_n \frac{\zeta_n}{\eta_n} = 0$ [for instance, one can take $\eta_n = 2^n \zeta_n$], let

$$B = \bigcap_{n=1}^{\infty} \eta_n V_n \quad \text{and} \quad \frac{\zeta_n}{\eta_n} \leq \mu \quad (\text{for all } n \geq 1)$$

for some $\mu \geq 0$. Then B is clearly a bounded, closed, disked subset of X such that $A \subset \mu B$. It is clear that

$$\mathscr{P}|_A \leq \gamma_B(\cdot)\text{-top}|_A .$$

In order to verify the result, it suffices to show that for any $\lambda > 0$ there exists an $k \geq 1$ such that

$$A \cap V_k \subset \lambda B. \tag{1}$$

In fact, since $\lim_n \frac{\zeta_n}{\eta_n} = 0$, there exists a natural number $N \geq 2$ such that $\frac{\zeta_n}{\eta_n} \leq \lambda$ (for all $n \geq N$), hence

$$A \subset \bigcap_{n=N}^{\infty} \lambda\eta_n V_n \tag{2}$$

[since each V_n is circled]. As $\bigcap_{j=1}^{N-1} \lambda\eta_j V_j$ is a $\mathscr{P}$-neighbourhood of 0, there exists an $k \geq 1$ such that

$$V_k \subset \bigcap_{j=1}^{N-1} \lambda\eta_j V_j . \tag{3}$$

It then follows from (2) and (3) that

$$A \cap V_k \subset \bigcap_{j=1}^{\infty} \lambda\eta_j V_j = \lambda B ,$$

which proves our assertion (1).

(11.2) **Corollary.** Let $(X, \mathscr{P})$ be a metrizable LCS. For any totally bounded (resp. compact) subset A of $(X, \mathscr{P})$, there exists a $\mathscr{P}$-bounded, closed, absolutely convex subset B of X such that A is a totally bounded (resp. compact) subset of the normed space $(X(B), \gamma_B)$.

Proof. Since totally bounded sets are bounded, the result follows immediately from (11.1).

The preceding two results can be extended to the case of a sequence of bounded (resp. totally bounded) sets as follows:

(11.a) Bounded sets in a metrizable LCS: Let $(X, \mathscr{P})$ be a metrizable LCS and $\{A_n\}$ a sequence of $\mathscr{P}$-bounded subsets of X.

(i) (Mackey countability condition). There exists $\lambda_n > 0$ such that $\bigcup_{n=1}^{\infty} \lambda_n A_n$ is bounded in X.

(ii) There exists a $\mathscr{P}$-bounded, closed, disked subset B of X such that

$$\mathscr{P}|_{A_j} = \gamma_B(\cdot)\text{-top}|_{A_j} \quad (\text{for all } j \geq 1).$$

If, in addition, each A_n is assumed to be totally bounded (resp compact), then all A_i are totally bounded (resp. compact) in the normed space $(X(B), \gamma_B)$.

By a similar argument given in the proof of (11.1), one can verify the following:

(11.3) **Proposition.** Let $(X, \mathscr{P})$ be a metrizable LCS. For any null sequence $\{x_n\}$ in $(X, \mathscr{P})$, there exists a $\mathscr{P}$-bounded, closed, disked subset B of X such that $\{x_n\}$ is a null sequence in the normed space $(X(B), \gamma_B)$.

Proof. The result follows from (11.2) by taking $A = \{x_n : n \geq 1\} \cup \{0\}$ (the compact set), but we give here a more direct proof.

We first claim that there exists a sequence $\{\lambda_n\}$ of positive numbers such that

$$\lim_n \lambda_n = \infty \text{ and } \mathscr{P}\text{-}\lim_n \lambda_n x_n = 0.$$

Indeed, let $\{V_n : n \geq 1\}$ be a local base for $\mathscr{P}$ consisting of disks which satisfy $V_{n+1} + V_{n+1} \subset V_n$ $(n \geq 1)$. For any $k \geq 1$, there is a positive integer n_k with $n_k > n_{k-1}$ (where $n_0 = 0$) such that

$$x_n \in k^{-1}V_k \quad (\text{for all } n \geq n_k)$$

[since $\mathscr{P}\text{-}\lim_n x_n = 0$], thus we get the required sequence $\{\lambda_n\}$ by setting

$$\lambda_n = k \quad \text{for } n_k \leq n < n_{k+1} .$$

Suppose now that B is the $\mathscr{P}$-closed absolutely convex hull of $\{\lambda_n x_n : n \geq 1\}$. Then B is bounded and $\gamma_B(x_n) \leq \frac{1}{\lambda_n}$, hence $\lim_n \gamma_B(x_n) = 0$.

(5.2) and (5.3) can be extended to the case of metrizable LCS as shown by the following:

(11.b) <u>Compact sets in metrizable LCS:</u> Let $(X, \mathscr{P})$ be a metrizable LCS.

(i) For any totally bounded subset B of X, there exists a null sequence $\{x_n\}$ in X such that $B \subset \overline{\Gamma}(\{x_n\})$ (the $\mathscr{P}$-closed disked hull of the set $\{ x_n : n \geq 1 \}$).

(ii) For any null sequence $\{x_n\}$ in X there exist positive numbers λ_n such that

$$\lim_n \lambda_n = +\infty \quad \text{and} \quad \mathscr{P}\text{-}\lim_n \lambda_n x_n = 0.$$

(iii) If, in addition, X is complete (i.e., an F–space), then for any relatively compact subset K of $(X, \mathscr{P})$, there exists a $\mathscr{P}$-compact, absolutely convex subset C with $K \subset C$ such that the $\mathscr{P}$-closure $\overline{K}$ of K is compact in the B–space $(X(C), \gamma_C)$.

(11.c) <u>Compact operators</u>: Let X and Y be TVS and $T \in L(X, Y)$. T is said to be <u>compact</u> if there exists a 0–neighbourhood V in X such that T(V) is relatively compact in Y.

Suppose now that X is a LCS, Y is an F–space and $T \in L(X,Y)$ is compact. Then there exists a Banach space G and compact operators $R \in L(X, G)$ and $S \in L(G, Y)$ such that $T = SR$.

Exercises

11–1. Let X be a (F)–space. For any totally bounded subset K of X. there exists a $\mathscr{P}$–compact disk B in X with $K \subset B$ such that the $\mathscr{P}$–closure $\overline{K}$ of K is compact in the B–space X(B). [Compare with (11.2).]

11–2. Prove (11.a).

11–3. A metrizable LCS $X_{\mathscr{P}}$ is normable if and only if it possesses a fundamental sequence of bounded sets. [Hint: Use (11.a)(ii).]

11–4. Prove (11.b) (i) and (ii).

11–5. Prove (11.c).

11–6. Let X and Y be metrizable LCS, and let $Q \in L(X,Y)$ be an open, continuous linear map. Show that:

(a) For any null sequence $\{y_n\}$ in Y, there exists a null sequence $\{x_n\}$ in X such that $y_n=Q(x_n)$, and hence deduce from Question 11–4 [i.e.,(11.b)(i)] that any totally bounded subset of Y is the image of some totally bounded subset of X under Q.

(b) Denote by $\tilde{X}$ the completion of X. Then the natural extension $\tilde{Q} : \tilde{X} \to \tilde{Y}$ is open. (Consequently, the separated quotient space of any (F)–space is an (F)–space)(see Junek [1983, p.62]).

12

The Bornological Space Associated with a Locally Convex Space

We start with the following:

(12.1) <u>Definition</u>. Let $(X,\mathscr{P})$ be a LCS and $\mathscr{U}$ a local base for $\mathscr{P}$. A subset V of X is said to be <u>bornivorous</u> if it absorbs any $\mathscr{P}$-bounded subset of X. Suppose now that

$$\mathscr{U}^{\times} = \{ V \subset X : V = \Gamma V \text{ and bornivorous } \}.$$

Then $\mathscr{U}^{\times}$ determines a unique locally convex topology, denoted by $\mathscr{P}^{\times}$, with $\mathscr{U}^{\times}$ as a local base. $\mathscr{P}^{\times}$ is referred to as the <u>bornological topology associated with</u> $\mathscr{P}$, and $(X,\mathscr{P}^{\times})$ is called the <u>bornological space associated with</u> $(X,\mathscr{P})$. A LCS $(X,\mathscr{P})$ is called a <u>bornological space</u> if $\mathscr{P}= \mathscr{P}^{\times}$; in this case, $\mathscr{P}$ is called a <u>bornological topology</u>.

<u>Remark</u> : (i) It is clear that $\mathscr{P}\leq \mathscr{P}^{\times}$ (hence $\mathscr{P}^{\times}$ is always Hausdorff), and that $\mathscr{P}^{\times}$ is the finest locally convex topology with the same bounded sets as $\mathscr{P}$.

(ii) As the family $\mathscr{B}_{\text{von}}$ of all closed, bounded disks in $(X,\mathscr{P})$ is a fundamental system of bounded sets in X, it follows that a subset V of X is bornivorous if and only if it absorbs any member in $\mathscr{B}_{\text{von}}$.

(12.2) Proposition. <u>Every metrizable</u> LCS $(X,\mathscr{P})$ <u>is bornological</u>.

<u>Proof.</u> Let $\{ V_n : n \geq 1 \}$ be a local base for $\mathscr{P}$ such that $V_{n+1} + V_{n+1} \subset V_n$, and let W be an absolutely convex, bornivorous subset of X. If W is not a $\mathscr{P}$-neighboourhood of 0 , then $V_n \not\subset 2^n W$ (for all $n \geq 1$). For any $n \geq 1$, let $x_n \in V_n \backslash 2^n W$. Then $\{x_n\}$ is a null sequence, hence $\mathscr{P}$-bounded and thus absorbed by W, which contracdicts $x_n \notin 2^n W$ (for all $n \geq 1$).

(12.3) **Lemma.** Let $(X, \mathscr{P})$ be a LCS. Then the following statements are equivalent:

(a) $(X, \mathscr{P})$ is bornological.

(b) Every seminorm on X, which is bounded on bounded subsets of X, is $\mathscr{P}$-continuous.

(c) For any LCS Y, any linear map $T : X \longrightarrow Y$ which sends bounded subsets of X into bounded subset of Y (such an T is said to be locally bounded) is continuous.

Proof. The equivalence of (a) and (b) is trivial.

(a) $\Rightarrow$ (c) Let U be an absolutely convex 0–neighbourhood in Y. Then $T^{-1}(U) = \Gamma T^{-1}(U)$ which is bornivorous [since T is locally bounded], hence a $\mathscr{P}$-neighbourhood of 0, thus T is continuous.

(c) $\Rightarrow$ (a): Let $\mathscr{P}^{\times}$ be the bornological topology associated with $\mathscr{P}$. Then the identity map $I_X : (X, \mathscr{P}) \longrightarrow (X, \mathscr{P}^{\times})$ is locally bounded, hence continuous, thus $\mathscr{P}^{\times} \leq \mathscr{P}$. As $\mathscr{P} \leq \mathscr{P}^{\times}$ is always true, we conclude that $(X, \mathscr{P})$ is bornological.

(12.4) **Proposition.** Every separated quotient space of a bornological space is bornological, and the locally convex direct sum of a family of bornological spaces is bornological.

Proof. (a) Let $(X, \mathscr{P})$ be a bornological space, let M be a closed vector subspace of X and $Q_M : X \longrightarrow X/M$ the quotient map. If U is absolutely convex and bornivorous in $(X/M, \hat{\mathscr{P}})$, then so is $Q_M^{-1}(U)$, hence $Q_M^{-1}(U)$ is a $\mathscr{P}$-neighbourhood of 0, thus U is a $\hat{\mathscr{P}}$-neighbourhood of 0 ; consequently, X/M is bornological.

(b) The proof is similar to that of (a), hence will be omitted .

Remark. If the inductive topology $\mathscr{T}_{\text{ind}}$ w.r.t. to a family of bornological spaces (under continuous linear maps) is Hausdorff, then $\mathscr{T}_{\text{ind}}$ is bornological.

(12.a) Bornological spaces : (i) The product space of a family of bornological spaces is, in general, not bornological, and any subspace of a bornological space is , in general, not bornological.

(ii) Dieudonné has shown that any finite–dimensional subspace of a bornological space is bornological, but it is not true for countable–codimensional subspaces as shown by Valdivia [1972]. Moreover, Valdivia [1971] has shown that any countable–codimensional subspace of a sequentially complete bornological space is bornological.

(12.5) **Proposition.** Let $(X,\mathscr{P})$ be a LCS and $\mathscr{P}^{x}$ the bornological topology associated with $\mathscr{P}$. Then there exists a family $\{(E_\alpha,\|\cdot\|_\alpha) : \alpha \in \Lambda\}$ of normed spaces and a family $\{ T_\alpha : \alpha \in \Lambda \}$ of linear maps $T_\alpha : E_\alpha \longrightarrow X$ with $X = \bigcup_{\alpha\in\Lambda} T_\alpha(E_\alpha)$ such that $\mathscr{P}^{x}$ is the inductive topology w.r.t. $\{(E_\alpha,\|\cdot\|_\alpha,T_\alpha) : \alpha \in \Lambda \}$.

Proof. Let $\mathscr{B}_{von}$ be the family of all closed, absolutely convex bounded subsets of X. For any $B \in \mathscr{B}_{von}$, let $J_B : X(B) \longrightarrow X$ be the canonical embedding. Then $J_B : (X(B),\gamma_B) \longrightarrow (X,\mathscr{P}^{x})$ is continuous [by (12.2) and (12.3)] and $X = \bigcup_{B \in \mathscr{B}_{von}} X(B)$ [since $\mathscr{B}_{von}$ is fundamental]. On the other hand, an absolutely convex subset of X is a $\mathscr{P}^{x}$–neighbourhood of 0 if and only if it absorbs any member in $\mathscr{B}_{von}$, thus $\mathscr{P}^{x}$ is the finest locally convex topology on X such that all $J_B : (X(B),\gamma_B) \longrightarrow (X,\mathscr{P}^{x})$ are continuous; in other words $\mathscr{P}^{x}$ is the inductive topology w.r.t. $\{(X(B),\gamma_B,J_B) : B \in \mathscr{B}_{von}\}$.

The preceding result, together with (9.b), enables us to make clear the essential difference between the bornological structure and the topological structure of a vector space; the former is a collection of internal pieces each of which is a normed space, while the latter is a collection of external hulls each of which is a normed space.

(12.b) Continuity of linear map on bornological spaces : Let $(X,\mathscr{P})$ be a bornological space, let Y be a LCS and $T \in L^{*}(X,Y)$. Then the following statements are equivalent.

(i) T is continuous.

(ii) $\{Tx_n\}$ is a null sequence in Y whenever $\{x_n\}$ is a null sequence in X.

(iii) T is locally bounded (i.e., T sends any bounded subset of X into a bounded subset of Y).

Exercises

12–1 Prove (12.b).

12–2 Let $X_{\mathscr{P}}$ be a LCS. A sequence $\{x_n\}$ in X is called a local–null (or b–null) sequence if there exist $\lambda_n > 0$ with $\lambda_n \to 0$ and a bounded subset B of X such that

$$x_n \in \lambda_n B \quad (\text{for all } n \geq 1).$$

Show that

(a) Any local–null sequence $\{x_n\}$ must be a $\mathscr{P}$-null sequence.

(b) A disk V in $X_{\mathscr{P}}$ is bornivorous if and only if V absorbs any local–null sequence. Consequently, $X_{\mathscr{P}}$ is bornological if and only if any disk in X which absorbs all local–null sequences in X is a $\mathscr{P}$-neighbourhood of 0.

12–3 Let $X_{\mathscr{P}}$ be a LCS and $\mathfrak{S}$ a family consisting of bounded subsets of X with $\cup\mathfrak{S} = X$. A subset V of X is said to be $\mathfrak{S}$–bornivorous if V absorbs any $B \in \mathfrak{S}$. $X_{\mathscr{P}}$ is called a $\mathfrak{S}$–bornological space if any $\mathfrak{S}$–bornivorous disk in X is a $\mathscr{P}$-neighbourhood of 0. A sequence $\{x_n\}$ in X is called a $\mathfrak{S}$–local–null sequence if there exist $\lambda_n > 0$ with $\lambda_n \to 0$ and $B \in \mathfrak{S}$ such that

$$x_n \in \lambda_n B \quad (\text{for any } n \geq 1).$$

(a) A disk V in X is $\mathfrak{S}$–bornivorous if and only if it absorbs all $\mathfrak{S}$–local–null sequences.

(b) The following statements are equivalent for $X_{\mathscr{P}}$:

(i) X is $\mathfrak{S}$–bornological.

(ii) Any seminorm p on X which is bounded on $B \in \mathfrak{S}$ is continuous.

(iii) For any LCS Y, any $T \in L^*(X,Y)$ which sends $B \in \mathfrak{S}$ into bounded subsets of Y is continuous.

(c) For any LCS Y, the following statements are equivalent for any $S \in L^*(X,Y)$:

(i) S sends $B \in \mathfrak{S}$ into bounded subsets of Y.

(ii) T sends any $\mathfrak{S}$–local–null sequence in X into a local–null sequence in Y.

(iii) T maps any $\mathfrak{S}$–local–null sequence in X into a null sequence in Y.

(iv) T maps any $\mathfrak{S}$–local–null sequence in X into a bounded subsets of Y.

13

Vector Bornologies

It is well–known that unit balls in normed spaces are neighbourhoods of 0 as well as bounded sets. In section 7, we have extended the notion of unit balls from the topological point of view. This section and next parallely deal with a generalization of the notion of unit balls from the boundedness, called vector bornologies, as defined by the following:

(13.1) **Definition.** Let X be a vector space. A family $\mathfrak{S}$ of subsets of X is called a vector bornology if it satisfies the following conditions:

(VB1) $X = \cup \{B : B \in \mathfrak{S}\}$.

(VB2) Any subset of any member in $\mathfrak{S}$ belongs to $\mathfrak{S}$.

(VB3) $B_1 + \cdots + B_n \in \mathfrak{S}$ whenever $B_i \in \mathfrak{S}$ $(i = 1, \cdots, n)$.

(VB4) $\lambda B \in \mathfrak{S}$ whenever $B \in \mathfrak{S}$ and $\lambda \in K$.

(VB5) The circled hull of any $B \in \mathfrak{S}$ belongs to $\mathfrak{S}$.

Members in $\mathfrak{S}$ are called bounded sets.

A vector bornololgy $\mathfrak{S}$ on X is said to be :

(a) separated if $\mathfrak{S}$ does not contain any vector subspace of X different from $\{0\}$;

(b) convex bornology if $\Gamma B \in \mathfrak{S}$ whenever $B \in \mathfrak{S}$.

A vector space equipped with a vector bornology (resp. convex bornology) is called a bornological vector space, abbreviated BVS (resp. convex bornological vector space, abbreviated CBVS). Hereafter we shall generally denote a BVS by $(X,\mathfrak{S})$ or $X^{\mathfrak{S}}$ or simply by X if the vector bornology $\mathfrak{S}$ does not require any special notation.

Any fundamental subfamily $\mathfrak{B}$ of a vector bornology $\mathfrak{S}$ on X is called a base for $\mathfrak{S}$ (namely, any member in $\mathfrak{S}$ is contained in some element of $\mathfrak{B}$).

If $\mathfrak{S}$ is a vector bornology (resp. convex bornology) on X, then the family {circled hull of $B : B \in \mathfrak{S}\}$ (resp. $\{\Gamma B : B \in \mathfrak{S}\}$) is a base for $\mathfrak{S}$. Moreover, it is not hard to show that the following holds.

(13.a) Bases of vector bornologies: Let X be a vector space.

(1) A family $\mathscr{B}$ of subsets of X is a base for a vector bornology (resp. convex bornology) if and only if $\mathscr{B}$ satisfies the following conditions:

(i) $X = \cup\{B : B \in \mathscr{B}\}$;

(ii) for any finite $B_1, \cdots, B_n \in \mathscr{B}$, there is an $B \in \mathscr{B}$ such that $B_1 + \cdots + B_n \subset B$;

(iii) for any $A \in \mathscr{B}$ and $\lambda \in K$, there is an $B \in \mathscr{B}$ such that $\lambda A \subset B$;

(iv) for any $A \in \mathscr{B}$, the circled hull of A (resp. ΓA) is contained in some member in $\mathscr{B}$.

(2) Let $\mathscr{B}$ be a base for a vector bornology (resp convex bornology) on X. Then the family, defined by

$$\mathscr{M}(\mathscr{B}) = \{D \subset X : D \subset B \quad \text{for some } B \in \mathscr{B}\},$$

is a vector bornology (resp. convex bornology) with $\mathscr{B}$ as a base ; $\mathscr{M}(\mathscr{B})$ is called the vector bornology (resp. convex bornology) generated by $\mathscr{B}$.

(13.2) **Definition.** Let $\mathfrak{S}_1$ and $\mathfrak{S}_2$ be two vector bornologies on X. We say that $\mathfrak{S}_1$ is coarser than $\mathfrak{S}_2$ (or $\mathfrak{S}_2$ is finer than $\mathfrak{S}_1$), denoted by $\mathfrak{S}_1 \leq \mathfrak{S}_2$, if $\mathfrak{S}_2 \subset \mathfrak{S}_1$.

Remark. Suppose that $\mathscr{B}_1$ and $\mathscr{B}_2$ are two bases for two vector bornologies on X, and that $\mathscr{M}(\mathscr{B}_1)$ is the vector bornology generated by $\mathscr{B}_1$ (for definition, see (13.a)(2)). If $\mathscr{M}(\mathscr{B}_1) \leq \mathscr{M}(\mathscr{B}_2)$, then we write $\mathscr{B}_1 \leq \mathscr{B}_2$ and also say that $\mathscr{B}_1$ is coarser than $\mathscr{B}_2$ (or $\mathscr{B}_2$ is finer than $\mathscr{B}_1$). It is clear that $\mathscr{B}_1 \leq \mathscr{B}_2$ if and only if for any $D \in \mathscr{B}_2$ there is a $B \in \mathscr{B}_1$ such that $D \subseteq B$.

(13.3) **Definition.** Let $X^{\mathfrak{S}}$ and $Y^{\mathfrak{M}}$ be BVS. A $T \in L^*(X,Y)$ is called a locally bounded map (w.r.t. $\mathfrak{S}$ and $\mathfrak{M}$) if

$$T(\mathfrak{S}) = \{T(B) : B \in \mathscr{B}\} \subseteq \mathscr{M}.$$

The set consisting of all locally bounded maps (w.r.t. $\mathfrak{S}$ and $\mathscr{M}$), denoted by $L^{\times}(X^{\mathfrak{S}},Y^{\mathscr{M}})$ (or simply $L^{\times}(X,Y)$), is a vector subspace of $L^{*}(X,Y)$).

A bijective linear map $T : X \longrightarrow Y$ is called a <u>bornological isomorphism</u> (w.r.t. $\mathfrak{S}$ and $\mathscr{M}$) if T and T^{-1} are locally bounded (w.r.t. $\mathfrak{S}$ and $\mathscr{M}$).

A linear functional f on X is said to be $\mathfrak{S}$–<u>bounded</u> (or <u>locally bounded for</u> $\mathfrak{S}$) if it is bounded on any $B \in \mathfrak{S}$. The set consisting of all $\mathfrak{S}$–bounded linear functionals on X, denoted by $X^{\times}$ or $(X^{\mathfrak{S}})^{\times}$, is a vector subspace of X^{*}; $X^{\times}$ is called the <u>bornological dual of</u> $X^{\mathfrak{S}}$.

<u>Remark</u>. Let $\mathscr{B}_X$ and $\mathscr{B}_Y$ be two bases for vector bornologies on X and Y resp. and $T \in L^{*}(X,Y)$. Then T is locally bounded (w.r.t. $\mathscr{B}_X$ and $\mathscr{B}_Y$) if and only if for any $D \in \mathscr{B}_X$ there is an $B \in \mathscr{B}_Y$ such that $T(D) \subset B$.

(13.4) <u>Examples</u>: (a) Let $(X,\mathscr{P})$ be a TVS.

(i) $\mathscr{M}_{von}$ (the family of all $\mathscr{P}$-bounded subsets of X) is a vector bornology, called the <u>von Neumann bornology</u> by Hogbe–Nlend [1977,p.21] and denoted by $\mathscr{M}^X_{von}(\mathscr{P})$ or $\mathscr{M}_{von}(\mathscr{P})$ (or $\mathscr{M}^X_{von}$ or $\mathscr{M}_{von}$); the family $\mathscr{B}_{von}$ of all closed, circled $\mathscr{P}$-bounded subsets of X is a base for $\mathscr{M}_{von}$.

(ii) $\mathscr{M}_p$ (the family of all $\mathscr{P}$-totally bounded subsets of X) is a vector bornology, called the <u>precompact bornology</u> and denoted by $\mathscr{M}^X_p(\mathscr{P})$ or $\mathscr{M}_p(\mathscr{P})$ ($\mathscr{M}^X_p$ or $\mathscr{M}_p$); the family $\mathscr{B}_p$ of all $\mathscr{P}$-closed circled, totally bounded subsets of X is a base for $\mathscr{M}_p$. Moreover, if $\mathscr{P}$ is Hausdorff, then $\mathscr{M}_p$ is the family of all precompact subsets of X.

(iii) $\mathscr{M}_c(\mathscr{P})$ (the family of all relatively $\mathscr{P}$-compact sets in X) is a vector bornology, called the <u>compact bornology</u> and denoted by $\mathscr{M}^X_c(\mathscr{P})$ or $\mathscr{M}_c(\mathscr{P})$(or $\mathscr{M}^X_c$ or $\mathscr{M}_c$), the family $\mathscr{B}_c$ of all compact circled sets is a base for $\mathscr{M}_c$.

(iv) The family $\mathscr{B}_{cd}$ of all compact disks in $X_{\mathscr{P}}$ is a base for a convex bornology $\mathcal{M}(\mathcal{B}_{cd})$ which is called the convex compact bornology and denoted by $\mathcal{M}_{cd}(\mathcal{P})$ or $\mathcal{M}^X_{cd}$ or $\mathcal{M}_{cd}$.

If $\mathcal{P}$ is Hausdorff, then all $\mathcal{M}_{von}$, $\mathcal{M}_P$, $\mathcal{M}_c$ and $\mathcal{M}_{cd}$ are separated.

(b) Suppose that $X_{\mathcal{P}}$ is a LCS.

(i) $\mathcal{M}_{von}$ and $\mathcal{M}_P$ are separated convex bornologies.

(ii) $\mathcal{M}_c$ is, in general, not a convex bornology (since the convex hull of a $\mathcal{P}$–compact subset of X is, in general, not $\mathcal{P}$–compact). If, in addition, $X_{\mathcal{P}}$ is quasi–complete then $\mathcal{M}_c$ is a convex bornology.

(iii) The family, defined by

$$\mathcal{B}_f = \{ \Gamma\{x_1,\cdots,x_n\} : \text{for any finite set } \{x_1,\cdots,x_n\} \text{ in } X, n \geq 1 \},$$

is a base for a convex bornology, which is called the finite–dimensional bornology and denoted by $\mathcal{M}_f(\mathcal{P})$ or $\mathcal{M}_f$ (i.e. $\mathcal{M}_f = \mathcal{M}(\mathcal{B}_f)$).

(iv) Recall that a bounded disk B in $X_{\mathcal{P}}$ is said to be infracomplete if $(X(B),\gamma_B)$ is a Banach space. The family

$$\mathcal{B}_{inf}(\mathcal{P}) = \{ B = \Gamma B \subset X : B \text{ is (bounded) infracomplete} \},$$

is a base for a convex bornology, which is called the infracomplete bornology (complete bornology in the terminology of Hogbe–Nlend [1977, p.42]), and denoted by $\mathcal{M}_{inf}(\mathcal{P})$ or $\mathcal{M}_{inf}$ (i.e. $\mathcal{M}_{inf} = \mathcal{M}(\mathcal{B}_{inf}(\mathcal{P}))$).

(v) Let Y be a TVS. A subset H of L(X,Y) (the vector space of all continuous linear maps) is said to be equicontinuous if for any 0–neighbourhood U in Y, the set, defined by

$$H^{-1}(U) = \bigcap_{T \in H} T^{-1}(U),$$

is a 0–neighbourhood in X. The family of all equicontinuous subsets of L(X,Y) is a vector bornology (convex bornology when Y is a LCS), which is called the equicontinuous bornology on L(X,Y) and denoted by $\mathcal{M}_{eq}(\mathcal{L})$.

(vi) On any LCS $X_{\mathcal{P}}$, we have

$$\mathcal{M}_f \subset \mathcal{M}_{cd} \subset \mathcal{M}_{inf} \subset \mathcal{M}_{von} \text{ and } \mathcal{M}_{cd} \subset \mathcal{M}_p$$

or, equivalently ,

$$\mathcal{M}_{von} < \mathcal{M}_{inf} < \mathcal{M}_{cd} < \mathcal{M}_f \text{ and } \mathcal{M}_p < \mathcal{M}_{cd}.$$

(13.5) **Definition.** Let $\mathfrak{S}$ be a vector bornology on X, let $\{x_n\}$ be a sequence in X and $A \subseteq X$.

(a) $\{x_n\}$ is called a b–null sequence (or Mackey–null sequence) for $\mathfrak{S}$ if there is a circled $B \in \mathfrak{S}$ and a sequence $\{\lambda_n\}$ in K such that

$$\lim_n \lambda_n = 0 \text{ and } x_n \in \lambda_n B \quad (\text{for all } n \geq 1).$$

(b) If $u \in X$ is such that $\{x_n - u\}$ is a b–null sequence for $\mathfrak{S}$, then we say that $\{x_n\}$ is b–convergent (for $\mathfrak{S}$) or b–convergent to u, denoted by

$$x_n \xrightarrow{M} u.$$

(c) A is said to be b–closed (or Mackey–closed) (for $\mathfrak{S}$) if the conditions

$$x_n \in A \text{ and } x_n \xrightarrow{M} u \text{ for } \mathfrak{S}$$

imply that $u \in A$.

Remarks : (i) It is clear that if $x_n \xrightarrow{M} x$, $y_n \xrightarrow{M} y$ and $\lambda_n \longrightarrow \lambda$ in K, then

$$x_n + y_n \xrightarrow{M} x + y \text{ and } \lambda_n x_n \xrightarrow{M} \lambda x.$$

It is also trivial that every b–convergent sequence for $\mathfrak{S}$ must be bounded, and that the image of a b–convergent sequence under a locally bounded map is again a b–convergent sequence.

(ii) The intersection of a family of b–closed subsets of $(X,\mathfrak{S})$ is b–closed (for $\mathfrak{S}$), hence the b–closure of A (for $\mathfrak{S}$), denoted by $\tilde{A}^b$, is defined to be the intersection of all b–closed sets (for $\mathfrak{S}$) containing A. Moreover, if A is a vector subspace. then so is $\tilde{A}^b$.

(iii) Let X and Y be BVS and $T \in L^{\times}(X,Y)$. If N is b–closed in Y, then $T^{-1}(N)$ is b–closed in X.

(13.b) For a vector bornology $\mathfrak{S}$ on X, the following statements are equivalent:

(a) $\mathfrak{S}$ is separated.

(b) Every b–convergent sequence in X (for $\mathfrak{S}$)has a unique limit.

(c) $\{0\}$ is b–closed (for $\mathfrak{S}$).

(13.6) **proposition.** Let $\mathfrak{S}$ be a vector bornology on X and $\{x_n\}$ a sequence in X. Then the following statements are equivalent.

(a) $\{x_n\}$ is a b–null sequence (for $\mathfrak{S}$).

(b) There exists a circled $B \in \mathfrak{S}$ and a decreasing sequence $\{\zeta_n\}$ of positive numbers such that

$$\lim \zeta_n = 0 \ \text{and} \ x_n \in \zeta_n B \quad (\text{for all } n \geq 1)\,.$$

(c) There exists a circled $B \in \mathfrak{S}$ such that for any $\epsilon > 0$ one can find a natural number $N(\epsilon)$ with

$$x_n \in \epsilon B \quad \text{for all } n \geq N(\epsilon).$$

If $\mathfrak{S}$ is a convex bornology, then one of (a), (b) and (c) is equivalent to the following:

(d) <u>There exists a disk</u> $B \in \mathfrak{S}$ <u>such that</u> $x_n \in X(B)$ <u>and</u> $\gamma_B(x_n) \longrightarrow 0$.

<u>Proof</u>. (a) ⇒ (b): Let $B \in \mathfrak{S}$ be circled and let $\lambda_n \in K$ be such that $\lim_n \lambda_n = 0$ and $x_n \in \lambda_n B$ (for all $n \geq 1$). For any positive integer k, there is an $N_k \geq 1$ such that $|\lambda_n| \leq \frac{1}{k}$ (for all $n \geq N_k$), hence

$$\lambda_n B \subset k^{-1} B \quad (\text{for all } n \geq N_k).$$

Without loss of generality one can assume that N_k is strictly increasing .For any $j \geq 1$ with $N_k \leq j < N_{k+1}$ $(k \geq 1)$, let

$$\zeta_j = \frac{1}{k}.$$

Then the sequence $\{\zeta_j\}$ has the required properties.

(b) ⇒ (c): Trivial.

(c) ⇒ (a): For any $n \geq 1$, let

$$\epsilon_n = \inf\{\epsilon > 0 : x_n \in \epsilon B\} \text{ and } \lambda_n = \epsilon_n + \frac{1}{n}.$$

Then $\lambda_n \longrightarrow 0$ and $x_n \in \lambda_n B$ (for all $n \geq 1$).

Finally, we assume that $\mathfrak{S}$ is a convex bornology, then the implication (d) ⇒ (a) holds with $\lambda_n = \gamma_B(x_n) + \frac{1}{n}$ (for all $n \geq 1$). On the other hand, (b) implies $x_n \in X(B)$ and $\gamma_B(x_n) \leq \zeta_n \longrightarrow 0$. Thus the implication (b) ⇒ (d) follows.

<u>Remark</u>. Because of (d), we say that a subset K of a separated CBVS $(X,\mathfrak{S})$ is b–<u>compact</u> (resp. b–<u>precompact</u>) if there is a disk $B \in \mathfrak{S}$ such that K is relatively compact (resp. totally bounded) in the normed space $(X(B),\gamma_B)$.

(13.7) <u>Definition.</u> Let $\{(X_i,\mathfrak{S}_i) : i \in \Lambda\}$ be a family of BVS (resp.CBVS) and $\pi_j : \prod_{i\in\Lambda} X_i \longrightarrow X_j$ the j–th projection. The family

$$\prod_{i\in\Lambda} \mathfrak{S}_i = \{A \subseteq \prod_{i\in\Lambda} X_i : \pi_i(A) \in \mathfrak{S}_i \quad (\text{for all } i \in \Lambda)\},$$

defines obviously a vector bornology (resp. convex bornology) on ΠX_i, called the product bornology , and the pair $(\Pi X_i, \Pi\mathfrak{S}_i)$ is called the bornological product space.

Remarks: (i) The product bornology admits a base consisting of sets of the form

$$\{\prod_{i\in\Lambda} B_i : B_i \in \mathfrak{S}_i \quad (i \in \Lambda)\}.$$

Moreover, $\prod_{i\in\Lambda} \mathfrak{S}_i$ is the coarsest vector (resp. convex) bornology on $\prod_{i\in\Lambda} X_i$ for which all projections π_i are locally bounded.

(ii) If each vector bornology $\mathfrak{S}_i$ on X is separated , then so is the product bornology $\prod_i \mathfrak{S}_i$.

Let $(X,\mathfrak{S})$ be a BVS (resp. CBVS) and let M be a vector subspace of X. Then the family

$$M \cap \mathfrak{S} = \{ M \cap B : B \in \mathfrak{S} \}$$

is a base for a vector (resp. convex) bornology on M, which is the coarsest vector (resp. convex) bornology on M such that the canonical embedding $J_M^X : M \longrightarrow (X,\mathfrak{S})$ is locally bounded. This vector (resp. convex) bornology on M is called the relative vector bornology induced by $\mathfrak{S}$, and denoted by $\mathfrak{S}_M$. The pair $(M,\mathfrak{S}_M)$ is called a bornological subspace of $(X,\mathfrak{S})$, and members in $\mathfrak{S}_M$ are said to be relatively bounded in M. It is clear that if $\mathfrak{S}$ is separated then so is $\mathfrak{S}_M$.

(13.8) **Definition**. Let $(X,\mathfrak{S})$ be a BVS (resp. CBVS), let M be a vector subspace of X and $Q_M : X \longrightarrow X/M$ the quotient map. Then the family

$$Q_M(\mathfrak{S}) = \{Q_M(B) : B \in \mathfrak{S}\},$$

is a base (since Q_M is onto) for a vector (resp. convex) bornology on X/M, called the quotient bornology and denoted by $\mathfrak{S}/M$. The pair $(X/M,\mathfrak{S}/M)$ is called the bornological quotient space.

Remark. The quotient bornology on X/M is the finest vector (resp. convex) bornology for which $Q_M : (X,\mathfrak{S}) \longrightarrow X/M$ is locally bounded.

(13.c) Let $(X,\mathfrak{S})$ be BVS and M a vector subspace of X. Then the quotient bornology $\mathfrak{S}/M$ on X/M is separated if and only if M is b–closed (for $\mathfrak{S}$).

(13.d) Complete convex bornologies : A disk B in a vector space X is said to be completant (or infracomplete) if the seminormed space $(X(B), r_B)$ is a Banach space. A CBVS $(X,\mathfrak{S})$ is called a complete convex bornological vector space (complete CBVS or CCBVS for short) if $\mathfrak{S}$ admits a basis consisting of completant disks; in this case, $\mathfrak{S}$ is referred to as a complete convex bornology.

(i) Every complete convex bornology must be separated.

(ii) Every b–closed vector subspace M of a CCBVS $(X,\mathfrak{S})$ must be complete w.r.t the relative bornology; and every vector subspace N of a separated CBVS $(X,\mathfrak{S})$, which is complete for the relative bornology, must be b–closed.

(iii) The bornological quotient space of a complete CBVS by a b–closed vector subspace must be a complete CBVS.

(iv) The bornological product space of a family of complete CBVS is a complete CBVS.

Exercises

13–1. Prove (13.a).

13–2. Let $\mathfrak{S}$ be a convex bornology on X and let $\mathscr{B}$ be a base for $\mathfrak{S}$ consisting of disks.

For any $B \in \mathscr{B}$, let r_B be the gauge of B defined on the vector subspace $X(B) = \bigcup_n nB$ of X, and let $J_B : X(B) \to X$ be the cannonical embedding. Show that:

(a) Each J_B is locally bounded with respect to the von Neumann bornology on the seminormed space $(X(B), r_B)$ and $\mathfrak{S}$.

(b) If $\mathfrak{M}$ is a convex bornology on X such that each J_B is locally bounded, then $\mathfrak{S} \subset \mathfrak{M}$.

13–3. Prove (13.b).

13–4. Prove (13.c).

13–5. Prove (13.d).

14

Initial Bornologies and Final Bornologies

This section is devoted to the fundamental methods for constructing new vector bornologies from given ones. These methods, which are dual to those of projective and inductive topologies (see §9), show that bornological structure is a collection of 'internal pieces' each of which is a normed space.

(14.1) **Definition**. Let X be a vector space, Let $\{ (Y_i, \mathscr{M}_i) : i \in \Lambda \}$ be a family of BVS (resp. CBVS) and $T_i \in L^*(X, Y_i)$ $(i \in \Lambda)$. Then the family

$$\mathfrak{S}_{ini} = \{A \subset X : T_i(A) \in \mathscr{M}_i \ (i \in \Lambda)\},$$

defines obviously a vector bornology (resp. convex bornology) on X, called the <u>initial bornology on</u> X <u>w.r.t.</u> $\{(Y_i, \mathscr{M}_i, T_i) : i \in \Lambda\}$.

Remarks: (i) The initial bornology $\mathfrak{S}_{ini}$ on X is the coarsest vector bornology for which all $T_i : X \longrightarrow (Y_i, \mathscr{M}_i)$ are locally bounded; moreover, if we let $T_i^{-1}(\mathscr{M}_i) = \{T_i^{-1}(D) : D \in \mathscr{M}_i\}$, then the family

(14.1.a) $$\mathscr{B}_{ini} = \bigcap_{i \in \Lambda} T_i^{-1}(\mathscr{M}_i)$$

is a base for $\mathfrak{S}_{ini}$.

(ii) If all $\mathscr{M}_i$ are separated, then $\mathfrak{S}_{ini}$ is separated if and only if $\bigcap_{i \in \Lambda} \mathrm{Ker}\, T_i = \{0\}$.

(iii) The initial bornology $\mathfrak{S}_{ini}$ has the following important property: <u>if</u> $(G, \mathfrak{m})$ <u>is a BVS and</u> $\varphi \in L^*(G, X)$, then φ <u>is locally bounded</u> (<u>for</u> $\mathfrak{m}$ <u>and</u> $\mathfrak{S}_{ini}$) <u>if and only if all</u> $T_i \circ \varphi : (G, \mathfrak{m}) \longrightarrow (Y_i, \mathscr{M}_i)$ <u>are locally bounded</u>.

(14.2) **Examples**: (a) <u>Bornological subspaces</u>: Let $(X, \mathfrak{S})$ be a BVS and M a vector subspace of X. Then the relative vector bornology $\mathfrak{S}_M$ induced by $\mathfrak{S}$ is the initial bornology w.r.t. the canonical imbedding $J_M : M \longrightarrow (X,\mathfrak{S})$.

(b) **Product bornologies:** Let $\{(X_i,\mathfrak{S}_i) : i \in \Lambda\}$ be a family of BVS and $\pi_i : \prod_{j\in\Lambda} X_j \longrightarrow X_i$ the projection. Then the product bornology $\prod_j \mathfrak{S}_j$ on $\prod_{j\in\Lambda} X_j$ is the initial bornology w.r.t. $\{(X_i,\mathfrak{S}_i,\pi_i) : i \in \Lambda\}$.

(c) **Least upper bound bornology:** Let X be a vector space, let $\{\mathfrak{S}_i : i \in \Lambda\}$ be a family of vector bornologies on X and $I_i : X \longrightarrow X$ the identity map $(i \in \Lambda)$. Then the initial bornology w.r.t. $\{(X,\mathfrak{S}_i,I_i) : i \in \Lambda\}$ is called the least upper bound bornology (the intersection of vector bornologies $\mathfrak{S}_i$ in the terminology of Hogbe–Nlend [1977, p.31]), and denoted by $\sup_i \mathfrak{S}_i$. It is clear that $\sup_i \mathfrak{S}_i$ is the coarsest vector bornology such that $\mathfrak{S}_j \leq \sup_i \mathfrak{S}_i$ (for all $j \in \Lambda$); moreover, (14.1.a) shows that the family $\bigcap_{i\in\Lambda} \mathfrak{S}_i$ is a base for $\sup_i \mathfrak{S}_i$.

(d) Let X be a vector space and $\mathscr{A}$ a family consisting of subsets of X. The least upper bound bornology of all vector bornologies (resp. convex bornologies) on X containing $\mathscr{A}$ is called the vector (resp. convex) bornology generated by $\mathscr{A}$ and denoted by $\mathscr{G}(\mathscr{A})$. It is clear that $\mathscr{G}(\mathscr{A})$ always exists since the power set 2^X of X is a convex bornology containing $\mathscr{A}$.

Initial bornologies can be identified with the relative bornologies induced by product bornologies as shown by the following:

(14.3) **Proposition.** Let X be a vector space, let $\{(Y_i,\mathscr{M}_i) : i \in \Lambda\}$ be a family of separated BVS and let $T_i \in L^*(X,Y_i)$ be such that $\bigcap_{i\in\Lambda} \operatorname{Ker} T_i = \{0\}$. Suppose that $\mathfrak{S}_{ini}$ is the initial bornology on X w.r.t. $\{(Y_i,\mathscr{M}_i,T_i) : i \in \Lambda\}$ and that $T : X \longrightarrow \prod_{i\in\Lambda} Y_i$ is defined by

$$Tx = [T_ix, i \in \Lambda] \quad (\text{for all } x \in X).$$

Then $(X,\mathfrak{S}_{ini})$ and $(TX, \prod_i \mathscr{M}_i|_{TX})$ are bornologically isomorphic.

Proof. T is clearly linear and one–one, and also satisfies

$$\pi_i \circ T = T_i \text{ and } T_i \circ T^{-1} = \pi_i|_{T(X)}$$

The result then follows from remark (iii) of (14.1) (since $\Pi \mathscr{M}_i$ is the initial bornology).

(14.4) **Definition.** Let Γ be a directed set, let $\{(Y_i, \mathscr{M}_i) : i \in \Gamma\}$ be a family of BVS (resp. CBVS), and for any $i,j \in \Gamma$ with $i \leq j$, let $T_{ij} : (Y_j, \mathscr{M}_j) \longrightarrow (Y_i, \mathscr{M}_i)$ be locally bounded. Then $\varprojlim T_{ij} Y_j$ (the projective spectrum of $\{Y_i : i \in \Gamma\}$ under T_{ij}) (see (9.4)), equipped with relative bornology induced by $\prod_i \mathscr{M}_i$ is called the <u>bornological projective limit of</u> $\{(Y_i, \mathscr{M}_i) : i \in \Gamma\}$ <u>under the locally bounded maps</u> T_{ij} and denoted by

$$\varprojlim T_{ij} (Y_j, \mathscr{M}_j).$$

Remark. For any $i \in \Gamma$, let π_{iR} be the restriction of the i–th projection π_i on $\varprojlim T_{ij} Y_j$. Then the vector (resp. convex) bornology on the bornological projective limit $\varprojlim T_{ij}(Y_j, \mathscr{M}_j)$ is the initial bornology on $\varprojlim T_{ij}$ w.r.t. $\{(Y_i, \mathscr{M}_i, \pi_{iR}) : i \in \Gamma\}$.

A similar argument given in the proof on (9.2) (e) shows that the bornological product space is an example of bornological projective limits.

(14.5) **Definition.** Let Y be a vector space, let $\{(X_i, \mathfrak{S}_i) : i \in \Lambda\}$ be a family of BVS (resp. CBVS) and $S_i \in L^*(X_i, Y)$ $(i \in \Lambda)$. Then the finest vector bornology (resp. finest convex bornology) on Y such that all S_i are locally bounded is called the <u>final vector bornology</u> (resp. <u>final convex bornology</u>) <u>on</u> Y <u>w.r.t.</u> $\{(X_i, \mathfrak{S}_i, S_i) : i \in \Lambda\}$ and denoted by $\mathscr{M}_{ind}$. Clearly

$$\mathscr{M}_{ind} = \mathscr{G}(\bigcup_{i \in \Lambda} S_i(\mathfrak{S}_i)),$$

that is, $\mathscr{M}_{ind}$ is the least upper bound bornology of all vector bornologies containing

$\bigcup_{i\in\Lambda} S_i(\mathfrak{S}_i)$.

Remark. The final vector bornology $\mathcal{M}_{ind}$ has the following remarkable property: Let $(Z, \mathfrak{m})$ be a BVS and $\psi \in L^*(Y, Z)$. Then ψ is locally bounded (for $\mathcal{M}_{ind}$ and $\mathfrak{m}$) if and only if all $\psi \circ S_i : (X_i,\mathfrak{S}_i) \longrightarrow (Z, \mathfrak{m})$ are locally bounded.

(14.6) Examples: (a) Bornological quotient spaces: Let $(X,\mathfrak{S})$ be a BVS (resp. CBVS), and M a vector subspace of X.Then the quotient bornology (resp. quotient convex bornology) $\mathfrak{S}/M$ on X/M is the final vector bornology (resp. final convex bornology) w.r.t. $(X,\mathfrak{S},Q_M)$.

(b) Bornological direct sums: Let $\{(X_i,\mathfrak{S}_i) : i \in \Gamma\}$ be a family of BVS (resp. CBVS) and let $J_i : X_i \longrightarrow \bigoplus_{j\in\Lambda} X_j$ be the i–th canonical injection. Then the final bornology on $\bigoplus_j X_j$ w.r.t. $\{(X_i, \mathfrak{S}_i, J_i) : i \in \Lambda\}$ is called the direct sum bornology and denoted by $\bigoplus_i \mathfrak{S}_i$. The space $(\bigoplus_i X_i, \bigoplus_i \mathfrak{S}_i)$ (or denoted by $\bigoplus_i (X_i, \mathfrak{S}_i)$) is called the bornological direct sum of $(X_i, \mathfrak{S}_i)$.

Remarks. (i) Let $\mathcal{F}(\Lambda)$ be the family of all non–empty finite subsets of Λ. Then the family

$$\{ \bigoplus_{i\in\alpha} B_i : B_i \in \mathfrak{S}_i \text{ and } \alpha \in \mathcal{F}(\Lambda) \}$$

is a base for the direct sum bornology $\bigoplus \mathfrak{S}_i$ [since J_i are injective and $\bigoplus_j X_j = \langle \bigcup_{i\in\Lambda} J_i(X_i) \rangle$ (the linear hull)].

(ii) It is easily seen that

$$\prod_{i\in\Lambda} \mathfrak{S}_i \leq \bigoplus_{i\in\Lambda} \mathfrak{S}_i \qquad \text{on } \bigoplus_i X_i.$$

These two vector bornologies are equal whenever Λ is finite.

(c) The finite–dimensional bornology: Let X be a vector space over K.Then

$X \sim K^{(\Lambda)}$ (algebraically) for some index set Λ (one can take card $\Lambda = \dim X$). Then X equipped with the finite–dimensional bornology (for definition, see (13.4) (b) (iii)) is bornologically isomorphic with the bornological direct sum $K^{(\Lambda)}$ of $(K, \mathcal{M}_{von}^{K})$. It is easily seen that it is the finest vector bornology on X and is always convex.

The following result, dual to (14.3),shows that final bornologies can be identified with a quotient bornology of some direct sum bornology.

(14.7) **Proposition.** Let Y be a vector space, let $\{(X_i, \mathfrak{S}_i) : i \in \Lambda\}$ be a family of BVS, let $S_i \in L^*(X_i, Y)$ $(i \in \Lambda)$ and suppose that

(1) $$Y = < \bigcup_{i \in \Lambda} S_i(X_i) > \quad \text{(the linear hull).}$$

Suppose that $\mathcal{M}_{ind}$ is the final vector bornology on Y w.r.t. $\{(X_i, \mathfrak{S}_i, S_i) : i \in \Lambda\}$, that $S : \oplus_j X_j \longrightarrow Y$ is defined by

(2) $$S(x) = \sum_i S_i(\pi_i(x)) \quad (\text{whenever } x = \sum_i J_i \pi_i(x) \in \oplus_j X_j),$$

and that $\breve{S} : \oplus_j X_j/_{\text{Ker } S} \longrightarrow Y$ is the injection (in fact, bijection by (1)) associated with S. Then $(\oplus_j X_j/_{\text{Ker } S}, \oplus_j \mathfrak{S}_j/_{\text{Ker } S})$ and $(Y, \mathcal{M}_{ind})$ are bornologically isomorphic under $\breve{S}$.

Proof. (2) implies that

$$S \circ J_i = S_i \quad (\text{for all } i \in \Lambda),$$

(where $J_i : X_i \longrightarrow \oplus_j X_j$ is the canonical injection), and the definition of $\breve{S}$ (on account of $Q_S = (\breve{S})^{-1} \circ S$ and $S \circ J_i = S_i$) implies that

$$(\breve{S})^{-1} \circ S_i = Q_S \circ J_i : X_i \longrightarrow \oplus_j X_j/_{\text{Ker } S} \quad (i \in \Lambda)$$

are locally bounded (where $Q_S : \underset{j}{\oplus} X_j \longrightarrow \underset{j}{\oplus} X_j/_{\mathrm{Ker}\,S}$ is the quotient map). The result then follows from the Remark of (14.5) and (14.6) (b).

(14.a) **Final bornologies:** Let $\{(X_i, \mathfrak{S}_i) : i \in \Lambda\}$ be a family of BVS, let M be a vector subspace of $\underset{j}{\oplus} X_j$, let $Q_M : \underset{j}{\oplus} X_j \longrightarrow \underset{j}{\oplus} X_j/_M$ be the quotient map and $Q_{M,i} = Q_M \circ J_i$ $(i \in \Lambda)$. Then:

(i) $\underset{j}{\oplus} X_j/_M = \sum_i Q_{M,i}(X_i)$ (= linear hull of $\underset{i \in \Lambda}{\cup} Q_{M,i}(X_i)$).

(ii) The quotient bornology $\underset{j}{\oplus} \mathfrak{S}_j/_M$ is the final vector bornology on $\sum_i Q_{M,i}(X_i)$ w.r.t. $\{(X_i, \mathfrak{S}_i, Q_{M,i}) : i \in \Lambda\}$.

(14.8) **Definition.** Let Γ be a direct set, let $\{(X_i, \mathfrak{S}_i) : i \in \Gamma\}$ be a family of BVS (resp. CBVS), and for any $i, j \in \Gamma$ with $i \leq j$, let

$$S_{ji} : (X_i, \mathfrak{S}_i) \longrightarrow (X_j, \mathfrak{S}_j) \quad \text{be locally bounded,}$$

and suppose that

$$H = \text{linear hull of } \cup \{\mathrm{Im}(J_i - J_j \circ S_{ji}) : i \leq j, \quad i, j \in \Gamma\}.$$

Then the space $(\oplus X_j/H, \oplus \mathfrak{S}_j/H)$ is called the <u>bornological inductive limit of</u> $\{(X_i, \mathfrak{S}_i) : i \in \Gamma\}$ <u>under the locally bounded maps</u> S_{ji}, and denoted by $\underrightarrow{\lim}\, S_{ji}(X_i, \mathfrak{S}_i)$, that is

$$\underrightarrow{\lim}\, S_{ji}(X_i, \mathfrak{S}_i) = (\oplus X_j/H, \oplus \mathfrak{S}_j/H) .$$

The vector bornology (resp. convex bornology) $\oplus \mathfrak{S}_j/H$ on $\underrightarrow{\lim}\, S_{ji} X_i$ is called the <u>inductive limit bornology w.r.t.</u> $\mathfrak{S}_i$ $(i \in \Gamma)$.

Remark. Let $Q_H : \underset{j}{\oplus} X_j \longrightarrow \oplus X_j/H$ be the quotient map and $Q_{H,i} = Q_H \circ J_i$

$(i \in \Gamma)$. Then (14.a) shows that the inductive limit bornology $\oplus\mathfrak{S}_j/H$ on $\varinjlim S_{ji}X_i$ w.r.t. $\mathfrak{S}_j$ is the final vector (resp. convex) bornology w.r.t. $\{ (X_i, \mathfrak{S}_i, Q_{H,i}) : i \in \Gamma \}$; moreover the following family

$$\bigcup_{i\in\Gamma} Q_{H,i}(\mathfrak{S}_i) = \bigcup_{i\in\Gamma} \{Q_{H,i}(A) : A \in \mathfrak{S}_i\}$$

is a base for $\oplus\mathfrak{S}_j/H$.

By a similar argument given in the proof of (9.6)(e), one can show that any bornological direct sum is an example of a bornological inductive limit.

(14.9) **Proposition.** Let $Y = \varinjlim S_{ji}(X_i, \mathfrak{S}_i)$ be a bornological inductive limit of separated BVS $(X_i,\mathfrak{S}_i)$ under the locally bounded maps $S_{ji}(i,j \in \Gamma)$. If all maps S_{ji} are injective, then Y is separated. In particular, every bornological direct sum of separated BVS is separated.

Proof. Every bounded subset of Y is a bounded subset of one of the BVS $(X_i,\mathfrak{S}_i)$ which is separated.

(14.10) **Theorem.** Every CBVS $(X,\mathfrak{S})$ is the bornological inductive limit of a family of seminormed spaces, and of normed spaces if $(X,\mathfrak{S})$ is separated.

Proof. Let $\mathscr{B}$ be a base for $\mathfrak{S}$ consisting of bounded disks. For any $A,B \in \mathscr{B}$ with $A \subset B$, let

$$J_{B,A} : X(A) \longrightarrow X(B) \quad \text{be the canonical embedding.}$$

Then $(X,\mathfrak{S}) = \varinjlim J_{B,A}(X(A), \mathscr{M}_{von})$.

Finally, if $(X,\mathfrak{S})$ is separated, then any $A \in \mathscr{B}$ does not contain a non–trivial vector subspace of X, hence γ_A must be a norm on $X(A)$.

(14.b) <u>Bornologically complementary subspaces:</u> Let $(X,\mathfrak{S})$ be a BVS, let M and N be vector subspaces of X such that $X = M + N$ and $M \cap N = \{0\}$. If

(14.b.1) $$(X,\mathfrak{S}) = (M,\mathfrak{S}_M) \oplus (N,\mathfrak{S}_N),$$

then we say that M and N are <u>bornologically complementary subspaces of</u> X, and that M (resp. N) is a <u>bornological complement of</u> N (resp. M). One can show that M and N are bornologically complementary subspaces of X if and only if the projection $P_M : X \longrightarrow M$ is locally bounded (for $\mathfrak{S}$ and $\mathfrak{S}_M$).

<u>Exercises</u>

14–1. Prove (14.a).

14–2. Prove (14.b).

14–3. Let $\{(X_i,\mathfrak{S}_i) : i \in \Lambda\}$ be a family of BVS, let $\prod_j \mathfrak{S}_j$ be the product bornology on $\prod_{j\in\Lambda} X_j$ and $\oplus_j \mathfrak{S}_j$ the direct sum bornology on $\oplus_{j\in\Lambda} X_j$. Show that:

(a) $\prod_j \mathfrak{S}_j$ has a base consisting of sets of the form :

$$B = \prod_{i\in\Lambda} B_i \text{ with } B_i \in \mathfrak{S}_i \quad (\text{for all } i \in \Lambda).$$

(b) $\oplus_j \mathfrak{S}_j$ has a base consisting of sets of the form:

$$B = \oplus_{i\in\Lambda} B_i \text{ (finite sum) with } B_i \in \mathfrak{S}_i \ (i \in \Lambda).$$

15

von Neumann Bornologies and Locally Convex Topologies Determined by Convex Bornologies

For any LCS $(X,\mathscr{P})$, it is known from (13.4)(b) that the von Neumann bornology $\mathscr{M}_{von}(\mathscr{P})$ and the infracomplete bornology $\mathscr{M}_{inf}(\mathscr{P})$ are separated, convex bornologies such that

$$\mathscr{M}_{von}(\mathscr{P}) \leq \mathscr{M}_{inf}(\mathscr{P}),$$

hence we have to distinguish the bornological convergence for $\mathscr{M}_{von}(\mathscr{P})$ as well as for $\mathscr{M}_{inf}(\mathscr{P})$, this leads naturally to the following:

(15.1) **Definition**. Let $(X,\mathscr{P})$ be a LCS with the von Neumann bornology $\mathscr{M}_{von}(\mathscr{P})$ and infracomplete bornology $\mathscr{M}_{inf}(\mathscr{P})$. Convergence of sequences for $\mathscr{M}_{von}(\mathscr{P})$ (resp. $\mathscr{M}_{inf}(\mathscr{P})$) is said to be b–convergence (resp. fast–convergence) in $(X,\mathscr{P})$.

Remark. A sequence $\{x_n\}$ is b–null sequence (resp. fast–null sequence) in $(X,\mathscr{P})$ if and only if there exists a bounded disk (resp. an infracomplete disk) B in $(X,\mathscr{P})$ such that $\{x_n\}$ is a null sequence in the normed space $(X(B),\gamma_B)$ (resp. in the Banach space $(X(B),\gamma_B)$).

(15.a). Let $(X,\mathscr{P})$ be a LCS and $\{x_n\}$ a sequence in X.

(i) Every b–null sequence in $(X,\mathscr{P})$ must be $\mathscr{P}$-convergent to 0 . The converse is true when $(X,\mathscr{P})$ satisfies the following property:

Every $\mathscr{P}$-compact subset K of X is compact in a normed space $(X(B),\gamma_B)$ for some $\mathscr{P}$-bounded disk B in X.

In particular, in a metrizable LCS, every $\mathscr{P}$-convergent sequence must be a b–convergent sequence in $(X, \mathscr{P})$.

(ii) The following statements are equivalent:

(1) $\{x_n\}$ is a fast–null sequence in $(X, \mathscr{P})$.

(2) There exists an infracomplete disk B in $(X, \mathscr{P})$ and a decreasing sequence $\{\zeta_n\}$ of positive mumbers such that

$$\lim_n \zeta_n = 0 \text{ and } x_n \in \zeta_n B \quad (\text{for all } n \geq 1).$$

(iii) There exists an infracomplete disk B in $(X, \mathscr{P})$ such that for any $\epsilon > 0$ one can find a natural number $N(\epsilon) \geq 1$ with

$$x_n \in \epsilon B \text{ for all } n \geq N(\epsilon).$$

(iv) There exists an infracomplete disk B in $(X, \mathscr{P})$ such that $\{x_n\}$ is a null sequence in the Banach space $(X(B), \gamma_B)$.

(15.2) <u>Definition.</u> Let $(X, \mathscr{P})$ be a LCS and $K \subset X$. We say that K is:

(a) b–<u>compact</u> (resp b–<u>precompact) in</u> $(X, \mathscr{P})$ if there exists a $\mathscr{P}$-bounded disk B in X such that K is compact (resp. totally bounded) in the normed space $(X(B), \gamma_B)$;

(b) <u>fast–compact</u> (resp. <u>fast–bounded</u>) in $(X, \mathscr{P})$ if there exists an infracomplete disk B in X such that K is compact (resp.bounded) in the Banach space $(X(B), \gamma_B)$.

<u>Remark.</u> Let $(X, \mathscr{P})$ be a LCS with the von Neumann bornology $\mathscr{M}_{von}(\mathscr{P})$ and infracomplete bornology $\mathscr{M}_{inf}(\mathscr{P})$.

(i) A subset K of X is b–compact (resp. b–precompact) in $(X, \mathscr{P})$ if and only if K is b–compact (resp. b–precompact) for $\mathscr{M}_{von}(\mathscr{P})$ (for definition, see the Remark of (13.6)).

(ii) A subset K of X is fast–compact (resp.fast–bounded) if and only if K is b–compact (resp. b–bounded) for $\mathscr{M}_{inf}(\mathscr{P})$.

(iii) We have the following implications:

fast–compactness $\Rightarrow$ b–compactness $\Rightarrow$ compactness in $(X, \mathscr{P})$.

The converses are, in general, not true. Moreover, if $(X, \mathscr{P})$ is an F–space (resp. metrizable), then $\mathscr{P}$–compact sets must be fast–compact (resp. b–compact) in $(X, \mathscr{P})$ (see (11.b)).

(iv) b–compactness and fast–compactness can be defined only on CBVS without any locally convex topologies (see the Remark of (13.6)).

It is well–known that the disked hull of a compact set in a LCS is, in general, not relatively compact; but it is true for fast–compactness as a part of the following interesting result shows.

(15.3) **Theorem.** Let $(X, \mathscr{P})$ be a LCS and $K \subset X$.

(a) If K is fast–compact in $(X, \mathscr{P})$ then so is $\overline{\Gamma K}$.

(b) If $\{x_n\}$ is a fast–null sequence in $(X, \mathscr{P})$, then $\overline{\Gamma}(\{x_n : n \geq 1\})$ is fast–compact in $(X, \mathscr{P})$.

(c) If K is fast–compact in $(X, \mathscr{P})$, then there exists a fast–null sequence $\{x_n\}$ in $(X, \mathscr{P})$ such that $K \subset \overline{\Gamma}(\{x_n : n \geq 1\})$.

(The bars denote the $\mathscr{P}$–closure).

Proof. (a) There exists an infracomplete disk B in X such that K is compact in the Banach space $(X(B), \gamma_B)$, hence the $\gamma_B(\cdot)$–closure of ΓK, denoted by $\widetilde{\Gamma K}$, is compact in $(X(B), \gamma_B)$ (by the completeness of $(X(B), \gamma_B)$). As the canonical embedding $J_B : (X(B), \gamma_B) \longrightarrow (X, \mathscr{P})$ is continuous, it follows that $\widetilde{\Gamma K} = \overline{\Gamma K}$, and hence that $\overline{\Gamma K}$ is fast–compact in $(X, \mathscr{P})$.

(b) Similar to that of part (a).

(c) There exists an infracomplete disk B in X such that K is compact in the Banach space $(X(B), \gamma_B)$. By (5.2), there exists a sequence $\{x_n\}$ in $X(B)$ such that

$$\gamma_B(x_n) \longrightarrow 0 \quad \text{and} \quad K \subset \tilde{\Gamma}(\{x_n : n \geq 1\}),$$

where $\tilde{\Gamma}(\{x_n : n \geq 1\})$ is the $\gamma_B(\cdot)$–closure of $\Gamma(\{x_n : n \geq 1\})$, hence $\{x_n\}$ is a fast–null sequence in $(X,\mathscr{P})$. On the other hand, the completeness of $(X(B),\gamma_B)$, together with $\gamma_B(x_n) \longrightarrow 0$, shows that $\tilde{\Gamma}(\{x_n : n \geq 1\})$ is compact in $(X(B),\gamma_B)$, so that $\tilde{\Gamma}(\{x_n : n \geq 1\}) = \overline{\Gamma}(\{x_n : n \geq 1\})$ (by the continuity of $J_B : (X(B),\gamma_B) \longrightarrow (X,\mathscr{P})$).

(15.4) **Definition.** A LCS $(X,\mathscr{P})$ is said to be infracomplete (bornologically complete in the terminology of Hogbe–Nlend [1977, p.46]) if the von Neumann bornology $\mathscr{M}_{von}(\mathscr{P})$ admits a base consisting of infracomplete disks in X. (i.e. $\mathscr{B}_{inf}(\mathscr{P})$ is a base for $\mathscr{M}_{von}(\mathscr{P})$).

(15.b) A characterization of infracomplete LCS: A sequence in a LCS $(X,\mathscr{P})$ is called a b–Cauchy sequence in $(X,\mathscr{P})$ if there exists a $\mathscr{P}$–bounded disk B in X such that $\{x_n\}$ is a Cauchy sequence in the normed space $(X(B),\gamma_B)$. One can show (see Hogbe–Nlend [1977,p.46]) that a LCS $(X,\mathscr{P})$ is infracomplete if and only if every b–Cauchy sequence in $(X,\mathscr{P})$ is $\mathscr{P}$–convergent. Consequently, every sequentially complete LCS must be infracomplete.

(15.5) **Definition**. A LCS $(X,\mathscr{P})$ is said to have a Schwartz bornology (or $(X,\mathscr{P})$ is a co–Schwartz space) if the von Neumann bornology $\mathscr{M}_{von}(\mathscr{P})$ admits a basis consisting of b–compact disks; in this case, $\mathscr{M}_{von}(\mathscr{P})$ is called a Schwartz bornology.

Remark. Similarly we say temporarily that a LCS $(X,\mathscr{P})$ has a fast–Schwartz bornology if the von Neumann bornology $\mathscr{M}_{von}(\mathscr{P})$ admits a basis consisting of fast–compact disks in $(X,\mathscr{P})$; in this case, $\mathscr{M}_{von}(\mathscr{P})$ is called a fast–Schwartz bornology. Clearly every fast–Schwartz bornology must be a Schwartz bornology.

(15.c) Schwartz bornologies. Suppose that a LCS $(X,\mathscr{P})$ has a Schwartz bornology. Then $(X,\mathscr{P})$ is infracomplete and bounded sets in X must be $\mathscr{P}$–totally bounded.

The construction of the associated bornological topology from bounded sets in Section 12 demonstrates a fundamental method for constructing locally convex topologies. The remainder of this section continues this idea for constructing locally convex topologies from given convex bornologies.

Let $\mathfrak{S}$ be a convex bornology on X with a base $\mathscr{B}$ consisting of disks. A set V in X is said to be $\mathfrak{S}$–bornivorous if it absorbs any element in $\mathscr{B}$ (or $\mathfrak{S}$). Suppose now that

$$\mathscr{U}_{\mathfrak{S}} = \{V \subset X : V = \Gamma V \text{ is } \mathfrak{S}\text{–bornivorous}\}.$$

It is easily seen from (7.4) that $\mathscr{U}_{\mathfrak{S}}$ determines a unique locally convex topology on X, denoted by $\mathfrak{S}_t$, such that $\mathscr{U}_{\mathfrak{S}}$ is a local base for $\mathfrak{S}_t$. It should be noted that $\mathfrak{S}_t$ is, in general, not Hausdorff.

(15.6) **Definition.** Let $\mathfrak{S}$ be a convex bornology on X. The locally convex topology (not necessarily Hausdorff) $\mathfrak{S}_t$ on X, constructed in the preceding, is called the associated topology of $\mathfrak{S}$. The family of all $\mathfrak{S}_t$–bounded subsets of X is called the associated bornology (or weak bornology) of $\mathfrak{S}$, and denoted by $(\mathfrak{S}_t)_{von}$ or $\mathscr{M}_{von}(\mathfrak{S}_t)$ or $\mathfrak{S}_{tb}$.

For example, if $(X,\mathscr{P})$ is a LCS, then the bornological topology $\mathscr{P}^{\times}$ associated with $\mathscr{P}$ is the associated topology of the von Neumann bornology $\mathscr{M}_{von}(\mathscr{P})$.

It is clear that

$$\mathfrak{S} \subset (\mathfrak{S}_t)_{von} \text{ and } (X,\mathfrak{S}_t)' = X^{\times},$$

and that $\mathfrak{S}_t$ is the finest locally convex topology on X such that elements in $\mathfrak{S}$ are $\mathfrak{S}_t$–bounded. Thus the associated topology $\mathfrak{S}_t$ is Hausdorff if and only if the bornological dual $X^{\times}$ of $(X,\mathfrak{S})$ separates points on X. This leads to the followng definition:

(15.7) **Definition.** A convex bornology $\mathfrak{S}$ on X is said to be:

(a) **regular** if $X^{\times} = (X,\mathfrak{S})^{\times}$ separates points on X;

(b) **topological** if $\mathfrak{S} = (\mathfrak{S}_t)_{von}$.

For any LCS $(X,\mathscr{P})$, the von Neumann bornology $\mathscr{M}_{von}(\mathscr{P})$ is topological as well as regular. Moreover, we have the following:

(15.d). Topological convex bornologies: A convex bornology $\mathfrak{S}$ on X is topological if and only if $\mathfrak{S}$ is the von Neumann bornology for some locally convex topology on X.

(15.8) **Definition.** Let $(X,\mathscr{P})$ be a LCS with the infracomplete bornology $\mathscr{M}_{inf}(\mathscr{P})$. If $\mathscr{P}$ coincides with the associated topology $(\mathscr{M}_{inf}(\mathscr{P}))_t$ of $\mathscr{M}_{inf}(\mathscr{P})$, then we say that $(X,\mathscr{P})$ is a fast–bornological space (ultrabornological or completely bornological space in the terminology of Hogbe–Nlend [1977, p.53]).

The following result gives some examples of fast–bornological spaces.

(15.9) **Proposition.** (a) Fast–bornological spaces are bornological.

(b) A bornological space $(X,\mathscr{P})$ for which $\mathscr{B}_{inf}$ is a base for the von Neumann bornology $\mathscr{M}_{von}(\mathscr{P})$ is fast–bornological ($\mathscr{B}_{inf}$ is the family of all infracomplete bounded disks in $(X,\mathscr{P})$.)

(c) Quasi–complete bornological spaces are fast–bornological. In particular, F–spaces and the Banach dual of a normed space are fast– bornological.

Proof. Part (b) follows from (15.8), while part (c) is a consequence of part (b) (since complete, bounded disks are infracomplete). As $\mathscr{M}_{inf}(\mathscr{P}) \subset \mathscr{M}_{von}(\mathscr{P})$, part (a) follows from (15.8) and the definition of bornological spaces.

(15.e) Criteria for fast–bornological spaces: Let $(X,\mathscr{P})$ be a LCS. A subset V of X is said to be fast–bornivorous if V absorbs any infracomplete (bounded) disk in X.

The following statements are equivalent for a LCS $(X, \mathcal{P})$.

(i) $(X, \mathcal{P})$ is fast–bornological.

(ii) For any LCS Y, any $T \in L^*(X,Y)$, which sends infracomplete (bounded) disks of $(X, \mathcal{P})$ into bounded subsets of Y, is continuous.

(iii) For any LCS Y, any $T \in L^*(X,Y)$, which sends compact convex subsets of $(X, \mathcal{P})$ into bounded sets in Y, is continuous.

(iv) Every seminorm on X, which is bounded on any infracomplete (bounded) disk in X, is continuous.

(v) Every fast–bornivorous disk in X is a $\mathcal{P}$–neighbourhood of 0.

Exercises

15–1. Prove (15.a).

15–2. Prove (15.b).

15–3. Prove (15.c).

15–4. Prove (15.d).

15–5. Prove (15.e).

15–6. Let $(X, \mathcal{P})$ be a LCS and M a vector subspace of X.

(a) If the LCS $(M, \mathcal{P}_M)$ is infracomplete, then M is b–closed in X.

(b) If $(X, \mathcal{P})$ is infracomplete and M is b–closed, then the subspace $(M, \mathcal{P}_M)$ and the quotient space $(X/_M, \dot{\mathcal{P}})$ are infracomplete.

16

Dual Pairs and the Weak Topology

We begin with the following:

(16.1) Definition. Let X and Y be vector spaces over $\mathbb{K}$, and suppose that there exists a bilinear functional ψ on $X \times Y$, satisfying the following separation conditions:

(S1) if $\psi(x_0, y) = 0$ (for all $y \in Y$) then $x_0 = 0$;

(S2) if $\psi(x, y_0) = 0$ (for all $x \in X$) then $y_0 = 0$.

Then the triple (X, Y, ψ) is called a dual pair (or a duality), and ψ is called the canonical bilinear functional of the duality. Usually, we write

$$\langle x, y\rangle = \psi(x, y) \quad \text{(for all } x \in X \text{ and } y \in Y\text{)},$$

while the triple (X, Y, ψ) is more conveniently denoted by $\langle X,Y\rangle$.

Remark. Because of the separation condition, the map

$$y \longrightarrow \langle \cdot, y\rangle : Y \longrightarrow X^*$$

is linear and injective; symmetrically the map

$$x \longrightarrow \langle x, \cdot\rangle : X \longrightarrow Y^*$$

is linear and injective. Thus we identify Y with a vector subspace of X^* which separates points on X, and symmetrically X with a vector subspace of Y^* which separates points on Y.

(16.2) Examples: (a) Let X be a vector space and X^* the algebraic dual of X. Then $\langle X, X^*\rangle$ is a dual pair under the natural bilinear form

(16.2.1) $$\langle u, x^*\rangle = x^*(u) \quad \text{(for all } u \in X \text{ and } x^* \in X^*\text{)}.$$

Moreover, if Y is a vector subspace of X^* which separates points on X, then the restriction on $X \times Y$ of the natural bilinear form (16.2.1) induces a duality between X and Y. In particular, if $(X, \mathscr{P})$ is a LCS with the topological dual X', then $\langle X, X' \rangle$ is a dual pair under the natural bilinear form:

$$\langle u, x' \rangle = x'(u) \quad (\text{for all } u \in X \text{ and } x' \in X').$$

(b) **Bornological duality.** Let $(X, \mathfrak{S})$ be a CBVS with the bornological dual $X^{\times}$. If $X^{\times}$ separates points on X, then $\langle X, X^{\times} \rangle$ is a duality under the natural bilinear form

$$\langle u, f \rangle = f(u) \quad (\text{for all } u \in X \text{ and } f \in X^{\times});$$

in this case, $\langle X, X^{\times} \rangle$ is referred to as the bornological duality.

(16.3) **Definition.** For a given dual pair $\langle X, Y \rangle$, the weak topology on X, denoted by $\sigma(X,Y)$, is defined by the family $\{p_y : y \in Y\}$ of seminorms on X, where

$$p_y(x) = |\langle x, y \rangle| \quad (\text{for all } x \in X).$$

Symmetrically we define the weak topology on Y, denoted by $\sigma(Y,X)$. Subsets of X which are closed (resp. open, compact, totally bounded etc.) for $\sigma(X,Y)$ are called weakly closed (resp. weakly open, weakly compact, weakly totally bounded, etc.)

By (S1), $\sigma(X,Y)$ is a Hausdorff locally convex topology on X which is the coarsest topology such that all linear functionals $\{\langle \cdot, y \rangle : y \in Y\}$ are continuous, hence Y can be identified with a vector subspace of $(X, \sigma(X, Y))'$; they are equal as shown by the following:

(16.4) **Lemma.** Let $\langle X, Y \rangle$ be a dual pair. Then

$$Y \simeq (X, \sigma(X, Y))' \quad (\text{algebraically isomorphic});$$

namely, an $f \in X^*$ is $\sigma(X,Y)$–continuous if and only if there exists a unique $y \in Y$ such that

(1) $$f(x) = \langle x, y\rangle \quad (\text{for all } x \in X).$$

(Hence we always identify Y with $(X, \sigma(X, Y))'$, and say that Y is the topological dual of $(X, \sigma(X, Y))$.)

Proof. Let $f \in X^*$ be $\sigma(X, Y)$–continuous. Then there exists a $\lambda > 0$ and $\{y_1, \cdots, y_n\} \subset Y$ such that

$$|f(x)| \leq \lambda \max_{1 \leq i \leq n} |\langle x, y_i\rangle| \quad (\text{for all } x \in X),$$

so that

(2) $$\langle x, y_i\rangle = 0 \quad (i = 1, \cdots, n) \Rightarrow f(x) = 0.$$

It then follows from a well–known result that there are $\mu_1, \cdots, \mu_n \in K$ such that

$$f(x) = \Sigma_{i=1}^n \mu_i \langle x, y_i\rangle = \langle x, \Sigma_{i=1}^n \mu_i y_i\rangle \quad (\text{for all } x \in X),$$

thus the result follows.

(16.5) **Corollary**. Let $(X, \mathscr{P})$ be a LCS with the topological dual X'. Then $X' = (X, \sigma(X, X'))'$.

(16.a) Let $\langle X, Y\rangle$ be a dual pair and Y_1 a vector subspace of Y. Then Y_1 is $\sigma(Y, X)$–dense in Y if and only if Y_1 separates points on X in the sense that

$$\langle x_0, y\rangle = 0 \quad (\text{for all } y \in Y_1) \Rightarrow x_0 = 0.$$

In particular, if $(X, \mathscr{P})$ is a LCS with the topological dual X', then X' is

$\sigma(X^*, X)$–dense in X^*.

(16.6) **Proposition.** *Let* $<X, Y>$ *be a duality. A subset* B *of* X *is* $\sigma(X, Y)$*–bounded if and only if* B *is* $\sigma(X, Y)$*–totally bounded.*

Proof. The sufficiency is trivial. To prove the necessity, let V be a $\sigma(X, Y)$–neighbourhood of 0, and let $y \in Y$ be such that

$$W = \{x \in X : |<x, y>| \leq 1\} \subseteq V.$$

Since B is $\sigma(X, Y)$–bounded, the set $\{<b, y> : b \in B\}$ is bounded in $\mathbb{K}$, hence relatively compact in $\mathbb{K}$; consequently, there exists a finite subset $\{ b_1, \cdots, b_n \}$ of B such that

$$\{<b, y> : b \in B\} \subset \bigcup_{i=1}^{n} (<b_i, y> + U_{\mathbb{K}}).$$

It then follows that

$$B \subset \bigcup_{i=1}^{n} (b_i + W) \subset \bigcup_{i=1}^{n} (b_i + V),$$

and hence that B is $\sigma(X, Y)$–totally bounded.

As a consequence of (16.6) and (2.11), we have the following:

(16.b) A Banach space $(E, \|\cdot\|)$ is finite–diminsional if and only if $\|\cdot\|$–top coincides with $\sigma(E, E')$.

(16.7) **Definition.** Let $<X, Y>$ be a dual pair and $\phi \neq B \subset X$. The set, defined by

$$B^o = \{y \in Y : \text{Re} <b,y> \leq 1 \ \ (\text{for all } b \in B)\},$$

is called the *polar* of B (*taken in* Y or w.r.t. $<X, Y>$); while the set, defined by

$$B^a = \{y \in Y : |<b, y>| \leq 1 \text{ (for all } b \in B)\},$$

is called the absolute polar of B (taken in Y or w.r.t. $<X, Y>$). (When there is need to specify the polar taken in Y , we shall write $B^o(Y)$ for B^o.) The bipolar of B, which is a subset of X , is defined to be the polar of B^o (taken in X), and denoted by B^{oo}.

(16.8) **Lemma.** Let $<X, Y>$ be a dual pair and $\phi \neq B \subset X$.

(a) B^o (resp. B^a) is a $\sigma(Y,X)$–closed convex (resp. $\sigma(Y,X)$–closed, disked) subset of Y containing 0.

(b) $(\lambda B)^o = \lambda^{-1}B^o$ (for all $\lambda \neq 0$).

(c) If $A \subset B \subset X$ then $B^o \subset A^o$.

(d) $B \subset B^{oo}$ and $B^o = B^{ooo}$.

(e) If $B_\alpha \subset X$ $(\alpha \in \Lambda)$, then

$$(\bigcup_{\alpha\in\Lambda} B_\alpha)^0 = \bigcap_\alpha B_\alpha^0 \quad \text{and} \quad (\bigcap B_\alpha)^0 \supset \overline{\text{co}}(\bigcup_\alpha B_\alpha^0)$$

where the bar denotes the $\sigma(Y, X)$–closure.

The proof is straighforward, hence will be omitted.

(16.9) **Theorem** (Bipolar theorem). Let $<X, Y>$ be a dual pair and $B \subset X$. Then

$$B^{oo} = \overline{\text{co}}(B \cup \{0\}).$$

(where the bar denotes the $\sigma(X,Y)$–closure).

Proof. Let $B_1 = \overline{\text{co}}(B \cup \{0\})$. Then $B_1 \subset B^{oo}$. If $B_1 \neq B^{oo}$, then we take $x_0 \in B^{oo} \setminus B_1$. As $Y = (X, \sigma(X, Y))'$, it follows from the strong separation theorem that there is an $0 \neq y \in Y$ such that

$$\sup\{\text{Re} <x,y> : x \in B_1\} < 1 < \text{Re} <x_0,y>;$$

in particular,

(1) $$\sup \{\mathrm{Re} \langle b,y\rangle : b \in B\} < 1 < \mathrm{Re} \langle x_0,y\rangle,$$

The left side of (1) shows that $y \in B^o$, while the right side of (1), together with $x_0 \in B^{oo}$ shows that $y \notin B^{ooo} = B^o$ (by (16.8) (d)). Thus we arrive at a contradiction.

(16.10) **Corollary.** *Let* $\langle X, Y\rangle$ *be a dual pair and* $B \subset X$. *Then*

$$B^o = (\overline{co}(B \cup \{0\}))^o = (co(B \cup \{0\}))^o.$$

Proof. As $B^o = B^{ooo}$, the result follows immediately from (16.9).

(16.c) Let $\langle X, Y\rangle$ be a dual pair and let $\{B_i : i \in \Lambda\}$ be a family of convex subsets of X with $o \in \bigcap_i B_i$. Then $(\bigcap_i B_i)^o = \overline{co}(\bigcup_i B_i^o)$ if and only if $\overline{\bigcap_i B_i} = \bigcap_i \overline{B_i}$.

(16.11) **Proposition.** *Let* $\langle X, Y\rangle$ *be a dual pair and* $B \subset X$.

(a) B *is* $\sigma(X, Y)$*–bounded if and only if* B^o *is an absorbing subset of* Y.

(b) B^o *is* $\sigma(Y,X)$*–bounded if and only if* B^{oo} *is an absorbing subset of* X.

(c) *Suppose that* $0 \in B = \overline{co}\, \overline{B}$. *Then* B *is* $\sigma(X,Y)$*–complete if and only if every* $f \in Y^*$ *which is bounded on* B^o *is represented by some element in* X.

Proof. (a) Follows from the following equivalent statements:

B is $\sigma(X,Y)$–bounded

$\Leftrightarrow$ for any $y \in Y$, $\sup \{|\langle b,y\rangle| : b \in B\} = \lambda_y < \infty$

$\Leftrightarrow$ $y \in \lambda_y B^o$

$\Leftrightarrow$ B^o is absorbing in Y.

(b) Follows from part (a) in view of the bipolar theorem.

(c) We first notice that $(Y^*, \sigma(Y^*, Y))$ is complete and that $\sigma(X,Y) = \sigma(Y^*,Y)|_X$, here we identify X as a vector subspace of Y^*. [In fact, Y^* is closed in the product space K^Y, while the product space K^Y is complete for the product topology]. Thus we have the following equivalent statements:

B is $\sigma(X,Y)$–complete

$\Leftrightarrow$ B is $\sigma(Y^*,Y)$–closed

$\Leftrightarrow$ $B = (B^o)^o(Y^*)$ (by the bipolar theorem and $o \in B = \overline{coB}$)

$\Leftrightarrow$ every $g \in Y^*$ which is bounded on B^o is represented by some element of X.

(16.12) Theorem (Smulia). Let <X, Y> be a dual pair, and let $B \subset X$ be such that $o \in B = \overline{co}\, B$ (the $\sigma(X,Y)$–closure). Then B is $\sigma(X,Y)$–compact if and only if B^o is absorbing in Y and every $g \in Y^*$ which is bounded on B^o is represented by some element in X.

Proof. Follows from the following equivalent statements:

B is $\sigma(X,Y)$–compact

$\Leftrightarrow$ B is $\sigma(X,Y)$–totally bounded and $\sigma(X,Y)$–complete

$\Leftrightarrow$ B is $\sigma(X,Y)$–bounded and $\sigma(X,Y)$–complete

$\Leftrightarrow$ B^o is absorbing in Y and every $g \in Y^*$ which is bounded on B^o is represented by some element in X.

(16.13) **Theorem.** (Alaoglu–Bourbaki). Let $(X, \mathscr{P})$ be a LCS. Then the polar of any 0–neighbourhood V in X is $\sigma(X',X)$–compact.

Proof. Without loss of generality one can assume that $V = \overline{co}\, V$, so that $V = V^{oo}$ (by (16.9)). Any $f \in X^*$ which is bounded on V must be continuous, hence it can be represented by some element of X', it follows from (16.11)(c) that V^o is $\sigma(X',X)$–complete. On the other hand, V is absorbing in X, hence V^o is $\sigma(X', X)$–bounded (by (16.6)(b)), thus V^o is $\sigma(X',X)$–totally bounded (by (16.6)).Consequently, V^o is $\sigma(X', X)$–compact.

Let $(X,\mathcal{P})$ be a LCS. Recall that a subset D of X' is equicontinuous if there exists an $V \in \mathcal{U}_X$ such that

$$V \subset \bigcap_{f \in D} \{x \in X : |f(x)| \leq 1\}$$

or, equivalently, $D \subset V^o$. Therefore, Alaoglu–Bourbaki's theorem asserts that any $\mathcal{P}$–equicontinuous subset of X' is relatively $\sigma(X',X)$–compact (and a fortiori $\sigma(X',X)$–bounded).

A normed space $(E, \|\cdot\|)$ is called a Banach dual space if there exists a normed space (F, q) such that $(E, \|\cdot\|)$ and the Banach dual (F', q') of (F, q) are isometric, that is, E = F' and $\|\cdot\|$ is the dual norm q' of q.

Every Banach dual space must be complete, but there are B–spaces which are not Banach dual spaces. On the other hand, if $(E, \|\cdot\|)$ is a Banach dual space, then Alaoglu–Bourbaki's theorem ensures that there is a Hausdorff locally convex topology $\mathcal{P}$ on E (that is, the topology $\sigma(E,F)$) such that the closed unit ball $U_E = \{x \in E : \|x\| \leq 1\}$ is $\mathcal{P}$–compact. The converse is true as shown by the following result.

(16.14) Theorem (Ng). A normed space $(E, \|\cdot\|)$ is a Banach dual space if and only if there exists a Hausdorff locally convex topology $\mathcal{P}$ on E such that the closed unit ball U_E in E is $\mathcal{P}$–compact.

Proof. It has only to verify the sufficiency. Consider the following vector subspace of E^* :

$$F = \{f \in E^* : f \text{ is } \mathcal{P}\text{-continuous on } U_E\}.$$

Then $(E, \mathcal{P})' \subseteq F \subseteq E' = (E, \|\cdot\|)'$ (the inclusion $F \subseteq E'$ follows from the $\mathcal{P}$-compactness of U_E). F is closed in the Banach dual E' of $(E, \|\cdot\|)$, hence (F, q) is a Banach space, where q the restriction on F of the dual norm of $\|\cdot\|$, i.e.,

$$q(f) = \|f\| = \sup\{|f(x)| : x \in U_E\} \quad (\text{for all } f \in F).$$

Let (F', q^*) be the Banach dual of (F,q), where q^* is the dual norm of q. It remains to show that $(E, \|\cdot\|)$ and (F', q^*) are isometric.

In fact, the map $J_F : E \longrightarrow F'$, defined by

$$(J_F x)f = f(x) \quad (\text{for all } f \in F),$$

is linear, injective (since F separates points of E) and

$$q^*(J_F x) = \sup\{|(J_F x)f| : f \in F,\ q(f) \leq 1\} \leq \|x\|.$$

Thus it has only to verify that

$$J_F(U_E) = U_{F'} = \{\psi : F' : q^*(\psi) \leq 1\}.$$

To this end, we first notice that $J_F(U_E)$ is $\sigma(F',F)$-compact (the restriction of J_F on U_E is continuous for $\mathcal{P}$ and $\sigma(F',F)$ and the definition of F); hence the bipolar theorem ensures that

$$J_F(U_E) = (J_F(U_E))^{\circ\circ},$$

where the polar is taken with respect to the duality <F,F'>. Observe that

$$\begin{aligned}(J_F(U_E))^{\circ} &= \{f \in F : |f(x)| \le 1 \ \text{(for all } x \in U_E)\} \\ &= F \cap \{x' \in E' : \|x'\| \le 1\} = \{f \in F : q(f) \le 1\} \\ &= \text{the closed unit ball } U_F \text{ in } F.\end{aligned}$$

Therefore $J_F(U_E) = U_F^{\circ} = U_{F'}$, which obtains our assertion.

The proof of the preceding result is taken from Ng [1971 b], but the main idea of the proof is essentially due to Edwards [1964]. While the usual form of Dixmier's theorem states as follows : A Banach space E is a Banach dual space if and only if there exists a vector subspace F of E', which separates points of E, such that U_E is $\sigma(E,F)$–compact.

Exercises

16–1. Prove (16.a).

16–2. Prove (16.b).

16–3. Let <X,Y> be a dual pair, let $A, B \subset X$ be such that $0 \in A \cap B$ and show that

$$(A+B)^{\circ} \subset A^{\circ} \cap B^{\circ};$$

if, in addition, B satisfies $\mu B \subset B$ for all $\mu \ge 0$, then

$$(A+B)^{\circ} = A^{\circ} \cap B^{\circ}.$$

16–4. Let <X,Y> be a dual pair and $\{B_\alpha : \alpha \in D\}$ a family of convex subsets of X, each containing 0. Show that:

(a) $(\bigcap_\alpha B_\alpha)^{\circ} = \overline{\mathrm{co}}(\bigcup_\alpha B_\alpha^{\circ})$ if and only if $\bigcap_\alpha \overline{B}_\alpha = \overline{\bigcap_\alpha B_\alpha}$

(b) Suppose that $\mathfrak{I}$ is a locally convex topology on X with $(X,\mathfrak{I})'=Y$. Then one of the following conditions ensures that $\bigcap_\alpha \overline{B}_\alpha = \overline{\bigcap_\alpha B_\alpha}$;

(i) each B_α is closed;

(ii) there exists an $\beta \in D$ such that B_β is a $\mathcal{T}$–neighbourbood of 0 and each B_α $(\alpha \neq \beta)$ is closed;

(iii) each B_α $(\alpha \in D)$ is a $\mathcal{T}$–neighbourhood of 0;

(iv) $\bigcap_\alpha B_\alpha$ is a $\mathcal{T}$–neighbourhood of 0.

In particular, if V and W are convex $\mathcal{T}$–neighbourhoods of 0 in X, then $(V \cap W)^\circ = co(V^\circ \cup W^\circ)$.

16–5. (Gelfand) Let E be a separable, B–space, let $K \subset E$ and let $\{f_n\}$ be a $\sigma(E',E)$–null sequence in E' (i.e., for any $x \in E$, $f_n(x) \longrightarrow 0$) such that

$$\lim_n f_n(x) = 0 \text{ uniformly on } K.$$

Show that K is relatively compact. [Hint : The proof is based on the following facts : (i) K is bounded; (ii) $\psi(K)$ is equicontinuous in $C(U_{E'})$, where $\psi : E \longrightarrow C(U_{E'})$, defined by $\psi(x)(f) = f(x)$ (for all $f \in U_{E'}$), is a metric injection. (And then apply Arzela–Ascoli's theorem).]

16–6. Let X be a LCS and $X_\sigma = (X,\sigma(X,X'))$. Show that

$$L(X_\sigma,F) = \mathfrak{F}(X,F) \text{ for any normed space } F,$$

where $\mathfrak{F}(X,F) = \{T \in L(X,F) : \dim T(X) < \infty\}$.

16–7. Let E be a B–space

(a) Let $\{x_n\}$ be a $\sigma(E,E')$–null sequence in E (i.e., for any $f \in E'$, $\lim_n f(x_n) = 0$). Show that if $\{ x_n : n \geq 1 \}$ is relatively compact (or relatively sequentially compact), then $\lim_n \|x_n\| = 0$.

(b) Let $B \subset E$ be bounded and $\sigma(E, E')$–closed. If B is sequentially $\sigma(E,E')$–compact, then for any bounded, $\sigma(E,E')$–closed subset A of E with $A \cap B = \emptyset$,

$$\text{dist}(A,B) > 0.$$

16–8. Prove that closed subspaces of a Banach dual space are still Banach dual spaces.

17

Elementary Duality Theory

For the study of duality theory, there are mainly three fundamental principles: The first one is to consider the duality between locally convex topology and bornology: namely to a given dual pair $<X, Y>$, any locally convex topology (resp. convex bornology) on either X or Y, consistent with $<X, Y>$, corresponds by polarity a convex bornology (resp. locally convex topology) on the other space; this idea is extremely convenient since, in practice, it is much easier to construct convex bornologies than locally convex topologies satisfying some given properties. The second one is to consider the duality between equicontinuity and boundedness for an $\mathfrak{S}$–topology; namely for a given LCS $(X,\mathscr{P})$, one compares two natural convex bornologies on X'; the equicontinuous bornology $\mathscr{M}_{equ}(\mathscr{P})$ and the von Neumann bornology $\mathscr{M}_{von}(\mathfrak{S}^o)$, this leads to the Banach–Steinhaus theorem and the notions of barrelledness and the infrabarrelledness. The final one is to consider the duality between locally bounded maps and the continuity of their dual maps; this is one of the fundamental operations in Analysis. This section is devoted to a study of these three fundamental principles.

(17.1) **Definition.** Let $<X, Y>$ be a dual pair, and let $\mathfrak{S}$ be a family of subsets of Y satisfying the following conditions:

(T1) each element in $\mathfrak{S}$ is $\sigma(Y,X)$–bounded.

(T2) $\cup\,\mathfrak{S} = Y$.

(T3) $\mathfrak{S}$ is directed upwards under the set inclusion " $\subset$ ".

Then the family

(17.1.1) $$\mathfrak{S}^o = \{\lambda D^o : D \in \mathfrak{S}, \lambda > 0\}$$

determines a Hausdorff locally convex topology, denoted by $\mathfrak{T}_{\mathfrak{S}}$, with $\mathfrak{S}^o$ as a local base at 0. This topology is referred to as the $\mathfrak{S}$–topology (or the topology of uniform convergence on sets in $\mathfrak{S}$) and $\mathfrak{S}$ is called a topologizing family for X (or a base of a compatible bornology on Y).

Remarks: (i) The Hausdorff property of $\mathfrak{T}_{\mathfrak{S}}$ is guaranteed by the condition (T2), which can be weakened by the following :

(T2*) $\cup\mathfrak{S}$ is total in Y in the sense that (the linear hull)

$$\langle \cup \mathfrak{S} \rangle \text{ is } \sigma(Y,X)\text{--dense in } Y.$$

If the family $\mathfrak{S}$, consisting of subsets of Y, satisfies only (T1) and (T2*), then $\mathfrak{S}$ still determines a Hausdorff locally convex topology $\mathfrak{T}_{\mathfrak{S}}$ on X such that the family

$$\{\lambda \bigcap_{i=1}^{n} B_i^o : \lambda > 0, B_i \in \mathfrak{S} \text{ and any finite } n\}$$

is a local base at 0 for $\mathfrak{T}_{\mathfrak{S}}$. On the other hand, the condition (T1) is essential since it guarantees that each D^o $(D \in \mathfrak{S})$ is absorbing.

(ii) Clearly the $\mathfrak{S}$–topology $\mathfrak{T}_{\mathfrak{S}}$ on X will not be changed if $\mathfrak{S}$ is replaced by one of the following families of subsets of Y:

(a) All subsets of any element in $\mathfrak{S}$;

(b) $\{\lambda B : \lambda \in K \text{ and } B \in \mathfrak{S}\}$;

(c) $\{\overline{\Gamma}(\bigcup_{i=1}^{n} B_i) : B_i \in \mathfrak{S} \text{ and any finite } n\}$.

Therefore, the $\mathfrak{S}$–topology on X is determined by a family $\{p_B : B \in \mathfrak{S}\}$ of seminorms, where

$$p_B(x) = \sup\{|\langle x,b'\rangle| : b' \in B\} \quad (\text{for all } x \in X).$$

An $f \in X^*$ is $\mathfrak{T}_{\mathfrak{S}}$–continuous if and only if f is bounded on some B^0 (where $B \in \mathfrak{S}$), thus

$$(X,\mathfrak{T}_{\mathfrak{S}})' = \langle \cup \{(B^0)^{\pi}(X^*) : B \in \mathfrak{S}\}\rangle = \langle \cup \{\overline{\Gamma B}^* : B \in \mathfrak{S}\}\rangle,$$

where $(B^0)^{\pi}(X^*)$ is the polar of B^0 w.r t. $\langle X,X^*\rangle$ and $\overline{\Gamma B}^*$ is the $\sigma(X^*,X)$–closed, disked hull of B in X^*, and $\langle\cdot\rangle$ denotes the linear hull.

(iii) A non–empty family $\mathscr{S}$ of $\sigma(Y,X)$–bounded subsets of Y is said to be <u>saturated</u> if it satisfies the following conditions:

(a) If $A \subset B$ with $B \in \mathscr{S}$ then $A \in \mathscr{S}$.

(b) $\lambda B \in \mathscr{S}$ whenever $\lambda \in K$ and $B \in \mathscr{S}$.

(c) $\overline{\Gamma}(\bigcup_{i=1}^{n} B_i) \in \mathscr{S}$ whenever $B_i \in \mathscr{S}$ and any finite n (the bar denotes the $\sigma(Y,X)$–closure).

The family of all $\sigma(Y,X)$–bounded subsets of Y is a saturated family, and the intersection of a collection of saturated families is saturated, it follows that a given family $\mathfrak{S}$ of $\sigma(Y,X)$–bounded subsets determines a smallest saturated family $\mathscr{A}(\mathfrak{S})$ containing $\mathfrak{S}$, which is called the <u>saturated hull of</u> $\mathfrak{S}$. We see from Remark (ii) that the $\mathfrak{S}$–topology on X coincides with the $\mathscr{A}(\mathfrak{S})$–topology.

(iv) A family $\mathscr{M}$ consisting of subsets of Y is callled a <u>bornology</u> if it satisfies the following conditions :

(B1) $\cup \mathscr{M} = X$.

(B2) $B_1 \cup \cdots \cup B_n \in \mathscr{M}$ whenever $B_i \in \mathscr{M}$ $(i = 1,\cdots,n)$.

(B3) $A \subset B \in \mathscr{M}$ implies that $A \in \mathscr{M}$.

Clearly vector bornologies are bornologies. A bornology $\mathscr{M}$ on Y is <u>compatible with</u> $\langle X, Y\rangle$ if members in $\mathscr{M}$ are weakly bounded. A subfamily $\mathscr{B}$ of a <u>bornology</u> $\mathscr{M}$ is called a <u>base for</u> $\mathscr{M}$ if for any $A \in \mathscr{M}$ there is some $D \in \mathscr{B}$ such that $A \subset D$. It is clear that a family $\mathfrak{M}$ consisting of subsets of Y is a base for a bornology if and only if $\mathfrak{M}$ satisfies the conditions (T2) and (T3). [Because of these reasons, together with (T1), $\mathfrak{S}$ is called a <u>base of a compatible bornology on</u> Y.] In this case,

$$\{B \subset Y : B \subset D \text{ for some } D \in \mathfrak{M}\} = \mathcal{M}(\mathfrak{M})$$

is the bornology with $\mathfrak{M}$ as a base. [$\mathcal{M}(\mathfrak{M})$ is called the bornology generated by $\mathfrak{M}$.] It is also clear that the $\mathfrak{S}$–topology on X coincides with the $\mathcal{M}(\mathfrak{S})$–topology (for any topologizing family $\mathfrak{S}$).

We list below some important examples of $\mathfrak{S}$–topologies.

(17.2) Examples and notations: Let $(X,\mathscr{P})$ be a LCS with X' as the topological dual, and $<X,X'>$ a dual pair.

(a) The weak topology $\sigma(X,X')$ on X: The family $\mathscr{F}(X')$ of all non–empty finite subsets of X' is a topologizing family for X, the $\mathscr{F}(X')$–topology on X is exactly the weak topology $\sigma(X,X')$ on X.

Dually, if $\mathscr{F}(X)$ is the family of all non–empty finite subsets of X, then $\mathfrak{T}_{\mathscr{F}(X)} = \sigma(X',X)$.

(b) The strong topology: The von Neumann bornology $\mathcal{M}_{von}(\mathscr{P})$ is a topologizing family for X', hence the $\mathcal{M}_{von}(\mathscr{P})$–topology is called the strong topology on X', and denoted by $\beta(X',X)$. The pair $(X', \beta(X', X))$ (or X'_β) is called the strong dual of X'. The topological dual of X'_β is called the bidual of X, and denoted by X" ; while the strong dual of X'_β is called the strong bidual of X, and denoted by $(X'', \beta(X'', X'))$ or X''_β. In particular, if $(E,\|\cdot\|)$ is a normed space, then the Banach dual $(E',\|\cdot\|)$ of $(E,\|\cdot\|)$ is the strong dual of $(E,\|\cdot\|)$.

Dually, as $\sigma(X',X)$ is a Hausdorff locally convex topology on X' such that $X = (X',\sigma(X', X))'$ (see (16.1)), it follows that the topology on X uniform convergence on $\sigma(X',X)$–bounded subsets of X' is called the strong topology on X, and denoted by $\beta(X,X')$.

(c) The strong–star topology: The family $\mathscr{M}_{von}(\beta(X,X'))$ of all $\beta(X,X')$–bounded subsets of X is a convex bornology, hence the $\mathscr{M}_{von}(\beta(X,X'))$–topology on X' is called the strong–star topology, and denoted by $\beta^*(X',X)$. Dually, the strong–star topology on X is denoted by $\beta^*(X,X')$. Clearly we have $\beta^*(X',X) < \beta(X',X)$.

(d) The topology of precompact convergence: The family $\mathscr{M}_p(\mathscr{P})$ of all $\mathscr{P}$–precompact subsets of X is a convex bornology, hence the $\mathscr{M}_p(\mathscr{P})$–topology on X' is called the topology of precompact convergence (or the topology polar to $\mathscr{P}$), and denoted by $\mathscr{P}^o$. As $\sigma(X,X')$–bounded sets are $\sigma(X,X')$–totally bounded (see(16.6)), it follows that $(\sigma(X,X'))^o = \beta(X',X)$.

(e) The topology of compact convex convergence: The family $\mathscr{B}_{cd}$ of all $\mathscr{P}$–compact disks in X is a base for a convex bornology on X, hence $\mathscr{B}_{cd}$ is a topologizing family for X'; the $\mathscr{B}_{cd}$–topology is called the topology of uniform convergence on $\mathscr{P}$–compact convex sets (or the Arens topology on X'), and denoted by $\kappa(X',X)$.

(f) The Mackey topology: The family of all $\sigma(X,X')$–compact disks in X is a topologizing family for X', it determines a Hausdorff locally convex topology on X, denoted by $\tau(X',X)$, which is the topology of uniform convergence on all $\sigma(X,X')$–compact disks, is called the Mackey topology on X'.

Dually, the Mackey topology on X, denoted by $\tau(X,X')$, is the topology of uniform convergence on all $\sigma(X',X)$–compact disks in X'. A LCS $(Y,\mathscr{T})$ is called a Mackey space if $\mathscr{T} = \tau(Y,Y')$.

(g) On X', we have

$$\sigma(X',X) < \kappa(X',X) < \beta(X',X),$$

and

$$\kappa(X',X) < \tau(X',X) < \beta(X',X).$$

If $(X,\mathscr{P})$ is quasi–complete, then $\kappa(X',X) = \mathscr{P}^{o}$.

(h) <u>The natural topology on a bornological dual</u>. Let $(X,\mathfrak{S})$ be a regular CBVS with its bornological dual $X^{\times}$. Then $<X, X^{\times}>$ is a dual pair, and each $f \in X^{\times}$ is bounded on each element in $\mathfrak{S}$; thus elements in $\mathfrak{S}$ are $\sigma(X,X^{\times})$–bounded; consequently, $\mathfrak{S}$ is a topologizing family for $X^{\times}$. The $\mathfrak{S}$–topology on $X^{\times}$ is called the <u>natural topology</u>.

(j) <u>Equicontinuous bornologies and ultra–strong topologies.</u> Let $\mathscr{U}_X$ be a local base at o for $\mathscr{P}$ consisting of $\sigma(X,X')$–closed disks in X. Then $\{V^o : V \in \mathscr{U}_X\}$ forms a fundamental system of $\mathscr{P}$–equicontinuous subsets of X' , thus $\{V^o : V \in \mathscr{U}_X\}$ is a base for a convex bornology on X', which is called the <u>equicontinuous bornology</u> and denoted by $\mathscr{E}_{\mathscr{P}}$. Clearly, each V^o $(V \in \mathscr{U}_X)$ is $\sigma(X',X)$–bounded, hence $\mathscr{E}_{\mathscr{P}}$ is a topologizing family for X. Furthermore, the bipolar theorem shows that $V = V^{oo}$ (for any $V \in \mathscr{U}_X$), hence $\mathscr{P}$ is the $\mathscr{E}_{\mathscr{P}}$–topology such that

$$(X,\mathfrak{T}_{\mathscr{E}_{\mathscr{P}}})' = (X,\mathscr{P})' = X'. \tag{17.2.1}$$

On the other hand, Alaoglu–Bourbaki's theorem shows that $\mathscr{E}_{\mathscr{P}}$ admits a base consisting of $\sigma(X',X)$–compact disks in X', hence $\mathscr{E}_{\mathscr{P}}$ is an infracomplete bornology on X'. The associated topology of $\mathscr{E}_{\mathscr{P}}$, denoted by $\beta_u(X',X)$ (which is always fast–bornological by (15.8)), is called the <u>ultra–strong topology</u> (see Hogbe–Nlend [1977,p.76]). The pair $(X',\beta_u(X',X))$(or X'_{β_u}) is called the <u>ultra–strong dual of</u> $(X,\mathscr{P})$. The topological dual of $(X',\beta_u(X',X))$ is called the <u>ultra–bidual of</u> $(X,\mathscr{P})$ and denoted by $X'^{\times}$. It is clear that a linear functional on X' is $\beta_u(X',X)$–continuous if and only if it is bounded on $\mathscr{P}$–equicontinuous subsets of X'. Therefore $X'^{\times}$ is the bornological dual of the CBVS

$(X', \mathcal{T}_{\mathcal{P}})$. It is also clear that

$$\beta(X',X) < \beta_u(X',X),$$

it then follows that $X'' \subset X'^{\times}$ If, in addition, X is a normed space then $X'' = X'^{\times}$.

Let $(X,\mathcal{P})$ be a LCS with the topological dual X'. It is known from (16.4) (resp. (17.2.1)) that the $\sigma(X,X')$–topology (resp. the original topology $\mathcal{P}$) is a Hausdorff locally convex topology on X such that

(1) $$X' = (X,\sigma(X,X'))' = (X,\mathcal{P})'.$$

It is well–known that there exists a normed space E such that

(2) $$E \subsetneqq (E',\beta(E',E))'.$$

These two facts lead to the following:

(17.3) <u>Definition.</u> Let <X, Y> be a dual pair. A locally convex topology $\mathfrak{T}$ on X is said to be <u>consistent with</u> <X, Y> if $Y = (X, \mathfrak{T})'$.

In view of the separation condition of duality, any topology on X, consistent with <X, Y>, must be Hausdorff. The weak topology $\sigma(X, Y)$ on X is the coarsest locally convex topology consistent with <X, Y>.

Let $(X, \mathcal{P})$ be a LCS. It is known from (17.2) (j) and (1) that $\mathcal{P}$ is the topology of uniform convergence on $\mathcal{P}$–equicontinuous subsets of X' (which are relatively $\sigma(X',X)$–compact) and $\mathcal{P}$ is consistent with $<X, X'>$. The converse holds as shown by the following important result.

(17.4) Theorem (Mackey–Arens). Let $<X, Y>$ be a dual pair. A locally convex topology $\mathscr{P}$ on X is consistent with $<X, Y>$ if and only if $\mathscr{P}$ is an $\mathfrak{S}$–topology for a family $\mathfrak{S}$ covering Y, consisting of relatively $\sigma(Y,X)$–compact, convex subsets of Y containing 0; in other words

$$\sigma(X, Y) \leq \mathscr{P} \leq \tau(X, Y).$$

Hence the Mackey topology $\tau(X,Y)$ is the finest locally convex topology on X consistent with $<X, Y>$.

Proof. The necessity follows from the bipolar theorem and Alaoglu–Bourbaki's theorem. To prove the sufficiency, we first notice that an $f \in X^*$ is $\mathfrak{T}_{\mathfrak{S}}$–continuous if and only if f is bounded on some B^o (with $B \in \mathfrak{S}$). So that

(1) $$(X, \mathfrak{T}_{\mathfrak{S}})' = < \cup \{(B^o)^{\pi}(X^*) : B \in \mathfrak{S}\}> \quad \text{(the linear hull)},$$

where $(B^o)^{\pi}(X^*)$ is the polar of B^o w.r.t $<X, X^*>$. As $\cup\mathfrak{S} = Y$, it follows that $\sigma(X,Y) \leq \mathfrak{T}_{\mathfrak{S}}$, and hence from (1) that $\mathfrak{T}_{\mathfrak{S}}$ is consistent with $<X,Y>$ if and only if

(2) $$(B^o)^{\pi}(X^*) \subset Y \quad \text{(for all } B \in \mathfrak{S}).$$

To prove (2), we may assume without loss of generality that each $B \in \mathfrak{S}$ is $\sigma(Y,X)$–closed. As usual, we identify Y with a vector subspace of X^*, hence $\sigma(Y, X) = \sigma(X^*, X)|_Y$, thus each $B \in \mathfrak{S}$ is $\sigma(X^*, X)$–closed in X^* (since B is $\sigma(Y, X)$–compact and $\sigma(X^*, X)$ is Hausdorff). It then follows from the bipolar theorem that $B = (B^o)^{\pi}(X^*)$, which obtains our assertion (2).

The closedness of a convex set and the notion of boundedness do not depend upon the consistent topology, but only on the duality. To do this, we need the following interesting result.

(17.5) **Lemma.** Let $(X, \mathcal{P})$ be a LCS. Then any barrel (i.e., $\sigma(X, X')$–closed, disked, absorbing set) in X absorbs every infracomplete disked subset of X.

Proof. Let V be a barrel in X and B an infracomplete disk in X. Then $(X(B), \gamma_B)$ is a Banach space and $\mathcal{P}|X(B)$ is coarser than the $\gamma_B(\cdot)$–topology, hence $V_B = V \cap X(B)$ is $\gamma_B(\cdot)$–closed, and a fortiori a barrel in $(X(B), \gamma_B)$. By (16.11)(a), the polar V_B^o of V_B, taken in $X(B)' = (X(B), \gamma_B)'$ (i.e. w.r.t. $<X(B), X(B)'>$), is $\sigma(X(B)', X(B))$–bounded, hence the uniform boundedness theorem shows that V_B^o is bounded in the Banach dual $X(B)'$ (on account of the completeness of $(X(B), \gamma_B)$); namely, there is a $\mu > 0$ such that

$$|f(y)| \leq \mu\, \gamma_B(y) \quad (\text{for all } f \in V_B^o \text{ and } y \in X(B)).$$

In particular, $\mu^{-1}B \subset V_B^{oo}$, it then follows from the bipolar theorem that $\mu^{-1}B \subset V_B^{oo} = V_B \subset V$ [since V_B is $\gamma_B(\cdot)$–closed and disked]; in other words, B is absorbed by V.

(17.6) **Theorem.** Let $<X, Y>$ be a duality.

(a) A convex set B in X is $\sigma(X,Y)$–closed if and only if it is $\tau(X,Y)$–closed; consequently, the $\tau(X,Y)$–closure and the $\sigma(X,Y)$–closure of any convex set in X are identical.

(b) (Mackey theorem). A subset B of X is $\tau(X,Y)$–bounded if and only if it is $\sigma(X,Y)$–bounded.

Proof. (a) As $Y = (X, \sigma(X,Y))' = (X, \tau(X,Y))'$, a similar argument, given in the proof of (8.5), shows that (a) is true.

(b) The necessity is trivial. To prove the sufficiency, one can assume, without loss of generality, that B is a disked, $\sigma(X, Y)$–bounded subset of X. To prove that B is $\tau(X,Y)$–bounded, it suffices to show, by the bipolar theorem, that the polar B^o of B

absorbs any $\sigma(Y,X)$–compact disk D in Y.

In fact, such a D is an infracomplete disk in Y, while B° is a barrel in $(Y, \sigma(Y,X))$ [see (16.11) (a)], thus B° absorbs D (by(17.5)).

(17.7) **Theorem.** (Banach–Mackey). For a dual pair $\langle X, Y\rangle$ the following assertions hold.

(a) Every infracomplete disk in $(X, \tau(X, Y))$ is $\beta(X, Y)$–bounded; consequently, $\tau(X, Y) < \overset{*}{\beta}(X, Y)$.

(b) A subset B of X is $\sigma(X, Y)$–bounded if and only if it is $\overset{*}{\beta}(X, Y)$–bounded.

Proof. (a) Let $\mathscr{B}_{von}(\sigma)$ be the family consisting of all $\sigma(Y,X)$–closed, $\sigma(Y, X)$–bounded disked subsets of Y. Then $\mathscr{B}^{0}_{von}(\sigma) = \{D^{\circ} : D \in \mathscr{B}_{von}(\sigma)\}$ is a local base at o for $\beta(X,Y)$, and each D° (where $D \in \mathscr{B}_{von}(\sigma)$) is a barrel in $(X, \sigma(X,Y))$ (by (16.11)(b)). Hence each infracomplete disk B in $(X, \sigma(X,Y))$ is absorbed by any D° (with $D \in \mathscr{B}_{von}(\sigma)$) (by (17.5)), thus B is $\beta(X,Y)$–bounded.

On the other hand, any $\sigma(Y,X)$–compact, disked subset B of Y is infracomplete, hence is $\beta(Y,X)$–bounded, thus $\tau(X,Y) < \overset{*}{\beta}(X,Y)$(by the definition of $\overset{*}{\beta}(X,Y)$).

(b) The sufficiency is obvious. To prove the necessity, one can assume, without loss of generality, that B is $\sigma(X,Y)$–closed and disked. Then B° is a $\beta(Y,X)$–neighbourhood of 0 in Y, hence B° absorbs all $\beta(Y,X)$–bounded subsets of Y or, equivalently, B is absorbed by any $\overset{*}{\beta}(X,Y)$–neighbourhood of 0. Thus B is $\overset{*}{\beta}(X,Y)$–bounded.

Part(a) of (17.7) shows that $\overset{*}{\beta}(X, Y)$ is not consistent with $\langle X, Y\rangle$; while part (b) of (17.7) shows that there exists a non–consistent topology on X, $\overset{*}{\beta}(X, Y)$ says, such that $\sigma(X, Y)$ and $\overset{*}{\beta}(X, Y)$ have the same bounded sets. However, this cannot occur for closedness of convex sets as shown by the following result.

(17.8) **Proposition.** *Let* $\langle X, Y\rangle$ *be a dual pair, and* $\mathfrak{S}$ *a topologizing family for* X. *Then the* $\mathfrak{S}$*–topology* $\mathfrak{T}_{\mathfrak{S}}$ *is consistent with* $\langle X, Y\rangle$ *if and only if each* $\mathfrak{T}_{\mathfrak{S}}$*–closed convex set in* X *must be* $\sigma(X,Y)$*–closed*.

Proof. The necessity follows from (17.6) (a). To prove the sufficiency, let $f \in X^*$ be $\mathfrak{T}_{\mathfrak{S}}$–continuous. Then the null space $f^{-1}(0)$ is a $\mathfrak{T}_{\mathfrak{S}}$–closed vector subspace of X, hence $\sigma(X,Y)$–closed (by the hypothesis); consequently, f is $\sigma(X,Y)$–continuous, thus $\mathfrak{T}_{\mathfrak{S}}$ is consistent with $\langle X, Y\rangle$.

It is known from (d) and (c) of (17.2) and (17.7) (a) that

$$\sigma(X,Y)^{\circ} = \beta(Y,X) > \beta^{*}(Y,X) > \tau(Y,X),$$

thus the polar topology is, in general, not consistent. However we mention the following interesting result.

(17.a) *The polar topology* : For any LCS $(X,\mathscr{P})$, the topology $\mathscr{P}^{\circ}$, polar to $\mathscr{P}$, is the finest $\mathfrak{S}$–topology on X' which coincides with $\sigma(X', X)$ on each $\mathscr{P}$–equicontinuous subset of X'; consequently, any $\mathscr{P}$–equicontinuous subset of X' is relatively $\mathscr{P}^{\circ}$–compact. (see Köthe [1969]).

(17.9) **Proposition.** *Let* $(X,\mathscr{P})$ *be a LCS and* $\mathfrak{S}$ *a topologizing family for* X'. *Then every* $\mathscr{P}$*–equicontinuous subset of* X' *is bounded for the* $\mathfrak{S}$*–topology*.

Proof. The result follows from (17.7)(a) (Since $\mathfrak{T}_{\mathfrak{S}} \le \beta(X',X)$ and every $\sigma(X',X)$–closed, $\mathscr{P}$–equicontinuous disk in X' is infracomplete), but we present here a somewhat more direct proof. Let $\mathcal{U}_X$ be a local basis for $\mathscr{P}$ consisting of $\sigma(X,X')$–closed, disks in X. It suffices to show that the polar of each $V \in \mathcal{U}_X$ is $\mathfrak{T}_{\mathfrak{S}}$–bounded or, equivalently, each $V \in \mathcal{U}_X$ absorbs any member in $\mathfrak{S}$; but this is obvious since members in $\mathfrak{S}$ are $\sigma(X,X')$–bounded (and hence $\mathscr{P}$–bounded).

Let $(X, \mathscr{P})$ be a LCS with its topological dual X'. Then the preceding result shows that $\mathscr{P}$-equicontinuous subsets of X' are $\beta(X',X)$–bounded as well as $\sigma(X',X)$. The converse is, in general, not true, Thus it is natural to deduce the following:

(17.10) Definition. A LCS $(X,\mathscr{P})$ is said to be barrelled (resp. infrabarrelled) if every $\sigma(X',X)$–bounded (resp. $\beta(X',X)$–bounded) subset of X' is $\mathscr{P}$-equicontinuous.

Remark. Clearly every barrelled space is infrabarrelled, any infrabarrelled space is a Mackey space (by (17.9)).

(17.a) Barrelled and infrabarrelled spaces (I): The following statements are equivalent for a LCS $(X, \mathscr{P})$.

(i) $(X,\mathscr{P})$ is barrelled (resp. infrabarrelled).

(ii) $\mathscr{P}= \beta(X, X')$ (resp. $\mathscr{P}= \beta^*(X,X')$).

(iii) Any barrel (resp. bornivorous barrel) in $(X,\mathscr{P})$ is a $\mathscr{P}$-neighbourhood of 0.

(iv) Every lower semi–continuous seminorm on X (resp. lower semi–continuous seminorm on X which is bounded on bounded subsets of X) is $\mathscr{P}$-continuous.

The remainder is devoted to a study of the third principle in the duality theory —— the duality between locally bounded maps and the continuity of their dual maps.

Let $<X,X'>$ and $<Y,Y'>$ be two dual pairs and $T \in L^*(X,Y)$. We define T^* on Y' by setting

$$T^*(y') = y' \circ T \quad (\text{for all } y' \in Y')$$

or, explicitly

$$<x, T^* y'> = <Tx, y'> \quad \text{(for all } x \in X).$$

Then T^* is called the algebraic adjoint of T. (Clearly each linear map has an algebraic adjoint). If T^* satisfies

$$T^*(Y') \subset X',$$

then we say that T^* is the dual (or adjoint) map of T w.r.t. $<X,X'>$ and $<Y,Y'>$, and denoted by T'. Of course there are linear maps which do not have any dual map. Thus the relation $T^*(Y') \subset X'$ is sometimes expressed by saying that the dual map of T exists. The existence of T' is characterized by the weak continuity of T as shown by the following result.

(17.11) **Lemma.** Let $<X, X'>$ and $<Y, Y'>$ be dual pairs and $T \in L^*(X, Y)$. Then $T^*(Y') \subset X'$ if and only if $T \in L(X_\sigma, Y_\sigma)$, where $X_\sigma = (X, \sigma(X, X'))$. In this case, $T' \in L(Y'_\sigma, X'_\sigma)$ and T is the dual map of T' (w.r.t. $<X, X'>$ and $<Y, Y'>$).

Proof. Necessity. Let $T^*(Y') \subset X'$ and let $\{x_\lambda\}$ be a net in X which is $\sigma(X, X')$–convergent to 0. Then we have, for any $y' \in Y'$, that

$$\lim_\lambda <Tx_\lambda, y'> = \lim_\lambda <x_\lambda, T^* y'> = 0,$$

that is, $\{Tx_\lambda\}$ is $\sigma(Y, Y')$–convergent to 0, thus $T \in L(X_\sigma, Y_\sigma)$.

Sufficiency. For any $y' \in Y'$, it has to show, by (16.4), that $T^* y'$ is $\sigma(X, X')$–continuous. Indeed, let $\{x_\lambda\}$ be a net in X such that $0 = \sigma(X, X')\text{–}\lim_\lambda Tx_\lambda$. Then $0 = \sigma(Y, Y')\text{–}\lim_\lambda Tx_\lambda$, hence $0 = \lim_\lambda <Tx_\lambda, y'> = \lim_\lambda <x_\lambda, T' y'>$, which obtains our assertion.

Finally, if $T \in L(X_\sigma, Y_\sigma)$, then T' exists and $T' \in L(Y'_\sigma, X'_\sigma)$. As $(X'_\sigma)' = X$, we conclude from the first part that $T' \in L(Y'_\sigma, X'_\sigma)$ if and only if $(T')^*(X) \subset Y$, and hence that $T = T'' = (T')^{*}|_{X}$.

The following result is useful for establishing the duality between locally bounded map and continuity of their adjoint map.

(17.12) **Lemma** (Polar formulas for image and inverse image). Let $<X, X'>$ and $<Y, Y'>$ be dual pairs and $T \in L(X_{\sigma}, Y_{\sigma})$.

(a) $(TA)^{o} = (T')^{-1}(A^{o})$ (for all $\phi \neq A \subset X$).

(b) $(T'N)^{o} = T^{-1}(N^{o})$ (for all $\phi \neq N \subset Y'$).

(c) $\overline{TX} = (\mathrm{Ker}\ T')^{o}$ and $\mathrm{Ker}\ T = (T'(Y'))^{o}$.

(Consequently, TX is $\sigma(Y, Y')$–dense in Y if and only if T' is one–to–one, while $T'(Y')$ is $\sigma(X', X)$–dense in X' if and only if T is one–to–one.)

(d) $(T^{-1}(W))^{o} = T'(W^{o})$ whenever W is a convex $\tau(Y, Y')$–neighbourhood of 0 in Y.

Proof. (a) Observe that

$$(TA)^{o} = \{y' \in Y' : \mathrm{Re}<Ta, y'> \leq 1 \quad (\text{for all} \in A)\}$$

$$= \{y' \in Y' : \mathrm{Re}<a, T'y'> \leq 1 \quad (\text{for all} \in A)\}.$$

Therefore, we have the following equivalent statements:

$$y' \in (TA)^{o} \Leftrightarrow T'y' \in A^{o} \Leftrightarrow y' \in (T')^{-1}(A^{o}),$$

thus (a) follows.

(b) The proof is similar to that of part (a), hence will be omitted.

(c) By the bipolar theorem and parts (a) and (b), we have

$$\overline{TX} = (TX)^{oo} = ((T')^{-1}(X^{o}))^{o} = ((T')^{-1}(\{0\}))^{o} = (\mathrm{Ker}\ T')^{o},$$

and

$$(T'(Y'))^{\circ} = T^{-1}((Y')^{\circ}) = T^{-1}(\{0\}) = \operatorname{Ker} T.$$

(d) The inclusion $T'(W^{\circ}) \subset (T^{-1}(W))^{\circ}$ is obvious. Conversely, if $x' \in (T^{-1}(W))^{\circ}$, then

$$\langle u, x' \rangle = 0 \quad (\text{for all } u \in \operatorname{Ker} T)$$

(since $\operatorname{Ker} T$ is a vector subspace with $\operatorname{Ker} T \subset T^{-1}(W)$) ,thus the map f, defined by

$$f(Tu) = \langle u, x' \rangle \quad (\text{for all } u \in X),$$

is a well–defined, continuous linear functional on the subspace $T(X)$ of $(Y, \tau(Y, Y'))$ because of

$$W \cap T(X) \subset \{y \in T(X) \ \operatorname{Re} f(y) \leq 1\}.$$

Now the Hahn–Banach extension theorem, together with Mackey–Arens′ theorem ensures that there is a $y' \in Y'$ such that

$$\langle Tu, y' \rangle = f(Tu) = \langle u, x' \rangle \quad (\text{for all } u \in X)$$

and

$$\operatorname{Re}\langle w, y' \rangle \leq 1 \quad (\text{for all } w \in W).$$

Therefore $y' \in W^{\circ}$ is such that $x' = T'y'$, thus $x' \in T'(W^{\circ})$.

(17.13) **Theorem.** Let $\langle X, X' \rangle$ and $\langle Y, Y' \rangle$ be dual pairs and $T \in L(X_{\sigma}, Y_{\sigma})$.

(a) Let $\mathfrak{S}$ (resp. $\mathcal{M}$) be topologizing family for X' (resp. Y') which consists

<u>of weakly closed disks and contains all non–empty subsets of any element in</u> $\mathfrak{S}$ (resp. $\mathcal{M}$) (in particular, saturated families). <u>Then</u>

$$(17.13.1) \qquad T(\mathfrak{S}) \subset \mathcal{M}$$

<u>if and only if</u> $T' : (Y', \mathfrak{T}_{\mathcal{M}}) \longrightarrow (X', \mathfrak{T}_{\mathfrak{S}})$ <u>is continuous</u>.

(b) <u>Let</u> $\mathfrak{S}'$ (resp. $\mathcal{M}'$) <u>be topologizing family for</u> X (resp. Y) <u>which consists of weakly closed disks and contains all non–empty subsets of any element in</u> $\mathfrak{S}'$ (resp. $\mathcal{M}'$) (in particular, saturated families). <u>If</u>

$$(17.13.2) \qquad T'(\mathcal{M}') \subset \mathfrak{S}',$$

<u>then</u> $T : (X, \mathfrak{T}_{\mathfrak{S}'}) \longrightarrow (Y, \mathfrak{T}_{\mathcal{M}'})$ <u>is continuous</u>. <u>The converse is true provided that</u>

$$(17.13.3) \qquad \sigma(Y, Y') \leq \mathfrak{T}_{\mathcal{M}'} \leq \tau(Y, Y').$$

<u>Proof.</u> (a) The necessity follows from the following fact:

$$TA \subset B \Rightarrow B^{o} \subset (TA)^{o} = (T')^{-1}(A^{o}).$$

To prove the sufficiency, let $A \in \mathfrak{S}$ be weakly closed and disked. Then A^{o} is a $\mathfrak{T}_{\mathfrak{S}}$–neighbourhood of 0 in X', hence there exists a weakly closed disk $B \in \mathcal{M}$ such that $T'B^{o} \subset A^{o}$ (since $B^{o} \subset (T')^{-1}A^{o}$). Now the bipolar theorem and (17.12) (b) show that

$$A = A^{oo} \subset (T'B^{o})^{o} = T^{-1}(B^{oo}) = T^{-1}B,$$

hence $T(A) \subset B$ and, thus (17.13.1) holds.

(b) Suppose that (17.13.2) holds and that $D \in \mathcal{M}'$. Then there exists a $C \in \mathfrak{S}'$ such that $T'(D) \subset C$, hence by (17.12) (b),

$$C^{o} \subset (T'(D))^{o} = T^{-1}(D^{o}).$$

This proves that $T : (X, \mathfrak{T}_{\mathfrak{S}'}) \longrightarrow (Y, \mathfrak{T}_{\mathcal{M}'})$ is continuous.

Conversely, let $T : (X, \mathfrak{T}_{\mathfrak{S}'}) \longrightarrow (Y, \mathfrak{T}_{\mathcal{M}'})$ be continuous and let $D \in \mathcal{M}'$. Then D^o is a convex $\mathfrak{T}_{\mathcal{M}'}$–neighbourhood of o in Y, hence there exists a weakly closed disk $C \in \mathfrak{S}'$ such that $C^o \subset T^{-1}(D^o)$. By (17.13.3), D^o is a convex $\tau(Y, Y')$–neighbourhood of 0, it then follows from (17.12) (d) that

$$C = C^{oo} \supset (T^{-1}(D^o))^o = T'(D^{oo}) \supset T'(D),$$

and hence that (17.13.2) holds.

(17.14) **Proposition.** Let $X_{\mathscr{P}}$ and $Y_{\mathscr{T}}$ be LCS with topological duals X' and Y' resp., let $X_\tau = (X, \tau(X,X'))$ and $X'_\beta = (X', \beta(X',X))$. Then

(1) $$L(X,Y) \subset L(X_\sigma, Y_\sigma) \subset L(X_\tau, Y_\tau),$$

(hence $L(X_\sigma, Y_\sigma) = L(X_\tau, Y_\tau)$) and

(2) $$T' \in L(Y'_\beta, X'_\beta) \quad (\text{for all } T \in L(X_\sigma, X_\sigma)).$$

<u>Proof.</u> Let $T \in L(X,Y)$. To prove that $T \in L(X_\sigma, Y_\sigma)$, it suffices to show, by (17.11), that $T^*(Y') \subset X'$. Indeed, for any $y' \in Y'$ and any net $\{x_\lambda\}$ in $X_{\mathscr{P}}$ with $0 = \mathscr{P}\text{-}\lim_\lambda x_\lambda$, we have $0 = \mathscr{T}\text{-}\lim_\lambda Tx_\lambda$ (since $T \in L(X,Y)$), hence $0 = \sigma(Y,Y')\text{-}\lim_\lambda Tx_\lambda$, and thus

$$0 = \lim_\lambda \langle Tx_\lambda, y' \rangle = \lim_\lambda \langle x_\lambda, T^* y' \rangle,$$

that is $T^* y'$ a $\mathscr{P}$-continuous linear functional on X.

Let $S \in L(X_\sigma, Y_\sigma)$, and let W be a convex $\tau(Y, Y')$–neighbourhood of 0. Then (17.12) (d) shows that $(S^{-1}(W))^o = S'(W^o)$. As W^o is a $\sigma(Y', Y)$–compact disk in Y', it follows from $S' \in L(Y'_\sigma, X'_\sigma)$ (see(17.11)) that $S'(W^o)$ is a $\sigma(X', X)$–compact disk in X', and hence that $S^{-1}(W)$ is a $\tau(X, X')$–neighbourhood of 0. Thus, $S \in L(X_\tau, Y_\tau)$.

To prove (2), we first note from (17.6) (b) that

$$\mathcal{M}^X_{\text{von}}(\sigma) = \mathcal{M}^X_{\text{von}}(\mathcal{P}).$$

The weak continuity of T ensures that $T(\mathcal{M}^X_{\text{von}}(\sigma)) \subset \mathcal{M}^Y_{\text{von}}(\mathcal{T})$, so that $T' \in L(Y'_\beta, X'_\beta)$ (by (17.13) (a)).

(17.15) **Proposition.** Let $(X, \mathcal{P})$ be a LCS, let M be a vector subspace of X and $N = \overline{M}$. Then:

(a) $(M, \mathcal{P}_M)' \simeq X'/M^\circ$ (algebraically isomorphic).

(b) $(X/N, \hat{\mathcal{P}})' \simeq N^\circ$.

Proof. (a) The canonical embedding $J_M : (M, \mathcal{P}_M) \longrightarrow (X, \mathcal{P})$ is continuous, hence its dual map J'_M exists (by (17.11)). Also the Hahn–Banach extension theorem shows that J'_M is onto. By (17.12)(c), we have $\text{Ker } J'_M = M^\circ$, it then follows that the injection $\hat{J}'_M$ associated with J'_M is an algebraic isomorphism from X'/M° onto $(M, \mathcal{P}_M)'$.

(b) The continuity of the quotient map $(X, \mathcal{P}) \longrightarrow (X/N, \hat{\mathcal{P}})$ ensures that its dual map Q'_N exists (by (17.11)). As Q_N is onto, it follow from (17.12)(c) that Q'_N is one–one. Clearly $Q'_N((X/N, \hat{\mathcal{P}})') = N^\circ$, we conclude that Q'_N is an algebraic isomorphism from $(X/N, \hat{P})'$ onto N°.

(17.c) Let $(X, \mathcal{P})$ be a LCS, let M be a vector subspace of X and $N = \overline{M}$. Then:

(i) $\sigma(X,X')|_M = \sigma(M,X'/M^\circ)$.

(ii) $\sigma(X,X')/N = \sigma(X/N, N^\circ)$ and $\tau(X,X')/N = \tau(X/N, N^\circ)$.

(17.d) A characterization of relatively weakly open maps: Let $\langle X, X' \rangle$ and $\langle Y, Y' \rangle$ be dual pairs and $T \in L(X_\sigma, Y_\sigma)$. Then $T'(Y')$ is $\sigma(X', X)$–closed in X' if and only if T is an open map from $(X, \sigma(X, X'))$ onto the subspace T(X) of $(Y, \sigma(Y, Y'))$ (called relatively weakly open). In particular, $T'(Y') = X'$ if and only if T is one–one and relatively weakly open; dually, $T(X) = Y$ if and only if T' is one–one as well as relatively weakly open.

(17.e) Dieudonné–Schwartz's homomorphism theorem: Let $(X, \mathscr{P})$ and $(Y, \mathscr{T})$ be F–spaces and $T \in L(X, Y)$. Then the following statements are equivalent:

(i) T is a homomorphism for $\mathscr{P}$ and $\mathscr{T}$ in the following sense: T is an open map from $(X, \mathscr{P})$ onto the subspace T(X) of $(Y, \mathscr{T})$.

(ii) T(X) is $\mathscr{T}$–closed in Y.

(iii) $T'(Y')$ is $\sigma(X', X)$–closed in X'.

(iv) T is a homomorphism for $\sigma(X, X')$ and $\sigma(Y, Y')$ (called a weak homomorphism).

Exercises

17–1. Let $\langle X,Y \rangle$ be a dual pair and let $\mathfrak{S}$ be a family covering Y, consisting of $\sigma(Y,X)$–bounded subsets of Y.

(a) Show that $(X,\mathfrak{T}_{\mathfrak{S}})' = \langle \cup\{(B^{\circ})^{\pi}(X^*) : B \in \mathfrak{S}\} \rangle$ (the linear hull in X^*), where $(B^{\circ})^{\pi}(X^*)$ is the polar of B° w.r.t. $\langle X,X^* \rangle$.

(b) Denoted by $\mathscr{A}(\mathfrak{S})$ the saturated hull of $\mathfrak{S}$, and let $\mathfrak{S}_1$ be a family covering Y, consisting of $\sigma(Y,X)$–bounded subsets of Y such that $\mathscr{A}(\mathfrak{S}) \subsetneq \mathfrak{S}_1$. Show that

$$\text{the } \mathfrak{S}\text{–topology} = \text{the } \mathscr{A}(\mathfrak{S})\text{–topology} \lneq \text{the } \mathfrak{S}_1\text{–topology}.$$

17–2. Let $(X, \mathscr{P})$ be a LCS, let $\mathscr{P}^{\circ}$ be the topology on X' of precompact convergence and D a $\mathscr{P}$–equicontinuous subset of X'.

(a) For any $f_{\circ} \in D$ and any $\mathscr{P}$–precompact subset A of X, show that there exists a finite subset $\{x_1,\cdots,x_n\}$ of X such that the following implication holds:

$$\sup_{1\le i\le n} |g(x_i)-f_o(x_i)| \le 1 \Rightarrow \sup_{u\in A} |g(u)-f_o(u)| \le 2 \text{ (for all } g \in D).$$

(b) Deduce from (a) that $\sigma(X',X) = \mathscr{P}^o$ on D.

17–3. Prove (17.b).

17–4. Prove (17.c).

17–5. Prove (17.d).

17–6. Prove (17.e).

17–7. (Weak* closed convex sets). Let E be a B–space.

(a) Show, by an example, that $\sigma(E',E'')$–closed convex sets in E' need not be $\sigma(E',E)$–closed [Hint: consider $E=L^1[0,1]$, and use Lusin's theorem to show that $L^\infty[0,1] = E'$ is the $\sigma(E',E)$–closure of C[0,1]; notice that C[0,1] is a $\beta(E',E)$–closed vector subspace of $E'=L^\infty[0,1]$.]

(b) (Krein–Smulian's theorem). Let $D \subset E'$ be convex. If D is almost $\sigma(E',E)$–closed in the sense that $D \cap (rU_{E'})$ is $\sigma(E',E)$–closed in $rU_{E'}$ (for any $r > 0$), then D is $\sigma(E',E)$–closed.

17–8. (Weak*–bounded sets) Let E be a B–space and $D \subset E'$. Show that the following statements are equivalent.

(i) D is bounded in $(E',\|\cdot\|)$.

(ii) D is $\sigma(E',E'')$–bounded.

(iii) D is $\sigma(E',E)$–bounded.

(iv) D is equicontinuous.

(v) D is relatively $\sigma(E',E)$–compact.

17–9. Let E be a normed space. Show by an example, that relatively $\sigma(E',E)$–compact sets in E' need not be bounded in $(E',\|\cdot\|)$. (Thus the completeness of E is essential for 17–9) [Hint : Consider $E = \mathbb{R}^{(\mathbb{N})}$ equipped with the ℓ^1–norm $\|\cdot\|_1$, and let $f_j \in E'$ be such that $f_j([\zeta_n]) = \zeta_j$ (for all $[\zeta_n] \in E$). The set $D = \{0,2f_1,2^2f_2,\cdots,2^nf_n,\cdots\}$ has the required properties.]

17-10 (Krein–Smulian's theorem on closed convex hulls of weak compact sets). Let E be a B–space. The closed convex hull (or the closed disked hull) of a $\sigma(E,E')$–compact subset of E is $\sigma(E,E')$–compact.

17–11 (Compactness and sequential compactness) Let E be a B–space.

(a) Show, by an example, that $\sigma(E',E)$–compact sets in E' need not be $\sigma(E',E)$–sequentially compact. [Hint : Consider the unit ball $U_{(L^\infty)'}$ and $f_n \in U_{(\ell^\infty)'}$, defined by $f_n([\zeta_j]) = \zeta_n$ (for all $[\zeta_j] \in \ell^\infty$); and then show that any subsequence of $\{f_n\}$ does not $\sigma((\ell^\infty)',\ell^\infty)$–converge.]

(b) If E is assumed to be separable, then $\sigma(E',E)$–compact sets in E' must be $\sigma(E',E)$–sequentially compact.

(c) (Eberlin–Smulian's theorem). A subset K of E is relatively $\sigma(E,E')$–compact if and only if it is relatively $\sigma(E,E')$–sequentially compact (i.e. any sequence in K has a subsequence which is $\sigma(E,E')$–convergent in E).

17–12. Let E be a normed space, and let $\sigma(E,E')$ be metrizable. Suppose that $\{u_n'\}$ is a sequence in E' such that the family

$$\{x \in E : |\langle x,u_n'\rangle| \leq r\} \quad \text{(where } r > 0 \text{ is rational)}$$

forms a subbase at 0 for $\sigma(E,E')$.

(a) Using Baire's category theorem to show that there is some $k \geq 1$ such that $\mu O_{E'} \subset \langle\{u_1' \cdots u_k'\}\rangle$ for some $\mu > 0$.

(b) Deduce from part (a) that dim E' $< \infty$, so that dim E $< \infty$.

17–13. Let E be a separable B–space, let $\{x_n : n \geq 1\}$ be a dense subset of $\{x \in E : \|x\|=1\}=S_E$, let $\{u_n : n \geq 1\}$ be a dense subset of U_E and let $f_n \in E'$ be such that

$$\|f_n\| = 1 \quad \text{and} \quad \langle x_n,f_n\rangle = \|x_n\| = 1 \quad \text{(for all } n \geq 1\text{)}.$$

(a) Show that the map $T : E \longrightarrow \ell^\infty$, defined by

$$Tx = [\langle x,f_n\rangle]_{n\geq 1} \quad \text{(for all } x \in E\text{)},$$

is a metric injection; in other words, E is isometric to a subspace of ℓ^∞.

(b) Show that the map $S : \ell^1 \to E$, defined by

$$S([\zeta_n]) = \Sigma_{n=1}^{\infty} \zeta_n u_n \quad \text{(for all } [\zeta_n] \in \ell^{\infty}\text{)},$$

is a surjective, continuous linear map, hence deduce that E is topologically isomorphic to a quotient space of ℓ^1.

17–14 Let E be a normed space.

(a) E is separable if and only if $\sigma(E',E)\big|_{U_{E'}}$ is metrizable.

(b) E' is separable if and only if $\sigma(E,E')\big|_{U_E}$ is metrizable.

18

Semi-Reflexive Spaces and Ultra-Semi-Reflexive Spaces

Let $X_{\mathscr{P}}$ be a LCS with bidual X'' and ultra-bidual $X'^{\times}$ resp., and let $\beta(X',X)$ and $\beta_u(X',X)$ be the strong topology and ultra-strong topology on X' resp. It is known that

$$\tau(X',X) < \beta(X',X) < \beta_u(X',X) \text{ and } X'' \subset X'^{\times}.$$

As X' separates points on X, it follows that the map $x \longrightarrow \hat{x}$ $(x \in X)$, defined by

$$\hat{x}(f) = <x, f> \quad (\text{for all } f \in X'),$$

is an injective linear map from X into X'' (and surely into $X'^{\times}$), which is called the evaluation map, and denoted by K_X. As usual, X is identified with a vector subspace of X'', thus

$$X \subset X'' \subset X'^{\times} \quad \text{and} \quad \beta^{*}(X,X') = \beta(X'',X')|_X.$$

It is natural to ask under what conditions on X or X', X (or precisely $K_X(X)$) coincides with X'' (resp. with $X'^{\times}$). This leads to the following:

(18.1) **Definition.** A LCS $(X,\mathscr{P})$ is said to be :

(a) semi-reflexive if $X = X''$ (more precisely, $K_X X = X''$);

(b) reflexive if $X = X''$ and $\mathscr{P} = \beta(X'',X')$.

Remark. It should be noted that the notion of semi-reflexivity does not depend upon the original topology $\mathscr{P}$, but only on the dual pair $<X,X'>$.

As $X'' = (X',\beta(X',X))'$, it follows that X is semi-reflexive if and only if $\beta(X',X)$ is consistent with $<X,X'>$. Moreover, we have the following:

(18.2) **Theorem.** <u>For a LCS</u> $(X,\mathscr{P})$, <u>the following statements are equivalent.</u>

(a) X <u>is semi–reflexive</u>.

(b) $\beta(X',X) = \tau(X',X)$.

(c) <u>Every</u> $\mathscr{P}$<u>–bounded subset of</u> X <u>is relatively</u> $\sigma(X,X')$<u>–compact</u>.

(d) $(X,\sigma(X,X'))$ <u>is quasi–complete</u>.

(e) $(X',\tau(X',X))$ <u>is a barrelled space</u>.

<u>**Proof.**</u> It is clear that (a) and (b) are equivalent. As $\sigma(X,X')$–bounded subsets of X are $\sigma(X,X')$–totally bounded, and a subset of X is $\sigma(X,X')$–compact if and only if it is bounded and complete for $\sigma(X,X')$, it follows that (c) and (d) are equivalent. We complete the proof by showing that (b) ⇒ (e) ⇒ (c) ⇒ (b).

(b) ⇒ (e): Let D be a barrel in $(X',\tau(X',X))$. Then the polar D^{o} of D w.r.t. $\langle X,X'\rangle$ is $\sigma(X,X')$–bounded, hence $D = D^{oo}$ is a $\beta(X',X)$–neighbourhood of 0 (by the bipolar theorem), thus D is a $\tau(X,X')$–neighbourhood of 0 (by the hypothesis). Therefore $(X',\tau(X',X))$ is barrelled.

(e) ⇒ (c): Let B be a $\mathscr{P}$–bounded disk in X. Then the polar B^{o} of B w.r.t. $\langle X,X'\rangle$ is a barrel in $(X',\tau(X',X))$ [since $\tau(X',X)$ is consistent with $\langle X,X'\rangle$], hence B^{o} is a $\tau(X',X)$–neighbourhood of 0. Consequently, B^{oo} is $\sigma(X,X')$–compact (by Alaoglu–Bourbaki's theorem), and <u>a fortiori</u> B is relatively $\sigma(X,X')$–compact.

(c) ⇒ (b): The assumption (c), together with the definition of Mackey topology implies that $\beta(X',X) \leq \tau(X',X)$, so that they are equal [since we always have $\tau(X',X) \leq \beta(X',X)$].

(18.a)<u>Semi–reflexive spaces</u>: (i) Any closed subspace of a semi–reflexive space is semi–reflexive.

(ii) The quotient space of a semi–reflexive space by a closed subspace need not be semi–reflexive (see Köthe [1969,§31.5]).

(iii) The product and the locally convex direct sum of a family of semi–reflexive spaces are semi–reflexive.

(iv) If the strong dual $(X',\beta(X',X)))$ of a LCS $X_{\mathscr{P}}$ is semi–reflexive then $\tau(X,X') = \tau(X'',X')|_{X'}$

(v) A Mackey space $(X,\mathscr{P})$ is infrabarrelled provided that its strong dual $(X',\beta(X',X))$ is semi–reflexive. The converse is, in general, not true. (Consider the space ℓ^1.)

(18.3) **Theorem.** For a LCS $(X,\mathscr{P})$, the following statements are equivalent.

(a) $(X,\mathscr{P})$ is reflexive.

(b) $(X,\mathscr{P})$ is semi–reflexive and infrabarrelled.

(c) $(X,\mathscr{P})$ is semi–reflexive and barrelled.

(d) $\mathscr{P}= \tau(X,X')$, $(X,\tau(X,X'))$ and $(X',\tau(X',X))$ are barrelled spaces.

(e) $\mathscr{P}= \tau(X,X')$, $(X,\mathscr{P})$ and $(X',\beta(X',X))$ are semi–reflexive.

(f) $\mathscr{P}= \tau(X,X')$, and $(X',\tau(X',X))$ is reflexive.

Proof. (a) ⇒ (b) ⇒ (c): Trivial [since $\beta^*(X,X') = \beta(X'',X')|_X$ and $X = X''$].

(c) ⇒ (d) ⇒ (e): Follows from (18.2) and the fact that barrelled spaces are Mackey spaces.

(e) ⇒ (a) : By (18.2) (b), the semi–reflexivity of $(X',\beta(X',X))$ implies that $\tau(X'',X') = \beta(X'',X')$; while the semi–reflexivity of X implies that $X = X''$, thus X is semi–reflexive and $\tau(X,X') = \tau(X'',X') = \beta(X'',X') = \beta(X,X')$; consequently, $(X,\mathscr{P})$ is reflexive.

(f) ⇔ (d) : Follows from the following equivalent statements:

$(X' ,\tau(X',X))$ is reflexive

⇔ $(X' ,\tau(X',X))$ is semi–reflexive and barrelled

⇔ $(X,\tau(X,X'))$ and $(X' ,\tau(X',X))$ are barrelled sapces (by (18.2)(e)).

(18.4) **Corollary.** Every metrizable, semi–reflexive LCS (and surely semi–reflexive normed space) must be reflexive.

<u>Proof.</u> Metrizable LCS are bornological and surely infrabarrelled, the result then follows from (18.3).

The Banach dual of any normed space is complete, hence every reflexive normed space must be complete. But this is not true for reflexive LCS as stated by the following:

(18.b) Komura [1964] has given examples showing that reflexive LCS are not complete; this also shows that the strong dual of a barrelled space is, in general, not complete.

(18.5) **Corollay.** <u>For a Banach space</u> E, <u>the following statements are equivalent</u>.

(a) E <u>is reflexive</u>.

(b) $\sigma(E',E)|_{U_{E'}} = \sigma(E',E'')|_{U_{E'}}$

(c) E' <u>is reflexive</u>.

(d) <u>A convex subset of</u> E' <u>which is</u> $\sigma(E',E'')$–<u>closed must be</u> $\sigma(E',E)$–<u>closed.</u>

<u>Proof.</u> (a) ⇒ (b): Trivial since $E = E''$.

(b) ⇒ (c): By Alaoglu–Bourbaki's theorem, $U_{E'}$ is $\sigma(E',E)$–compact, hence $\sigma(E',E'')$–compact (by the hypothesis), thus E' is reflexive by (18.4) and (18.2).

(c) ⇒ (a): If E' is reflexive, then as we have already shown (see (b) ⇒ (c)) that E'' is reflexive. Since E (or precisely $K_E E$) is a <u>closed</u> vector subspace of E), it follows (see (18.a)) that E is reflexive.

(a) ⇒ (d): Trival.

(d) ⇒ (a): For any $u'' \in E''$, Ker u'' is a $\sigma(E',E'')$–closed vector subspace of E', hence Ker u'' is $\sigma(E',E)$–closed, thus u'' is a $\sigma(E',E)$–continuous linear functional on E. It then follows from (16.2) that there exists a $u \in E$ such that $u'' = K_E u$, and hence that E is semi–reflexive; consequently, E is reflexive.

(18.c) James' theorem: Let E be a Banach space. If E is reflexive, then the $\sigma(E,E')$–compactness of U_E shows that for any $f \in E'$ there is a $u_o \in U_E$ such that

$$\langle u_o,f\rangle = \|f\| = \sup_{x\in U_E} |\langle x,f\rangle|.$$

The converse, called James' theorem (if every $f \in E'$ attains its supremum on U_E then E is reflexive), is true as shown by James [1964].

It is well–known that every finite–dimensional Banach space is reflexive, and that there are infinite–dimensional Banach spaces which are, in general, not reflexive; however every point in E" is related to a point in E by means of finite–dimensional subspace of E' as shown by Helley's selection theorem. Actually we have a much more strong result as stated by the following important result, due to Lindenstrauss and Rosenthal [1969].

(18.6) **Therorem** (The principle of local reflexivity).Let F be a Banach space and let G be any finite–dimensional Banach space. For any operator $T \in L(G,F'')$, any finite–dimensional subspace N of F' and any $\epsilon > 0$, there exists an operator $S \in L(G,F)$ with $\|S\| \le \|T\| + \epsilon$ such that

$$\langle y',Tx\rangle = \langle y',K_F S(x)\rangle \text{ for all } x \in G \text{ and } y' \in N, \tag{18.6.1}$$

as well as

$$K_F S(x) = Tx \quad \text{for all } x \in G \text{ with } Tx \in K_F(F).$$

For a proof of this result, we refer to Lindenstrauss and Rosenthal [1969] or Pietsch [1980,p.384] or Johnson /Rosenthal/ Zippin [1971] or Deans [1973].

(18.d) (Junek [1983, p.87]). Let X be a semi–reflexive LCS. Then for any closed bounded disk B in X,

$$(X(B),\gamma_B) \equiv (X'_{B^0},\hat{p}_{B^0})'_\beta$$

[where γ_B is the gauge of B defined on $X(B)$, p_{B^0} is the gauge of the polar B^0 defined on X', $\hat{p}_{B^0}$ is the quotient norm defined on $X'_{B^0} = X'/p_B^{-1}(0)$ of p_{B^0} by $p_{B^0}^{-1}(0)$, $(X'_{B^0},\hat{p}_{B^0})'_\beta$ is the Banach dual of $(X'_{B^0},\hat{p}_{B^0})$ (see also (10.b)).]

(18.7) **Definition.** A LCS $(X, \mathscr{P})$ is said to be ultra–semi–reflexive (completely reflexive in the terminology of Hogbe–Nlend [1977]) if $X = X'^{\times}$.

A LCS $(X,\mathscr{P})$ is ultra–semi–reflexive if and only if the ultra–strong topology $\beta_u(X',X)$ on X' is consistent with $\langle X,X'\rangle$. Clearly, ultra–semi–reflixive spaces must be semi–reflexive, but the converse is, in general, not true (see Hogbe–Nlend [1977, p.135]). However, reflexive Banach spaces must be ultra semi– reflexive.

It is known (see (18.b)) that reflexive LCS are, in general, not complete; but the situation is completely different for ultra–semi–reflexive spaces as shown by the following:

(18.8) **Theorem.** (Schwartz, see Hogbe–Nlend [1977, p.89]). Every ultra–semi–reflexive LCS is complete.

Proof. Let $\{x_i, i \in D\}$ be a $\mathscr{P}$–Cauchy net in X. For any $f \in X'$, $\{ f(x_i), i \in D \}$ is a Cauchy net in **K**, hence converges to an element $\psi(f) \in$ **K**. This defines a linear functional $\psi : f \longrightarrow \psi(f)$ on X'. We show that $\psi \in X'^{\times}$ or, equivalently, ψ is bounded on any $\mathscr{P}$–equicontinuous subset of X'.

In fact, let V be a closed, disked $\mathscr{P}$–neighbourhood of 0. There exists an $i_0 \in D$ such that

$$x_i - x_j \in V \quad (\text{for all } i, j \geq i_0),$$

hence

$$\sup\{|f(x_i) - f(x_j)| : f \in V^0\} \leq 1 \quad (\text{for all } i, j \geq i_0).$$

Passing to the limit on j, we obtain

(1) $$\sup\{|f(x_i) - \psi(f)| : f \in V^0\} \leq 1 \quad (\text{for all } i \geq i_0).$$

As V is absorbing, it follows that V° is $\sigma(X', X)$–bounded, and hence that there exists an $\mu > 0$ such that

$$\sup\{|f(x_{i_0})| : f \in V^{\circ}\} \leq \mu ;$$

we conclude from (1) that ψ is bounded on V°. Thus $\psi \in X'^{\times}$.

Now the ultra–semi–reflexivity of X ensures that there exists a unique $u \in X$ such that $\psi = K_X u$. A standard argument shows that $u = \mathscr{P}\text{-}\lim_i x_i$. Hence $(X, \mathscr{P})$ is complete.

(18.e) The strong dual of an ultra–semi–reflexive LCS is fast–bornological.

Exercises

18–1. Prove (18.a).

18–2. Prove (18.d).

18–3. Prove (18.e).

18–4. Prove (18.f).

18–5. Let E be a reflexive B–space. Show that any $0 \neq f \in E'$ attains its supremum on U_E in the sense : there is an $u_o \in U_E$ such that $\langle u_o, f\rangle = \sup_{x \in U_E} |f(x)| = \|f\|$.

[The converse called James' theorem, is true, shown by James.]

18–6. (Montel spaces) A LCS $X_{\mathscr{P}}$ is called a <u>semi–Montel space</u> if every bounded subset of X is relatively $\mathscr{P}$–compact. Semi–Montel, infrabarrelled spaces are called <u>Montel spaces</u>.

(a) Show that a LCS $X_{\mathscr{P}}$ is a semi–Montel space if and only if $X_{\mathscr{P}}$ is quasi–complete and $\beta(X',X) = \mathscr{P}^{o}$ (the topology of uniform convergence on $\mathscr{P}$–precompact sets.).

(b) Show that the strong dual of a Montel space is a Montel space.

(c) Show that the strong dual of a semi–Montel, bornological space is a semi–Montel space.

(d) Show that subspaces of a semi–Montel space are semi–Montel spaces, and that the product and the locally convex direct sum space of a family of semi–Montel spaces are semi–Montel spaces.

(e) Suppose that $X_{\mathscr{P}}$ is a semi–Montel space. Show that $\mathscr{P}$ and $\sigma(X,X')$ coincide on each $\mathscr{P}$–bounded set in X; consequently, a sequence $\{x_n\}$ in X is $\mathscr{P}$–convergent if and only if it is $\sigma(X,X')$–convergent.

19

Some Recent Results on Compact Operators and Weakly Compact Operators

We start with the following:

(19.1) <u>Definition.</u> Let E and F be normed spaces and $T \in L(E,F)$. We say that T is:

(a) a <u>finite operator</u> (or an <u>operator of finite rank</u>) if

$$\dim T(E) < \infty;$$

(b) <u>compact</u> (resp. <u>precompact</u>) if $T(U_E)$ is relatively compact(resp. totally bounded) in F;

(c) <u>weakly compact</u> if $T(U_E)$ is relatively $\sigma(F, F')$–compact in F;

(d) <u>completely continuous</u> if every $\sigma(E, E')$–convergent sequence $\{x_n\}$ in E is mapped into a norm convergent sequence $\{Tx_n\}$ in F.

Denote by $\mathscr{F}(E,F)$ (resp. $\mathscr{L}^p(E,F)$, $\mathscr{L}^c(E,F)$, $\mathscr{L}^{wc}(E,F)$, and $\mathscr{L}^{cc}(E,F)$) the vector space of all finite operators (resp. precompact, compact, weakly compact and completely continuous operators) from E into F.

<u>Remarks:</u> (i) It is clear that

$$\mathscr{F}(E,F) \subset \mathscr{L}^c(E,F) \subset \mathscr{L}^p(E,F),$$
$$\mathscr{L}^c(E,F) \subset \mathscr{L}^{wc}(E,F) \text{ and } \mathscr{L}^c(E,F) \subset \mathscr{L}^{cc}(E,F).$$

Moreover, if F is complete then $\mathscr{L}^c(E,F) = \mathscr{L}^p(E,F)$.

(ii) As the unit ball U_E in E is a 0–neighbourhood as well as a bounded set, the notion of compact (resp. weakly compact) can be generalized to the case of LCS in two directions from the global (i.e., neighbourhoods) and local (i.e., bornologies) property as follows: Let $(X, \mathscr{P})$ and $(Y, \mathscr{T})$ be LCS and $T : X \longrightarrow Y$ a linear map. We say that

T is:

(a) precompact (resp. compact, weakly compact and bounded) if there exists a $\mathscr{P}$-neighbourhood V of 0 in X such that T(V) is totally bounded (resp. relatively $\mathscr{T}$-compact, relatively $\sigma(Y,Y')$-compact and bounded) in Y.

(b) locally precompact (resp. locally compact, locally weakly compact and locally bounded) if for any $\mathscr{P}$-bounded subset B of X, T(B) is precompact (resp. relatively $\mathscr{T}$-compact, relatively $\sigma(Y,Y')$-compact and bounded) in Y.

Denote by $\mathscr{L}_L^p(X,Y)$ (resp. $\mathscr{L}_L^c(X,Y)$, $\mathscr{L}_L^{wc}(X,Y)$, and $\mathscr{L}_L^b(X,Y)$) the vector space of all precompact (resp. compact, weakly compact and bounded) operators from X into Y, and by $L^{lp}(X,Y)$ (resp. $L^{lc}(X,Y)$, $L^{lwc}(X,Y)$, and $L^{lb}(X,Y)$ (or $\mathscr{L}^x(X,Y)$)) the vector space of all locally precompact (resp. locally weakly compact and locally bounded) operators from X into Y. Then we have :

$$\mathscr{L}_L^c(X,Y) \subset \mathscr{L}_L^p(X,Y) \cap \mathscr{L}_L^{wc}(X,Y) \subset \mathscr{L}_L^b(X,Y) \subset L(X,Y);$$

$$\mathscr{L}_L^b(X,Y) = L(E,F) \text{ whenever } Y \text{ is normable; and}$$

$$\mathscr{L}_L^\alpha(X,Y) \subset L^{l\alpha}(X,Y) \text{ whenever } \alpha \text{ represents c, p, wc and b.}$$

Moreover, the composite of two operators in which one of them belongs to $\mathscr{L}_L^\alpha$ is in $\mathscr{L}_L^\alpha$ (where α represents c, p, wc and b).

(19.a) Finite representation: Let E and F be normed spaces and $T \in L(E,F)$. Then T is a finite operator if and only if T is of the form

$$\text{(19.a.1)} \qquad Tx = \Sigma_{i=1}^n \langle x, u_i' \rangle y_i \text{ (for all } x \in E) \text{ with } u_i' \in E' \text{ and } y_i \in F,$$

thus we write $T = \Sigma_{i=1}^n u_i' \otimes y_i$. Of course, the above finite representation (19.a.1) is not unique, but it can be assumed that both sets $\{u_i' : i = 1,\cdots,n\}$ and $\{y_i : i = 1,\cdots,n\}$ are linearly independent sets of n elements. The number n is uniquely determined by T and called the rank of T (see Schaefer [1966, EX. (III 18) p.119]).

(19.b) Compact operators between Banach spaces: Let E and F be Banach spaces. Then $\mathscr{L}^c(E,F)$ is a closed vector subspace of $(L(E,F),\|\cdot\|)$. Moreover, the composite of operators in which one of them is compact must be compact.

Finite operators are compact operators, the converse is, in general, not true. However we have the following interesting result.

(19.2) **Proposition.** Let E and F be Banach spaces and $T \in L(E,F)$ a compact operator. If $T(E)$ is closed in F then T is a finite operator.

Proof. The closedness of $T(E)$ in F, together with the completeness of F, shows that T is an open map from E onto the Banach space $T(E)$ (by Banach's open mapping theorem), hence $T(U_E)$ is a relative 0–neighbourhood in $T(E)$ which is relatively compact (by the assumption on the compactness of T), thus $\dim T(E) < \infty$.

The first duality theorem for compact operators is the following:

(19.3) **Theorem (Schauder).** Let E and F be Banach spaces and $T \in L(E,F)$. Then T is compact if and only if $T' : F' \longrightarrow E'$ is compact.

Proof. **Necessity.** We shall employ Arzela–Ascoli's theorem to prove that $T'(U_{F'})$ is relatively sequentially compact in the Banach space E' or, equivalently, for any sequence $\{y_n'\}$ in $U_{F'}$, there exists a subsequence $\{y_{n_k}'\}$ of $\{y_n'\}$ and $u' \in E'$ such that

$$\lim_{k\to\infty} \|T' y_{n_k}' - u'\| = 0. \tag{1}$$

In fact, by assumption, $K = \overline{T(U_E)}$ is compact in F. On the other hand, $U_{F'} \subset (C(K),\|\cdot\|_\infty)$ has the following properties:

(i) $U_{F'}$ <u>is uniformly bounded on</u> K (i.e. $U_{F'}$ <u>is bounded in the Banach space</u> $(C(K),\|\cdot\|_\infty)$). Indeed, for any $x \in U_E$ with $z = Tx \in T(U_E)$,

$$|g(z)| \leq \|g\| \|z\| = \|g\| \|Tx\| \leq \|g\| \|T\| \|x\|$$
$$\leq \|T\| \quad (\text{for all } g \in U_{F'}),$$

hence

$$\sup_{g \in U_{F'}} \|g\|_\infty = \sup \{\sup_{z \in K} |g(z)| : g \in U_{F'}\} \leq \|T\| .$$

(ii) $U_{F'}$ <u>an equicontinuous subset of</u> $C(K)$: For any $g \in U_{F'}$ we have

$$|g(z_1) - g(z_2)| \leq \|g\| \|z_1 - z_2\|$$
$$\leq \|z_1 - z_2\| \quad (\text{for all } z_1, z_2 \in T(U_E)),$$

thus $U_{F'}$ is equicontinuous.

By Arzelá–Ascoli's theorem, $U_{F'}$ is relatively compact in $(C(K),\|\cdot\|_\infty)$, thus for any sequence $\{y'_n\}$ in $U_{F'}$, there exists a subsequence $\{y'_{n_k}\}$ of $\{y'_n\}$ such that

$$\sup_{x \in U_E} |y'_{n_k}(Tx) - y'_{n_j}(Tx)| \longrightarrow 0 \quad (\text{as } k, j \longrightarrow \infty).$$

It then follows that

$$\|T'y'_{n_k} - T'y'_{n_j}\| = \sup_{x \in U_E} \{|T'y'_{n_k}(x) - T'y'_{n_j}(x)|\}$$
$$= \sup_{x \in U_E} \{|y'_{n_k}(Tx) - y'_{n_j}(Tx)|\} \longrightarrow 0 \quad (\text{as } k,j \longrightarrow \infty);$$

in other words, $\{T'y'_{n_k}\}$ is a Cauchy sequence in the Banach dual E', thus there is an $u' \in E'$ such that $\lim_k \|T'y'_{n_k} - u'\| = 0$. Therefore T' is compact.

<u>Sufficiency.</u> We first observe that $K_E U_E \subset U_{E''}$, it then follows from

$K_F T = T''K_E$ that

$$K_F T(U_E) = T''K_E(U_E) \subset T''(U_{E''}).$$

Suppose now that T' is compact. Then $T'' : E'' \longrightarrow F''$ is compact (by the first half of the theorem), thus $T''(U_{E''})$ is totally bounded in F'' ; consequently, $K_F(TU_E)$ is totally bounded in F''. As $K_F : F \longrightarrow F''$ is a metric injection, we conclude that TU_E is totally bounded in the Banach space F, and hence that T is compact.

Remark. Let E and F be normed spaces and $T \in L(E,F)$. One can show that T is precompact if and only if $T' \in \mathscr{L}^c(F',E')$.

(19.c) A generalization of Schauder's theorem (see Köthe [1979, p.202]): Let X and Y be F–spaces and $T \in L(X,Y)$. Then T is locally compact if and only if $T' \in L(Y'_\beta, X'_\beta)$ is locally compact.

Using Schauder's duality theorem for compact operators and Grothendieck's structure theorem for compact sets in Banach spaces, we are able to verify the following remarkable result.

(19.4) **Theorem** (Terzioglu [1971],Randtke [1972]). Let E and F be Banach spaces and $T \in L(E,F)$. Then the following statements are equivalent:

(a) T is compact.

(b) There exists a null sequence $\{f_n\}$ in E' such that

(1) $$\|Tx\| \leq \sup_n |\langle x,f_n\rangle| \quad (\text{for all } x \in E).$$

(c) There exists an $[\lambda_n] \in c_0$ and an equicontinuous sequence $\{u_n'\}$ in E' such that

$$\|Tx\| \le \sup_n |\lambda_n <x, u_n'>| \quad (\underline{\text{for all}}\ x \in E).$$

(d) <u>There exists a closed subspace</u> H <u>of</u> c_0 , $R \in \mathscr{L}^c(E,H)$ and $S \in L(H,F)$ <u>such that</u> $T = SR$.

<u>Proof.</u> (a) ⇒ (b): By (19.3), $T' : F' \longrightarrow E'$ is compact, hence $T'(U_{F'})$ is relatively compact in E', thus Grothendieck's structure theorem for compact sets (see (5.2)) shows that there exists a null sequence $\{f_n\}$ in E' such that each element $u' \in T'(U_{F'})$ has a representation

$$u' = \Sigma_{n=1}^{\infty} \mu_n f_n \text{ with } \Sigma_n |\mu_n| \le 1.$$

For any $x \in E$, we have

$$\|Tx\| = \sup \{|<Tx, y'>| : y' \in U_{F'}\} = \sup \{|<x, T'y'>| : y' \in U_{F'}\}$$

$$\le \Sigma_{k=1}^{\infty} |\mu_k| \sup_n |<x, f_n>| \le \sup_n |<x, f_n>| .$$

(b) ⇔ (c): Trivial.

(b) ⇒ (d): Suppose that $\{f_n\}$ is a null sequence in E' such that (1) holds. Then $\{f_n\}$ is bounded, $\sigma(E',E)$–null sequence, hence the map $R : E \longrightarrow c_0$, defined by,

$$Rx = [<x, f_n>]_{n \ge 1} = \Sigma_n <x, f_n> e_n \quad (\text{for all } x \in E),$$

is an operator with $\|R\| = \sup_n \|f_n\|$ such that

(2) $$\|Tx\| \le \sup_n |<x,f_n>| = \|Rx\|_{\infty} \quad (\text{for all } x \in E)$$

[see(3.j) (ii)]. It is not hard to show that the set, defined by

$$K = \{[\zeta_n] \in c_0 : |\zeta_n| \le \|f_n\| \ (\text{for all } n \ge 1)\},$$

is compact in c_0 (since $\|f_n\| \longrightarrow 0$). As $R(U_E) \subset K$, it follows that $R \in \mathscr{L}^c(E,c_0)$.

On the other hand, (2) ensures that $\mathrm{Ker}\, R \subset \mathrm{Ker}\, T$, hence there is an operator $S : R(E) \longrightarrow F$ such that

$$(3) \qquad T = SR \text{ and } \|SR(x)\| = \|Tx\| \le \|Rx\|_\infty$$

(by (1) and(2)). Let H be the closure of $R(E)$ in c_0. Then $R : E \longrightarrow H$ is still compact. On the other hand, the completeness of F ensures that $S \in L(R(E), F)$ has a unique continuous extension to H, which is denoted again by S. By (3), we have $T = SR$.

(d) ⇒ (a): Trivial.

<u>Remark.</u> If E and F are assumed only normed spaces, one can show that T is precompact if and only if the statement (b) (or (c)) is true.

We say that an operator $T \in L(E,F)$ admits a <u>compact factorization through a Banach space</u> (resp. a <u>LCS</u>) Z if there are compact operators $R \in L(E,Z)$ and $S \in L(Z,Y)$ such that

$$T = SR.$$

As an immediate consequence of (19.4), we obtain the following:

(19.5) **Corollary** (Figiel [1973]). <u>Let</u> E <u>and</u> F be <u>Banach spaces</u>. <u>An operator</u> $T \in L(E,F)$ <u>admits a compact factorization through a Banach space if and only if it admits a compact factorization through a closed subspace of</u> c_0.

From the equivalence of (a) and (d) of (19.4), it is natural to ask in what cases a compact operator $T : E \longrightarrow F$ admits a compact factorization through the whole space c_0. The answer is given by Terzioglu [1972], as stated by the following:

(19.6) **Theorem** (Terzioglu [1972]). Let E and F be Banach spaces and $T \in L(E,F)$ a compact operator. Then T admits a compact factorization through the whole space c_0 if and only if T is an ∞–nuclear operator in the following sense: there exists an $[\zeta_n] \in c_0$, an equicontinuous sequence $\{f_n\}$ in E' and a summable sequence $\{y_n\}$ in F [for definition see (3.h)] such that

$$Tx = \sum_n \zeta_n \langle x, f_n \rangle y_n \quad \text{(for all } x \in E).$$

For a proof, see Terzioglu [1972] or Köthe [1979, p.277].

(19.d) Precompact operators between LCS (Randtke [1972, (2.10)] and Wong [1979, (1.1.2) and (1.1.6)]): Let X and Y be LCS, and $T \in L(X,Y)$. If there exist an $[\zeta_n] \in c_o$, an equicontinuous sequence $\{f_n\}$ in X' and a bounded disk B in Y such that

(d.1) $$r_B(Tx) \le \sup_n |\zeta_n \langle x, f_n \rangle| \quad \text{(for all } x \in X)$$

(where r_B is the gauge of B defined on Y(B)), then T is precompact.

The converse is true provided that Y is metrizable; in this case (i.e., Y is metrizable), it is equivalent to the following:

(*) There exist normed spaces E and F, a precompact operator $\bar{T} \in L(E,F)$ and operators $Q \in L(X,E)$ and $J \in L(E,Y)$ such that $T = J\bar{T}Q$.

In order to extend the results of (19.4) and (19.6), we require the following terminology and results.

(19.7) **Definition.** Let X and Y be LCS. A seminorm p on X is said to be precompact if there exists an $[\zeta_n] \in c_0$ and an equicontinuous sequence $\{f_n\}$ in X' such that

$$p(x) \le \sup_n |\zeta_n \langle x, f_n \rangle| \quad \text{(for all } x \in X).$$

We say that an operator $T \in L(E,F)$ is:

(a) a quasi–Schwartz (or precompact–bounded in the terminology of Wong [1979, p.24]) if there exists a precompact seminorm p on X such that $\{Tx : p(x) \leq 1\}$ is bounded in Y;

(b) b–precompact if there exists a 0–neighbourhood V in X such that T(V) is b–precompact in Y (for definition, see (15.2)(a)).

We denote by $\mathscr{L}_L^{pb}(X, Y)$ the vector space of all b–precompact operators form X into Y.

Remarks:(i) The notions of precompact seminorms and quasi–Schwartz operators are due to Randtke [1972], while the concept of b–precompact operators is due to Wong [1982 a].

(ii) It is clear that every precompact seminorm on a LCS must be continuous.

(iii) The composite of two operators in which one of them is quasi–Schwartz (resp. b–precompact) is quasi–Schwartz (resp. b–precompact). Moreover, $\mathscr{L}_L^{pb}(X,Y) \subset \mathscr{L}_L^{p}(X,Y)$, and they are equal when Y is metrizable [see (11.2) and (15.2) (a)]. Thus the notion of b–precompact operators is a natural generalization of that of precompact operators. We shall show (below (19.8)) that b–precompact operators are exactly quasi–Schwartz operators.

(19.e) Characterization of precompact seminorms (Randtke [1972, (2.4)]): Let $(X, \mathscr{P})$ be a LCS, let p be a continuous seminorm on X, and let X_p be the quotient space $X/p^{-1}(0)$ equipped with the quotient norm p of p by $p^{-1}(0)$. Then the following statements are equivalent.

(i) p is precompact.

(ii) $Q_p : (X, \mathscr{P}) \longrightarrow X_p$ is a precompact operator.

(iii) There exists a continuous seminorm q on X with $p \leq q$ such that the canonical map $Q_{p,q} : X_q \longrightarrow X_p$ is precompact.

(19.8) **Theorem** (Randtke [1972], Wong [1982a]). Let X and Y be LCS and $T \in L(X,Y)$. Then the following statements are equivalent.

(a) T is quasi–Schwartz.

(b) There exist normed spaces E and F, a precompact operator $\tilde{T} \in L(E,F)$ and operators $Q \in L(X,E)$ and $J \in L(F, Y)$ such that

$$T = J\tilde{T}Q.$$

(c) T is a b–precompact operator.

(d) T admits a quasi–Schwartz factorization through a vector subspace H of c_0 in the following sense: there exists quasi–Schwartz operators $T_1 \in L(X, H)$ and $T_2 \in L(H, Y)$ such that

$$T = T_2 T_1 .$$

Proof. (a) ⇒ (b): Let p be a precompact seminorm on X such that $B = \{Tx : p(x) \leq 1\}$ is a bounded disk in Y and $V = \{x \in X: p(x) < 1\}$. Then $p^{-1}(0) \subset \operatorname{Ker} T$ [by the boundedness of B], hence there is an $S \in L(X_p, Y(B))$ (in fact, $\|S\| \leq 1$ by $T(V) \subset B$) such that $J_B S Q_p = T$. Now the precompactness of p ensures [see (19.e)] that there is a continuous seminorm q on X with $p \leq q$ such that $Q_{p,q} : X_q \longrightarrow X_p$ is precompact. Therefore we have

$$T = J_B S Q_{p,q} Q_q \text{ with } \tilde{T} = S Q_{p,q} \in \mathscr{L}^p(X_q, Y(B)),$$

hence the implication follows.

(b) ⇒ (c): By (11.2), every precompact set in a metrizable LCS $(Z, \mathfrak{T})$ must be b–precompact in $(Z, \mathfrak{T})$, the implication follows from Remark (iii) of (19.7).

(c) ⇒ (d): There exists a disked 0–neighbourhood V in X and a bounded disk B in Y such that $T(V)$ is precompact in the normed space $Y(B) = (Y(B), \gamma_B)$. We claim that the precompact operator $T : X \longrightarrow (Y(B), \gamma_B)$ admits a precompact factorization through a vector subspace of c_0.

In fact, by (d.1) of (19.d), there exists an $[\zeta_n] \in c_0$ and an equicontinuous sequence $\{f_n\}$ in X' such that

(1) $$\gamma_B(Tx) \leq \sup_n \{|\zeta_n^2 \langle x, f_n \rangle|\} \quad (\text{for all } x \in X).$$

Now we define a map $T_1 : X \longrightarrow c_0$ by setting

(2) $$T_1 x = [\zeta_n \langle x, f_n \rangle]_{n \geq 1} \quad (\text{for all } x \in X),$$

and let $H = T_1 X$. Then T_1 is precompact operator from X into a vector subspace H of c_0 [see (3.j)(ii)]. The diagonal translation $D : c_0 \longrightarrow c_0$, defined by

(3) $$D([\eta_n]) = [\zeta_n \eta_n]_{n \geq 1} \quad (\text{for all } [\eta_n] \in c_0),$$

is a compact operator such that $\operatorname{Ker}(D \circ T_1) \subset \operatorname{Ker} T$ (by (1)), hence there exists a linear map $S : D(H) \longrightarrow Y(B)$ such that $T = SDT_1$. For any $x \in X$, we have from (2) and (3) that $(DT_1)x = [\zeta_n^2 \langle x, f_n \rangle]_{n \geq 1}$, it then follows from (1) that

$$\gamma_B((SDT_1)x) = \gamma_B(Tx) \leq \sup_n \{|\zeta_n^2 \langle x, f_n \rangle|\} = \|DT_1 x\|_\infty,$$

and hence that S is continuous. Consequently, $SD : H \longrightarrow Y(B)$ is a precompact operator. This proves our assertion.

Finally, since H and $Y(B)$ are normed spaces, it follows that $T_1 : X \longrightarrow H$ and $SD : H \longrightarrow Y(B)$ are quasi–Schwartz operators, and hence that $T_2 = J_B SD : H \longrightarrow Y$ is a quasi–Schwartz operator such that

$$T = SDT_1 = (J_B SD)T_1 = T_2 T_1,$$

thus T is the composite of two quasi–Schwartz operators T_1 and T_2.

(d) ⇒ (a): Trivial.

The equivalence of (a) and (b) is due to Randtke [1972.(2.1)], while the equivalences between (a) , (c) and (d) are due to Wong [1982a,Theorems 2 and 5].

It is clear that b–precompact sets in a LCS Y are precompact, but the converse is, in general, not true (except for the metrizable case of Y [see (11.2) and (15.2) (a)]). Thus the notion of quasi–Schwartz operators is a natural generalization of that of precompact operators; consequently, the preceding result is a generalization of (19.4).

From (19.8), it is natural to ask in what case a quasi–Schwartz operator has a quasi–Schwartz factorization through the whole space c_0 . To answer this question, we require the following:

(19.9) **Definition (Randtke[1972]).** A sequence $\{y_n\}$ in a LCS Y is said to be strongly summable if it satisfies the following two conditions:

(i) for any $[\lambda_n] \in \ell^\infty$ the series $\sum_n \lambda_n y_n$ converges in Y;

(ii) $\{ \sum_n \lambda_n y_n : \|[\lambda_n]\|_\infty \leq 1, [\lambda_n] \in \ell^\infty \}$ is bounded in Y.

Let X be a LCS. An operator $T \in L(X,Y)$ is called a Schwartz operator if there exists an $[\xi_n] \in c_0$, an equicontinuous sequence $\{f_n\}$ in X' and a strongly summable sequence $\{y_n\}$ in Y such that

$$Tx = \sum_n \zeta_n \langle x,f_n \rangle y_n \quad (\text{for all } x \in X).$$

Remark. It can be shown (see Wong [1980b, Lemma 1]) that if Y is a sequentially complete LCS and $\{z_n\}$ is a weakly summable sequence in Y, then for any $[\lambda_n] \in c_0$, the sequence $\{\lambda_n z_n\}$ in Y must be strongly summable. Thus the notion of Schwartz operators is a natural extension of that of ∞–nuclear operators (see (19.6)).

(19.10) **Theorem** (Wong [1980b]). Let X and Y be LCS and $T \in L(X,Y)$ a quasi–Schwartz operator. Then T is a Schwartz operator if and only if T admits a quasi–Schwartz factorization through the whole space c_0. (Consequently, Schwartz operators must be compact.)

For a proof of this result, we refer to Wong [1980b, Theorem 3].

Combining (19.8),we see that the preceding result is a generalization of (19.6).

We conclude this section with some criteria for weak compactness of operators.

(19.11) **Theorem.** Let E and F be Banach spaces and $T \in L(E,F)$. Then the following statements are equivalent.

(a) T is weakly compact.

(b) $T' : (F',\sigma(F',F)) \longrightarrow (E',\sigma(E',E''))$ is continuous.

(c) $T''(E'') \subset K_F(F)$.

Proof: (b) $\Leftrightarrow$ (c): Follows (16.4) and (17.11).

(a) $\Rightarrow$ (c): We first show that if $B \subset F$ is relatively $\sigma(F,F')$–compact, then

(1) $$\operatorname{clu}_{\sigma(F'',F')} K_F(B) \subset K_F(F),$$

where $\operatorname{clu}_{\sigma(F'',F')} K_F(B)$ denotes the $\sigma(F'',F')$–closure of $K_F(B)$.

In fact, as the $\sigma(F,F')$–closure of B, denoted by $\overline{B}^{\sigma}$, is $\sigma(F,F')$–compact and $K_F : (F,\sigma(F,F')) \longrightarrow (F'',\sigma(F'',F'))$ is continuous, it follows that $K_F(\overline{B}^{\sigma})(\subset K_F(F))$ is $\sigma(F'',F')$–compact in F'', and surely $\sigma(F'',F')$–closed; consequently,

$$\operatorname{clu}_{\sigma(F'',F')} K_F(B) \subset \operatorname{clu}_{\sigma(F'',F')} K_F(\overline{B}^{\sigma}) = K_F(\overline{B}^{\sigma}) \subset K_F(F),$$

which obtains our assertion (1).

Now the weak compactness of T implies that $T(U_E)$ is relatively $\sigma(F,F')$–compact, hence (1) and $T''K_E = K_F T$ imply that

(2) $$\mathrm{clu}_{\sigma(F'',F')}(T''K_E)U_E = \mathrm{clu}_{\sigma(F'',F')}K_F(T(U_E)) \subset K_F(F).$$

Notice that $U_{E''} = (U_E^{\circ})^{\circ}(E'') = \mathrm{clu}_{\sigma(E'',E')}K_E(U_E)$ and that

$$T'' : (E'',\sigma(E'',E')) \longrightarrow (F'',\sigma(F'',F')) \text{ is continuous,}$$

[this is equivalent to say that $T''(\mathrm{clu}_{\sigma(E'',E')}A) \subset \mathrm{clu}_{\sigma(F'',F')}T''A$ (for any $A \subset E''$)], we then conclude from (2) that

$$T''U_{E''} \subset \mathrm{clu}_{\sigma(F'',F')}T''(K_E(U_E)) \subset K_F(F),$$

and hence from $E'' = \bigcup_n nU_{E''}$ that $T''E'' \subset K_F(F)$.

(c) ⇒ (a): By (16.4), $T'' : (E'',\sigma(E'',E')) \longrightarrow (F,\sigma(F,F'))$ is continuous, hence $T''(U_{E''})$ is $\sigma(F,F')$–compact (since $U_{E''}$ is $\sigma(E'',E')$–compact (by Alaoglu–Bourbaki's theorem)). On the other hand, since $\sigma(F,F') = \sigma(F'',F')|_F$ and

$$K_F(TU_E) = T''K_E(U_E) \subset T''U_{E''} \subset K_F(F),$$

we conclude that $T(U_E)$ is relatively $\sigma(F,F')$–compact.

The equivalence of (a) and (c) of (19.11) is due to Gantmacher and Nakamura.

Let E and F be Banach spaces and $T \in L(E,F)$. If either E or F is reflexive, then T must be weakly compact. On the other hand,the composite of two operators in which one of them is weakly compact must be weakly compact; consequently, if $T \in L(E,F)$ admits a factorization through a reflexive Banach space, then T must be weakly compact. The converse is true as shown by the following important result.

(19.12) **Theorem** (Davis/Figiel/Johnson/Pelczynski: [1974]). Let E and F be Banach spaces and $T \in L(E,F)$. Then T is weakly compact if and only if T admits a factorization through a reflexive Banach space; in other words, there exists a reflexive Banach space G and operators $S \in L(G,F)$ and $R \in L(E,G)$ such that $T = SR$.

For a proof, see Davis/ Figiel/ Johnson/Pelczynski [1974]) or Pietsch [1980, pp.55–57].

(19.f) Grothendieck's characterization for locally weakly compact operators (see Köthe [1979, p.204]): Let X and Y be LCS and $T \in L(X, Y)$. Then the following two statements are equivalent.

(i) T is locally weakly compact.

(ii) $T''X'' \subset K_Y(Y)$.

Moreover, (i) (or (ii)) implies the following;

(iii) T' sends equicontinuous subsets of Y' into relatively $\sigma(X',X'')$–compact subsets of X'.

If Y is complete, then (iii) implies (i).

(It should be noted that (19.f) is actually a generalization of (19.11).)

(19.g) A characterization for completely continuous operators (see Pietsch [1980 ,p.61]): Let E and F be Banach spaces and $T \in L(E,F)$. Then T is completely continuous if and only if for any Banach space E_0 and any weakly compact operator $R : E_0 \longrightarrow E$ the operator $TR : E_0 \longrightarrow F$ is compact.

(19.h) Criteria for compactness, weak compactness and complete continuity of operators with ℓ^1 or c_0 or ℓ^∞ as domains: In the following, F is always assumed to be a Banach space.

(A) Let $T \in L(\ell^1,F)$ and $y_n = Te_n$ for all $n \geq 1$. (Hence $\{y_n\}$ is a bounded sequence in F such that $T[\zeta_n] = \Sigma_{n=1}^{\infty} \zeta_n y_n$ (for all $[\zeta_n] \in \ell^1$) and $\|T\| = \sup_n \|y_n\|$ [see (3.j) (i)']. Consider the following statements.

(i) $\{y_n\}$ is a $\sigma(F, F')$–null sequence.

(ii) $T : (\ell^1, \sigma(\ell^1, c_0)) \longrightarrow (F, \sigma(F, F'))$ is continuous.

(iii) T is weakly compact.

Then (i) $\Leftrightarrow$ (ii) $\Rightarrow$ (iii).

(The implication (i) $\Rightarrow$ (ii) is essentially contained in Pietsch [1980, (3.2.3)].)

(B) Let $T \in L(c_0,F)$ and $y_n = Te_n$ for all $n \geq 1$. [Hence $\{y_n\}$ is a weakly summable sequence in F such that $T([\zeta_n]) = \Sigma_n \zeta_n y_n$ and $\|T\| = \sup\{\Sigma_n |<y_n,g>| : g \in U_{F'}\}$ by (3.j)(ii)′.] Then the following statements are equivalent.

(i) $\{y_n\}$ is summable in Y.

(ii) T is weakly compact.

(iii) T is compact.

(iv) T is completely continuous.

(The equivalence of (i) and (iii) is due to Randtke [1973] (also see Dazord [1976, p.109])).

(C)(Traves [1967, pp.454–457]). Let $T \in L(\ell^\infty,F)$ and $y_n = Te_n$ for all $n \geq 1$. Then the following statements are equivalent.

(i) $\{y_n\}$ is a summable sequence in F.

(ii) $T : (\ell^\infty,\sigma(\ell^\infty,\ell^1)) \longrightarrow (F,\sigma(F,F'))$ is continuous.

(iii) $T' : (F',\sigma(F',F)) \longrightarrow (\ell^1,\sigma(\ell^1,\ell^\infty))$ is continuous.

(iv) T is compact.

(19.j) <u>Criteria for compactness, weak compactness and complete continuity of operators with ranges contained in ℓ^1 or c_0 or ℓ^∞</u>: In the following, E will be always assumed to be a Banach space.

(A) (Randtke [1974]).Let $S \in L(E,\ell^1)$ and $f_n = S'e_n$ (for all $n \geq 1$).[Hence $\{f_n\}$ is weak* summable in E' for which $Sx = \Sigma_n <x,f_n>e_n$ and $\|S\| = \sup\{\Sigma_n |<x,f_n>| : x \in U_E\}$ (see (3.j) (i))]. Then the following statements are equivalent:

(a) $\{f_n\}$ is summable in E'.

(b) S is weakly compact.

(c) S is compact.

(B) (Pietsch [1980,p.61]and Dazord [1976,p.108]).Let $S \in L(E,c_0)$ and $f_n = S'e_n$ (for all $n \geq 1$). [Hence $\{f_n\}$ is a $\sigma(E',E)$–null sequence in E' such that $S(x) = \Sigma_n \langle x,f_n \rangle e_n$ and $\|S\| = \sup_n \|f_n\|$ (see (3.j)(ii)).] Consider the following statements:

(Wi) $\{f_n\}$ is a $\sigma(E',E'')$–null sequence in E'.

(Wii) S is weakly compact.

(Ci) $\{f_n\}$ is a null sequence in the Banach dual E'.

(Cii) S is compact.

Then (Wi) $\Leftrightarrow$ (Wii) and (Ci) $\Leftrightarrow$ (Cii).

As an application of (19.j)(B) and (19.k)(B), we obtain immediately (19.6).

Exercises

19–1. Prove (19.a).

19–2. Prove (19.b).

19–3. Prove (19.c).

19–4. Let E and F be B–space and $T \in L(E,F)$.

(a) Let M be a closed subspace of E and assume that $TJ_M^E : M \rightarrow F$ is a topological injection. Show that the followiing statements are equivalent:

(i) TJ_M^E is a finite operator.

(ii) TJ_M^E is compact.

(iii) $\dim TJ_M^E(M) < \infty$.

(iv) $\dim M < \infty$.

(b) Let N be a closed subspace of F and assume that $Q_N^F : E \rightarrow {}^F/_N$ is open. Show that the following statements are equivalent:

(i) $Q_N^F T$ is a finite operator.

(ii) $Q_N^F T$ is compact.

(iii) $\dim {}^F/_N < \infty$.

19–5. Prove (19.d).

19–6. Prove (19.e).

19–7. Let E and F be normed space and $T \in L(E,F)$.

(a) Show that T is compact if and only if for any bounded sequence $\{x_n\}$ is E, the sequence $\{Tx_n\}$ in Y contains a convergent subsequence.

(b) Using part (a) (or otherwise) to show that compact operators are completely continuous operators.

19–8. (Criterion of b–compactness) Let Y be a LCS. Show that the following statements are equivalent:

(a) Every compact disk in Y is b–compact.

(b) For any LCS X, it follows from $T \in \mathscr{L}^c(X,Y)$ that $T' : Y'_\beta \to X'_\beta$ is compact.

(c) The same as (b) but X being a B–space.

19–9. Let Y be a sequentially complete LCS and $\{y_n\}$ a weakly summable sequence in Y. Show that for any $[\lambda_n] \in c_o$, the sequence $\{\lambda_n y_n\}$ is strongly summable (Wong [1980b,lemma 1]).

19–10. Prove (19.f).

19–11. Prove (19.g).

19–12. Prove (19.h).

19–13. Prove (19.j).

19–14. Show that the following two statements are equivalent for a Banach space E :

(i) Every $\sigma(E',E)$–null sequence in E' is $\sigma(E',E'')$–null.

(ii) Every $S \in L(E,c_o)$ is weakly compact.

[Hint: Use (3.j)(ii) and (19.j)(B).](Raebiger [1984/85]).

19–15. Let E and F be B–spaces and $T \in L(E,F)$.

(a) Suppose that T is weakly compact. Then $\operatorname{Im} T$ is reflexive if and only if $\operatorname{Im} T$ is closed in F. In particular, if T has a bounded inverse, then E is reflexive (compare with (19.2)). [Hint : Use (17.c).]

(b) Suppose that T has a bounded inverse. Show that T is weakly compact if and only if E is reflexive.

(c) Let M be a closed subspace of E and assume that $TJ_M^E : M \to \operatorname{Im} T$ is a topological injection. Show that the following statements are equivalent:

(i) TJ_M^E is weakly compact.

(ii) TM is reflexive.

(iii) M is reflexive.

(Compare with 19–4(a)).

(d) Let N be a closed subspace of F and assume that $Q_N^F T : E \to {}^F/_N$ is open. Show that $Q_N^F T$ is weakly compact if and only if ${}^F/_N$ is reflexive. (Compare with 19–4 (b)).

20

Precompact Seminorms and Schwartz Spaces

It is known (see (17.b) and (12.3)) that a LCS $(X, \mathscr{P})$ is barrelled (resp. infrabarrelled, bornological) if and only if each lower semi–continuous seminorm on X (resp. lower semi–continuous seminorm on X which is bounded on bounded sets in X, seminorm on X which is bounded on bounded sets in X) is $\mathscr{P}$–continuous. This section continues this idea of using seminorms satisfying some expected properties to investigate Schwartz spaces. This class of locally convex spaces can be regarded as the 'best spaces in Analysis' since Schwartz spaces are closer to finite–dimensional spaces than Banach spaces are [bounded sets are precompact] on one hand, as well as are closer to Banach spaces than other classes of spaces are [complete Schwartz spaces are ultra–semi–reflexive].

Recall that a seminorm p on a LCS $(X, \mathscr{P})$ is <u>precompact</u> if there exists an $[\zeta_n] \in c_0$ and an equicontinuous sequence $\{u_n'\}$ in X' such that

$$p(x) \leq \sup_{n\geq 1} |\zeta_n <x,u_n'>| \quad \text{(for all } x \in X).$$

Precompact seminorms in $(X, \mathscr{P})$ are continuous, but not conversely. Hence a LCS $(X, \mathscr{P})$ is called a <u>Schwartz space</u> (by Randtke [1972]) if every $\mathscr{P}$–continuous seminorm on X is precompact. $(X, \mathscr{P})$ is called a <u>co–Schwartz space</u> if its strong dual X_β' is a Schwartz space.

In terms of criteria for precompact seminorms, we are able to present the following :

(20.1) **Theorem** (Grothendieck, Terzioglu [1969], Randtke [1972]). <u>The following statements are equivalent for a LCS</u> $(X, \mathscr{P})$:

(a) $(X, \mathscr{P})$ <u>is a Schwartz space</u>.

(b) $Q_p \in \mathscr{L}_L^P(X, X_p)$ <u>for any continuous seminorm</u> p <u>on</u> X.

(c) <u>For any continuous seminorm</u> p <u>on</u> X <u>there is a continuous seminorm</u> q <u>on</u> X <u>with</u> $p \leq q$ <u>such that</u> $Q_{pq} \in \mathscr{L}_L^P(X_q, X_p)$.

(d) <u>For any continuous seminorm</u> p <u>on</u> X <u>there is a continuous seminorm</u> q <u>on</u> X <u>with</u> $p \leq q$ <u>such that the canonical embedding</u> $X'(V_p^0) \longrightarrow X'(X_q^0)$ <u>is compact</u>, <u>where</u> $V_p = \{x \in X : p(x) \leq 1\}$.

(e) $\mathscr{L}(X,F) = \mathscr{L}_L{}^P(X,F)$ <u>for any normed</u> <u>(or Banach)</u> <u>space</u> F.

(f) (i) <u>Bounded subsets of</u> X <u>are precompact</u>, <u>and</u>

(ii) $(X,\mathscr{P})$ <u>is a quasi–normable space</u> (by Grothendieck [1973,p.176]) <u>in the following sense</u> : for any $V \in \mathscr{U}_X$ there is an $W \in \mathscr{U}_X$ such that for any $\lambda > 0$ it is possible to find a bounded set B in X with $W \subset B + \lambda V$.

Proof. (a) ⇔ (b) : As $p = \hat{p} \circ Q_p$, the equivalence follows from the definition of precompact seminorms and (19.d) (or (19.e)).

(b) ⇒ (c) : Let $U \in \mathscr{U}_X$ be such that $Q_p(U)$ is precompact in the normed space X_p. Then $\text{Ker}\ Q_U \subset \text{Ker}\ Q_p$, hence there exists a unique map $\hat{Q}_p \in L(X_U,X_p)$ such that $Q_p = \hat{Q}_p Q_U$ ($\hat{Q}_p$ is called the map obtained from Q_p by passing to the quotient). Since $\hat{p} \circ Q_p = p$ and $Q_p \in L^P(X,X_p)$, it follows that $\hat{Q}_p$ is a precompact operator. Now, let q be the gauge of $W = U \cap V_p$ (where $V_p = \{x \in X : p(x) \leq 1\}$). Then q is a continuous seminorm on X such that $p \leq q$. As $W \subset U$ we have $Q_{U,W} \in L(X_W,X_U)$, so that $Q_{p,q} = \hat{Q}_p Q_{U,W}$ is precompact.

(c) ⇒ (b) : Trivial.

(c) ⇔ (d) : Follows from Schauder's duality theorem (19.3).

(b) ⇒ (e) : Let $T \in L(X,F)$ and p the gauge of $T^{-1}(U_F)$. Then p is a continuous seminorm on X such that

$$\text{Ker}\ Q_p = p^{-1}(0) \subset \text{Ker}\ T$$

[by the boundedness of U_F], hence there is an $\tilde{T} \in L(X_p,F)$ such that $T = \tilde{T} Q_p$, and thus $T \in \mathscr{L}_L{}^P(X,F)$ [by the hypothesis].

(e) ⇒ (b) : Obvious.

(b) ⇒ (f) : For any bounded disk A in X and any $V \in \mathscr{U}_X$, the canonical map $K_{V,A} : X(A) \longrightarrow X_V$ is precompact [since $K_{V,A} = Q_V \circ J_A$], hence A must be precompact.

To prove (ii), let $W \in \mathscr{U}_X$ be such that $Q_V(W)$ is precompact in the normed space X_V. As $Q_V(V)$ is the unit ball in X_V, for any $\lambda > 0$ there exists a finite subset $\{x_1, \cdots, x_m\}$ of W such that $Q_V(W) \subset \bigcup_{j=1}^{m}(Q_V(x_j) + \frac{\lambda}{2}Q_V(V))$, hence

$$W \subset \bigcup_{j=1}^{m}(x_j + \frac{\lambda}{2}V) + p_V^{-1}(0) \subset \bigcup_{j=1}^{m}(x_j + \lambda V)$$

[since $p_V^{-1}(0) \subset \delta V$ for any $\delta > 0$].

(f) ⇒ (b) : For any $V \in \mathscr{U}_X$, by (ii) there is a $W \in \mathscr{U}_X$ such that for any $\lambda > 0$ it is possible to find a bounded subset B of X with $W \subset B + \frac{\lambda}{2}V$. As B is precompact [by (i)], there is a finite subset $\{x_1, \cdots, x_m\}$ of X such that $B \subset \bigcup_{j=1}^{m}(x_j + \frac{\lambda}{2}V)$, hence $W \subset \bigcup_{j=1}^{m}(x_j + \lambda V)$, in other words, $Q_V : X \longrightarrow X_V$ is precompact. This completes the proof.

As $\{B^0 : B \in \mathscr{B}_{von}^X\}$ is a local base at 0 for $\beta(X',X)$, it follows from (20.1)(c) that X is a co–Schwartz space if and only if for any $B \in \mathscr{B}_{von}^X$ there is an $A \in \mathscr{B}_{von}^X$ with $A^0 \subset B^0$ such that

(20.A) $$Q_{B^0,A^0} : X'_{A^0} \longrightarrow X'_{B^0} \text{ is precompact.}$$

Moreover, we have the following :

(20.2) **Theorem** (Terzioglu [1969], Randtke [1972]). For a LCS X the following statements are equivalent :

(a) X is a co–Schwartz space.

(b) For any bounded disk B in X there is a bounded disk A in X with $B \subset A$ such that $J_{A,B} \in \mathscr{L}_L^P(X(B), X(A))$.

(c) (i) Bounded subsets of X are precompact, and

(ii) any precompact subset of X is b–precompact.

(d) Bounded subsets of X are b–precompact.

Proof. The equivalence of (b), (c) and (d) is trivial.

(b)⟹(a): Follows from Schauder's duality theroem(19.3)(or the Remark of 19.3)).

(a)⟹(b): We first observe that for any bounded disk C in X, the evaluation map $K_C : X(C) \to X(C)''$ and $\widehat{J_C'} : X'/\mathrm{Ker}\, J_C' \to X(C)'$ are metric isomorphisms.

Now if A and B are bounded disks in X such that (20.A) holds, then it is easily shown that the following diagrams commute:

$$\begin{array}{ccc} X(A)' & \xrightarrow{J'_{A,B}} & X(B)' \\ \widehat{J'_A} \uparrow & & \uparrow \widehat{J'_B} \\ X'_{A^\circ} & \xrightarrow[Q_{B^\circ,A^\circ}]{} & X'_{B^\circ} \end{array} \qquad \begin{array}{ccc} X(B) & \xrightarrow{J_{A,B}} & X(A) \\ K_B \downarrow & & \downarrow K_A \\ X(B)'' & & X(A)'' \\ (\widehat{J'_B})' \downarrow & & \downarrow (\widehat{J'_A})' \\ (X'_{B^\circ})' & \xrightarrow[(Q_{B^\circ,A^\circ})']{} & (X'_{A^\circ})' \end{array}$$

Thus if suffices to show that

$$(\widehat{J'_B})' \circ K_B : X(B) \to (X'_{B^\circ})' \quad \text{and} \quad (\widehat{J'_A})' \circ K_A : X(A) \to (X'_{A^\circ})'$$

are metric injections. Indeed, let r_B be the gauge of B defined on X(B) and let $(\widehat{p}_{B^\circ})^*$ be the dual norm of the quotient norm $\widehat{p}_{B^\circ}$. For any $x \in X(B)$, we have

$$\begin{aligned} (\widehat{p}_{B^\circ})^*(\hat{J}'_B(K_B x)) &= \sup\{|\langle Q_{B^\circ}(u'), (\hat{J}'_B)'(K_B x)\rangle| : u' \in B^\circ\} \\ &= \sup\{|\langle x, \hat{J}'_B(Q_{B^\circ} u')\rangle| : u' \in B^\circ\} \\ &= \sup\{|\langle x, J'_B u'\rangle| : u' \in B^\circ\} \\ &= \sup\{|\langle x, u'\rangle| : u' \in B^\circ\} = r_B(x), \end{aligned}$$

which obtain our assertion.

As an immediate consequence of (20.1)(f) and (20.2)(c), we obtain:

(20.3) Corollary. A normed space which is either a Schwartz space or a co–Schwartz space is finite–dimensional.

(20.a) (Randtke [1972])(A) For a LCS X, the following statements are equivalent :

(i) X is a Schwartz space.

(ii) $\mathscr{L}_L^{pb}(X,Y) = \mathscr{L}_L^{b}(X,Y)$ for any LCS Y.

(iii) $\mathscr{L}^{pb}(X,Y) = \mathscr{L}_L^{p}(X,Y)$ for any LCS Y.

(iv) $\mathscr{L}(X,F) = \mathscr{L}_L^{pb}(X,F)$ for any normed (or Banach) space F.

(v) For any LCS (or normed space) Y and any continuous bilinear form ψ on $X \times Y$, there exist a precompact seminorm p on X and a continuous seminorm q on Y such that

$$|\psi(x,y)| \leq p(x)q(y) \quad \text{for all } (x,y) \in X \times Y.$$

In particular, if X is a Schwartz space and Y is a LCS, then for any continuous bilinear form ψ on $X \times Y$, there is a unique $T \in \mathscr{L}_L^{pb}(X,Y'_\beta)$ such that

$$\psi(x,y) = \langle y,Tx \rangle \quad \text{for all } (x,y) \in X \times Y.$$

(B) If Y is a co–Schwartz space then

$$\mathscr{L}_L^{pb}(X,Y) = \mathscr{L}_L^{p}(X,Y) = \mathscr{L}_L^{b}(X,Y) \quad \text{for any LCS } X.$$

(20.b) Let $X_{\mathscr{P}}$ be a LCS and let $\mathfrak{S}$ be a topologizing family for X' consisting of closed bounded disks in X such that $X = \cup\mathfrak{S}$. Then $(X',\mathfrak{T}_{\mathfrak{S}})$ is a Schwartz space if and only if for any $B \in \mathfrak{S}$ there is an $A \in \mathfrak{S}$ such that

$$B \subset A \quad \text{and} \quad J_{A,B} \in \mathscr{L}_L^{p}(X(B),X(A)).$$

A LCS X is called a :

(i) semi–Montel space if every bounded subset of X is relatively $\mathscr{P}$–compact;

(ii) Montel space if it is semi–Montel and infrabarrelled;

(iii) countably infrabarrelled if every $\beta(X',X)$–bounded subset of X' which is the countable union of $\mathscr{P}$-equicontinuous sets in X' is $\mathscr{P}$-equicontinuous;

(iv) σ-barrelled (resp. σ-infrabarrelled) if every $\sigma(X',X)$-bounded (resp. $\beta(X',X)$-bounded) sequence in X' is $\mathscr{P}$-equicontinuous;

(v) (DF)-space if it is countably infrabarrelled and possesses a fundamental sequence of bounded sets.

Semi-Montel (resp. Montel) spaces are semi-reflexive (resp. reflexive) for which $\beta(X',X)$ coincides with $\mathscr{P}$ (the topology of precompact convergence). The strong dual of an F-space (resp. (DF)-space) is a (DF)-space (resp. F-space).

(20.4) Proposition (Properties of bounded sets). Let $X_{\mathscr{P}}$ be either a Schwartz space or a co-Schwartz space. Then bounded subsets of X are $\mathscr{P}$-precompact; consequently, the following statements hold :

(a) $\beta(X',X) = \mathscr{P}^{\circ}$, hence $\mathscr{P}$-equicontinuous subsets of X' are relatively $\beta(X',X)$-compact.

(b) X is quasi-complete if and only if it is a semi-Montel space.

(c) If X is assumed to be quasi-complete, then every $\sigma(X,X')$-null sequence in X is $\mathscr{P}$-null.

(d) Suppose that X is a (DF)-space. Then

(i) $\mathscr{P}$ coincides with the topology of uniform convergence on relatively $\beta(X',X)$-compact subsets of X';

(ii) $(X',\beta(X',X))$ is a Fréchet and Montel space;

(iii) $X_{\mathscr{P}}$ is infrabarrelled.

Proof. By (20.1)(f)(i) or (20.2)(c)(i), bounded subsets of X are $\mathscr{P}$-precompact, hence $\beta(X',X) = \mathscr{P}$.

(a) It can be shown (see Ex. 17.2) that for any $V \in \mathscr{U}_X$, $\mathscr{P}|_{V^\circ} = \sigma(X',X)|_{V^\circ}$, the conclusion then follows from Alaoglu-Bourbaki's theorem.

(b) Trivial.

(c) Let X be quasi–complete and let $\{x_n\}$ be a $\sigma(X,X')$–null sequence in X. Then $B = \overline{\Gamma}(\{x_n: n \geq 1\})$ is bounded and closed, hence $\mathcal{P}$–compact <u>and a fortiori</u> $\mathcal{P} = \sigma(X,X')$ on B. Thus $\{x_n\}$ is a $\mathcal{P}$–null sequence.

(d) Suppose now that X is a (DF)–space. Then X'_β is a F–space with $\beta(X',X) = \mathcal{P}^{\circ}$, hence the topology on X of uniform convergence on all relatively $\beta(X',X)$–compact subsets of X, denoted by $\mathfrak{T}_{\beta c}$, is locally convex (by the completeness of X'_β).

(i) By part (a), $\mathcal{P} \leq \mathfrak{T}_{\beta c}$. Conversely, let D be a relatively $\beta(X',X)$–compact subset of X'. Then there exists a $\beta(X',X)$–null sequence $\{u'_n\}$ in X' such that $D \subset \overline{\Gamma}^{\beta}(\{u'_n : n \geq 1\})$ (the $\beta(X',X)$–closure) [by Grothendieck's theorem (11.b)]. Clearly, $\{u'_n : n \geq 1\}$ is a countable $\beta(X',X)$–bounded subsets of X', it is $\mathcal{P}$–equicontinuous [by the countable infrabarrelledness of X], hence there is an $V \in \mathcal{U}_X$ such that $\{u'_n: n \geq 1\} \subset V^{\circ}$, and thus $D \subset \overline{\Gamma}^{\beta}(\{u'_n : n \geq 1\}) \subset V^{\circ}$ [since V° is a $\beta(X',X)$–compact disk]; this implies that D is $\mathcal{P}$–equicontinuous; in other words, $\mathfrak{T}_{\beta c} \leq \mathcal{P}$.

(ii) Since X'_β is metrizable, it suffices to show that every $\beta(X',X)$–bounded sequence $\{u'_n\}$ in X' contains a convergent subsequence; but this is trivial since such a sequence $\{u'_n\}$ is equicontinuous [by the countable infrabarrelledness of X], hence $\{u'_n : n \geq 1\}$ is relatively $\beta(X',X)$–compact, and thus it contains a convergent subsequence.

(iii) Follows from (ii) and (i).

Part (c) of (20.4) is due to Terzioglu [1969]; while other parts are quite well–known.

(20.c) (Separability for equicontinuous and bounded sets).

(i) If $X_{\mathcal{P}}$ is a Schwartz space then any $\mathcal{P}$–equicontinuous subset of X' is $\beta(X',X)$–separable.

(ii) If $X_{\mathcal{P}}$ is a co–Schwartz space then any bounded set in X is $\mathcal{P}$–separable.

(20.5) Proposition (Properties on the spaces X_V). <u>Let</u> $X_{\mathcal{P}}$ <u>be a Schwartz space</u>. <u>Then the following assertions hold</u> :

(a) <u>For any</u> $V \in \mathcal{U}_X$, <u>the normed space</u> X_V <u>is separable</u>, <u>hence metrizable Schwartz spaces are separable</u>.

(b) For any bounded subset B of X, $Q_W(B)$ is separable in the normed space $(X_W, \hat{p}_W)$ (for any $W \in \mathcal{U}_X$).

(c) For any equicontinuous subset D of X', there is a $W \in \mathcal{U}_X$ such that D is relatively compact and separable in the B–space $X'(W^\circ)$.

(d) Suppose that X is σ–barrelled. For any $\sigma(X',X)$–null sequence $\{u_n'\}$ in X', there is an $V \in \mathcal{U}_X$ such that $\{u_n'\}$ is a null–sequence in the B–space $X'(V^\circ)$, and a fortiori, a $\beta(X',X)$–null sequence.

Proof. (a) By (20.1)(b), $Q_V \in \mathcal{L}_L^p(X,X_V)$, hence there is a $W \in \mathcal{U}_X$ such that $Q_V(W)$ is precompact in the normed space X_V, thus $Q_V(W)$ is separable in X_V, consequently, X_V is separable by the surjectivity of Q_V.

Assume that $X_{\mathcal{P}}$ is metrizable, and that $\{V_n : n \geq 1\}$ is a countably local base. As each normed space X_{V_n} is separable, there is a countable set $\{x_k^{(n)} : k \geq 1\}$ such that $\{Q_{V_n}(x_k^{(n)}): k \geq 1\}$ is dense in the normed space X_{V_n}, we then conclude from

$$p_{V_n}(u) = \hat{p}_{V_n}(Q_{V_n}(u)) \quad \text{(for all } u \in X \text{ and } n \geq 1)$$

that the countable set $\bigcup_{n=1}^{\infty} \{x_k^{(n)} : k \geq 1\}$ is dense in $X_{\mathcal{P}}$.

(b) For any $W \in \mathcal{U}_X$, since $Q_W \in \mathcal{L}_L^p(X,X_W)$, it follows from (20.1)(f)(i) that $Q_W(B)$ is precompact in the normed space X_W.

(c) Let $V \in \mathcal{U}_X$ be such that $D \subset V^\circ$. By (20.1)(d), there is a $W \in \mathcal{U}_X$ with $V \subset W$ such that the canonical map $X'(V^\circ) \longrightarrow X'(W^\circ)$ is compact, hence V° is compact in the B–space $X'(W^\circ)$, thus D is relatively compact and separable in the B–space $X'(W^\circ)$.

(d) $\{u_n'\}$ is a $\sigma(X',X)$–bounded sequence in X', hence it is $\mathcal{P}$–equicontinuous [by σ–barrelledness]. By part (c), there is an $V \in \mathcal{U}_X$ such that $\{u_n' : n \geq 1\}$ is relatively compact in the B–space $X'(V^\circ)$, hence

$$\sigma(X',X) = r_{V^\circ}(\cdot)\text{–topology on } \{u_n' : n \geq 1\},$$

thus $\lim_n r_{V^\circ}(u_n') = 0$.

Parts (a) and (d) are due to Terzioglu [1969], while the other parts are quite well–known.

The following result, due to Terzioglu [1969], should be compared with part (d) of (20.5).

(20.6) Proposition. Let $X_{\mathscr{P}}$ be a σ–barrelled, co–Schwartz space. Then any $\sigma(X',X)$–null sequence $\{u_n'\}$ in X' is a $\beta(X',X)$–null sequence.

Proof. There exists a $V \in \mathscr{U}_X$ such that $\{u_n':n\geq 1\} \subset V^{\circ}$ [by the σ–barrelledness of X]. Since $\beta(X',X) = \mathscr{P}^{\circ}$ [by (20.4)(a)] and $\sigma(X',X)\big|_{V^{\circ}} = \mathscr{P}^{\circ}\big|_{V^{\circ}}$, we conclude that $\{u_n'\}$ is $\beta(X',X)$–null sequence.

(20.d) (A Characterization of co–Schwartz spaces). A LCS X is a co–Schwartz space if and only if it satisfies the following :

(i) for any $V \in \mathscr{U}_X$ and any bounded disk B in X, the canonical map $K_{V,B}$: $X(B) \longrightarrow X_V$ is precompact, and

(ii) any precompact subset of X is b–precompact.

Quasi–complete Schwartz spaces are semi–Montel spaces, while complete Schwartz spaces are ultra–semi–reflexive as shown by the following interesting result.

(20.7) Theorem (Schwartz). Every complete Schwartz space $X_{\mathscr{P}}$ is ultra–semi–reflexive. Consequently $\beta(X',X) = \beta_u(X',X)$ and the strong dual X'_β of X is fast–bornological.

Proof. Denote by $\mathfrak{T}_{\mathscr{U}_X^{\circ}}$ the topology on $X'^{\times}$ of uniform convergence on all V° with $V \in \mathscr{U}_X$. Then $\mathfrak{T}_{\mathscr{U}_X^{\circ}}\big|_X = \mathscr{P}$, hence the $\mathscr{P}$-completeness of X ensures that X is $\mathfrak{T}_{\mathscr{U}_X^{\circ}}$–closed in $X'^{\times}$.

In order to show that $X = X'^{\times}$, it suffices to show that X is $\mathfrak{T}_{\mathscr{U}_X^{\circ}}$–dense in $X'^{\times}$ or, equivalently, every $\mathfrak{T}_{\mathscr{U}_X^{\circ}}$–continuous linear functional on $X'^{\times}$ vanishing on X is identically zero [by the strong separation theorem]. To this end, it has only to show $X' = (X'^{\times}, \mathfrak{T}_{\mathcal{U}_X^\circ})'$ or, equivalently [by Mackey–Arens' theorem], each V° $(V \in \mathcal{U}_X)$ is $\sigma(X',X'^{\times})$–compact [since $<X',X'^{\times}>$ is a dual pair].

In fact, for any $V \in \mathcal{U}_X$, by (20.5)(c), there is an $W \in \mathcal{U}_X$ such that V° is compact in the B–space $X'(W^\circ)$. On the other hand, the definition of $X'^{\times}$ shows that W° is $\sigma(X',X'^{\times})$–bounded, so that $\sigma(X',X'^{\times})|_{X'(W^\circ)} \leq r_{W^\circ}(\cdot)$–topology, consequently, V° is $\sigma(X',X'^{\times})$–compact, which proves our assertion.

Finally, X'_β must be fast–bornological (see (18.e)).

It is known [see (17.2) (j)] that the strong dual of any normed space is the ultra–strong dual, hence the preceding result shows that complete Schwartz spaces are closer to B–spaces than other classes of space.

(20.8) Proposition. (Terzioglu [1969]). For a F–space $X_{\mathcal{P}}$ the following statements are equivalent.

(a) X is a co–Schwartz space.

(b) Bounded subsets of X are $\mathcal{P}$–precompact.

(c) X is a Montel space.

In particular, Schwartz F–spaces are co–Schwartz spaces.

Proof. In a metrizable LCS, precompact sets are b–precompact [see (11.2)], the equivalence of (a) and (b) then follows from (20.2)(d).

Finally, the implication (a) $\Rightarrow$ (c) follows from (20.4)(b), while (c) $\Rightarrow$ (b) is trivial.

Dually, we have the following :

(20.9) Proposition. For a (DF)–space $X_{\mathcal{P}}$ the following statements are equivalent.

(a) X is a Schwartz space.

(b) Bounded subsets of X are $\mathcal{P}$–precompact and $X_{\mathcal{P}}$ is infrabarrelled.

(c) (i) Bounded subsets of X are $\mathscr{P}$–precompact, and

(ii) for any $\beta(X',X)$–null sequence $\{u_n'\}$ in X' there exists a $V \in \mathscr{U}_X$ such that $\{u_n'\}$ is a null sequence in the B–space $X'(V^{\circ})$.

In particular, co–Schwartz (or Montel), (DF)–spaces are Schwartz spaces.

Proof. We first observe that in all statements, X_β' is a F–space such that $\beta(X',X) = \mathscr{P}^{\circ}$ [since bounded sets are $\mathscr{P}$–precompact], hence W° is $\beta(X',X)$–compact (for any $W \in \mathscr{U}_X$).

(a)$\Longrightarrow$(b): Follows from (20.4)(d).

(b)$\Longrightarrow$(c): We have to prove (ii). Let $\{u_n'\}$ be a $\beta(X',X)$–null sequence in X'. Then $\{u_n'\}$ is a b–null sequence [by the metrizability of X_β'], hence there is a $\beta(X',X)$–closed bounded disk D in X' such that $\lim_n r_D(u_n') = 0$. Now the infrabarrelledness of X ensures that there is an $V \in \mathscr{U}_X$ such that $V^{\circ}=D$, hence $\{u_n'\}$ is a null sequence in the B–space $X'(V^{\circ})$.

(c)$\Longrightarrow$(a): For any $W \in \mathscr{U}_X$, as W° is $\beta(X',X)$–compact, there exists a $\beta(X',X)$–null sequence $\{u_n'\}$ in X' such that each $x' \in W^{\circ}$ is of the form

$$(1) \qquad \Sigma_{n=1}^{\infty}\lambda_n u_n' \quad (\text{for } \beta(X',X)) \text{ with } \Sigma_n|\lambda_n| \le 1.$$

By (ii), there is a $V \in \mathscr{U}_X$ such that $\{u_n'\}$ is a null sequence in the B–space $X'(V^{\circ})$. We claim that

$$(2) \qquad x' = \Sigma_{n=1}^{\infty}\lambda_n u_n' \quad (\text{in } X'(V^{\circ})),$$

and that W° is compact in the B–space $X'(V^{\circ})$.

In fact, there is a $N\ge1$ such that $r_{V^{\circ}}(u_n') \le 1$ $(n \ge N)$, hence

$$r_{V^{\circ}}(\lambda_n u_n' + \cdots + \lambda_{n+m}u_{n+m}') \le \Sigma_{j=1}^{m}|\lambda_{n+j}| \text{ (for all } n\ge N,\ m\ge0),$$

i.e., the sequence $\{\Sigma_{j=1}^{n}\lambda_j u_j',\ n\ge1\}$ is a Cauchy sequence in the B–space $X'(V^{\circ})$, thus (2) holds by the completeness of $X'(V^{\circ})$. On the other hand, since $\lim_n r_{V^{\circ}}(u_n') = 0$ and

$X'(V^{\circ})$ is a B–space, it follows that $\overline{\Gamma}^{\|\cdot\|}(\{u_n':n\geq 1\})$ (the $r_{V^{\circ}}(\cdot)$–closure) is compact in $X'(V^{\circ})$, and hence from (1) and (2) that $W^{\circ} \subset \overline{\Gamma}^{\|\cdot\|}(\{u_n':n\geq 1\})$, thus W° is compact in the B–space $X'(V^{\circ})$. This proves our assertion.

Now the compactness of W° in the B–space $X'(V^{\circ})$ implies that the canonical map $X'(W^{\circ}) \to X'(V^{\circ})$ is compact, so that X is a Schwartz space [by (20.1)].

Finally, suppose that $X_{\mathscr{P}}$ is a (DF)–space. If X is either a co–Schwartz space or a Montel space, then all conditions in (b) are satisfied by X, hence X is a Schwartz space.

The equivalence of (a) and (c) is due to Terzioglu [1969], while the equivalence of (a) and (b) is well–known [see Grothendieck [1973,p.177]].

Examples. (a) Fréchet co–Schwartz spaces are, in general, not Schwartz.

(b) Schwartz (DF)–spaces are, in general, not co–Schwartz spaces.

(c) Montel spaces are, in general, not Schwartz spaces.

Proof. It is known from Köthe [1969, §31.5] that there is a Fréchet Montel space G in which there exists a closed subspace N such that

(*) $\qquad {}^{G}/_{N}$ is not a Montel space.

For these two spaces G and N, it follows from (20.8) that G is a co–Schwartz space. We claim that G is not a Schwartz space; hence G is a co–Schwartz, Montel space which is not a Schwartz space (i.e., (a) and (c) hold).

In fact, if G is a Schwartz space, then ${}^{G}/_{N}$ is also a Schwartz F–space [see (20.10) below], hence ${}^{G}/_{N}$ is a Montel space [by (20.4)], which contradicts (*)

To prove (b), assume on the contrary, that Schwartz (DF)–spaces are co–Schwartz. Then, we first apply this assumption to the space G, G'_{β} is a Schwartz, (DF))–space, hence G'_{β} is a co–Schwartz space, (DF)–space; in other words, $(G'',\beta(G'',G'))$ is a Schwartz, F–space, hence G is a Schwartz space [see (20.10) below], which gives a contradiction.

(20.10) **Theorem** (Grothendieck [1973,p.182]). (a) Subspaces of a Schwartz space are Schwartz spaces; the product of an arbitrary family of Schwartz space is a Schwartz space.

(b) Every separated quotient space of a Schwartz space by a closed vector subspace is a Schwartz space; the locally convex direct sum of a countable family of Schwartz spaces is a Schwartz space.

Proof. (a) It is easily shown that if $T \in L(X,Y)$ and q (a continuous seminorm on Y) is such that either T or q is precompact, then $q \circ T$ is a precompact seminorm on X. Using this fact, (a) follows easily.

(b) Let $X_{\mathscr{P}}$ be a Schwartz space, let N be a closed subspace of X and $\hat{\mathscr{P}}$ the quotient topology on $X/_N$. For any disked $\hat{\mathscr{P}}$-neighbourhood $\hat{U}$ of $\hat{0}$ in $X/_N$, there exists, by (20.1), a $V \in \mathscr{U}_X$ such that for any $\lambda > 0$ it is possible to find a finite subset $\{x_1,\cdots,x_m\}$ of X such that

$$V \subset \bigcup_{j=1}^{m} (x_j+\lambda(Q_N)^{-1}(\hat{U})).$$

Now, $\hat{V}=Q_N(V)$ is a 0–neighbourhood in $(X/_N, \hat{\mathscr{P}})$ such that

$$\hat{V} \subset \bigcup_{j=1}^{m} (Q_N(x_j) + \lambda\hat{U}),$$

hence $(X/_N, \hat{\mathscr{P}})$ is a Schwartz space by (20.1).

The proof for countable direct sum will be based on two conditions of (20.1)(f). Let $X = \bigoplus_{i=1}^{\infty} X_i$, where each X_i is a Schwartz space. Since a subset B of X is bounded (resp. precompact) if and only if $\pi_i(B)$ is bounded (resp. precompact) in X_i and $\Delta = \{i \in N: \pi_i(B) \neq \{0\}\}$ is finite, it follows from (20.1)(f) that bounded subsets of X are precompact. In order to verify that X satisfies the condition (ii) of (20.1)(f), let W be an disked 0–neighbourhood in X and $J_n : X_n \to X$ the canonical embedding map. Then we identify X_n with the closed subspace $J_n(X_n)$ of X, hence each $W_n = J_n^{-1}(W) = W \cap X_n$ is a disked 0–neighbourhood V_n in X_n, and thus there exists a disked 0–neighbourhood V_n in X_n with $V_n \subset W_n$ such that V_n satisfies the condition (ii) of (20.1)(f) (with respect to W_n). The absolutely convex hull of $\bigcup_{n\geq 1}(n+1)^{-1}V_n$, denoted by V, is a disked 0–neighbourhood in X. Now for any $\lambda > 0$, we have

$$(n+1)^{-1}V_n \subset \lambda W_n \subset \lambda W \quad \text{for all } n > 0 \text{ with } n+1 \geq \lambda^{-1}.$$

For any n with $0 < n \leq \lambda^{-1}$, there exists a bounded subset B_n of X_n such that

$$(n+1)^{-1}V_n \subset V_n \subset B_n + \lambda^{-1}W_n.$$

The disked hull of $\cup\{B_n : 0 < n \leq \lambda^{-1}\}$, denoted by B, is clearly a bounded subset of E, and satisfies

$$(n+1)^{-1}V_n \subset B + \lambda W \quad \text{for all } n \geq 1.$$

Since $B+\lambda W$ is a disk, it follows that $V \subset B + \lambda W$, as asserted.

(20.e) (Universal Schwartz spaces (Randtke [1973 b]). Let $\mathfrak{T}$ be the locally convex topology on ℓ^∞, determined by the family of all $\|\cdot\|_\infty$–precompact seminorms on ℓ^∞. Then a LCS X is a Schwartz space if and only if it is topologically isomorphic to a subspace of the product space $((\ell^\infty)^\Lambda, \mathfrak{T}^\Lambda)$ for some index set Λ.

Examples. (d) Let $X_{\mathscr{P}}$ be a LCS. It is clear that any $\sigma(X,X')$–continuous seminorm on X is precompact, so that X_σ is always a Schwartz space.

(e) The spaces $C^m(\Omega)$ and $C^\infty(\Omega)$ of L. Schwartz (where $\Omega \neq \phi$ is open in $\mathbb{R}^n$). Let $\{K_n\}$ be a sequence of compact subsets of $\mathbb{R}^n$ with

$$\Omega = \bigcup_{m=1}^{\infty} K_m \quad \text{and} \quad K_m \subset \text{Int } K_{m+1} \quad (\text{for all } m \geq 1),$$

such that each compact subset of Ω is contained in some K_n [see (7.15)(c)].

Notation: For any $\alpha = (\alpha_1, \cdots \alpha_n) \in N^n$, D^α stands for the differential operator

$$D^\alpha = \left(\frac{\partial}{\partial t_1}\right)^{\alpha_1} \cdots \left(\frac{\partial}{\partial t_n}\right)^{\alpha_n} = \frac{\partial^{|\alpha|}}{\partial t_1^{\alpha_1} \cdots\cdots \partial t_n^{\alpha_n}},$$

where $|\alpha| = \Sigma_{i=1}^n \alpha_i$ is the order of this differential operator. [We have D^0 = identity and

$D^{\alpha}D^{\beta} = D^{\alpha+\beta}$.] $D^{\alpha}f$ (when it exists) is called the partial derivative of f with order $|\alpha|$.

Now for any $k \geq 1$, the set, defined by

$$C^k(\Omega) = \{f \in C(\Omega): D^{\alpha}f \in C(\Omega) \text{ for all } \alpha \in N^n \text{ with } |\alpha| \leq k\},$$

is a vector subspace of $C(\Omega)$. [Elements in $C^k(\Omega)$ are said to be of the class C^k] For any K_m, the functional $p_m^{(k)}$ on $C^k(\Omega)$, defined by

$$p_m^{(k)}(f) = \max\{|D^{\alpha}f(t)|: t \in K_m, |\alpha| \leq k\} \quad (\text{for any } f \in C^k(\Omega)),$$

is a seminorm on $C^k(\Omega)$, hence $\{p_m^{(k)} : m \geq 1\}$ determines a metrizable locally convex topology on $C^k(\Omega)$, denoted by $\mathscr{P}_{Sc}^{(k)}$; moreover, $(C^k(\Omega), \mathscr{P}_{Sc}^{(k)})$ is complete and

$$C^k(\Omega) \subset C^{k-1}(\Omega) \subset \cdots \subset C^{0}(\Omega) = C(\Omega) \quad (\text{for all } k \geq 1);$$

thus we set

$$C^{\infty}(\Omega) = \bigcap_{k=0}^{\infty} C^k(\Omega)$$

and let $J^{(k)}: C^{\infty}(\Omega) \to C^k(\Omega)$ be the canonical embedding. [Elements in $C^{\infty}(\Omega)$ are said to be of the class C^{∞} or infinitely differentiable.] On $C^{\infty}(\Omega)$, denote by $\mathscr{P}_{(Sc)}$ the projective topology with respect to $\{(C^k(\Omega), \mathscr{P}_{Sc}^{(k)}, J^{(k)}): k \geq 0\}$. Then $(C^{\infty}(\Omega), \mathscr{P}_{(Sc)})$ is a Fréchet space and $\mathscr{P}_{(Sc)}$ is determined by the family $\{p_m^{(k)}: m \geq 1, k \geq 0\}$ of seminorms. Now we show that $(C^{\infty}(\Omega), \mathscr{P}_{Sc})$ is a Schwartz space (hence a co–Schwartz space by (20.8)). To do this, it suffices to show that the quotient map

$$Q_m^{(k)}: C^{\infty}(\Omega) \to C^{\infty}(\Omega)/\text{Ker } p_m^{(k)} \quad \text{is precompact.}$$

In fact, for the compact set K_m, there is a $\delta > 0$ such that the compact set

(1) $$C = \{\zeta \in \mathbb{R}^n : |\zeta - t| = (\Sigma_{i=1}^{n}(\zeta_i - b_i)^2)^{\frac{1}{2}} \leq \delta \text{ for some } t \in K_m\}$$

is contained in Ω, hence the set

(2) $$V = \{f \in C^{\infty}(\Omega) : \max\{|D^{\alpha}f(t)| : t \in C, |\alpha| \leq k+1\} \leq 1\}$$

is a $\mathscr{P}_{(Sc)}$–neighbourhood of 0 [there is a K_i such that $C \subset K_i$, hence the unit ball determined by $p_i^{(k+1)}$ is contained in V]. Now we claim that $Q_m^{|k|}(V)$ is precompact in $C^\infty(\Omega)/_{\text{Ker } p_m^{(k)}}$. Indeed, for any $\alpha \in N^n$ with $|\alpha| \le k$, let $C_\alpha(K_m)$ be the Banach space $C(K_m)$, and let $\prod_{|\alpha| \le k} C_\alpha(K_m)$ be equipped with the product norm of the sup–norm on each $C_\alpha(K_m)$ [for definition, see (2.c) (ii)]. Then the map Ψ, defined by

$$(3) \qquad \Psi(Q_m^{(k)}(f)) = [D^\alpha f, |\alpha| \le k] \quad \text{for all } Q_m^{(k)}(f) \in C^\infty(\Omega)/_{\text{Ker } p_m^{(k)}},$$

is a metric injection from $C^\infty(\Omega)/_{\text{Ker } p_m^{(k)}}$ into $\prod_{|\alpha| \le k} C_\alpha(K_m)$ [since $\|\Psi(Q_m^{(k)}(f))\|_{\mathscr{P}\infty} = \max_{|\alpha| \le k} \|D^\alpha f\|_\infty = \max\{|D^\alpha f(t)| : t \in K_m, |\alpha| \le k\} = p_m^{(k)}(f) = \widehat{p}_m^{(k)}(Q_m^{(k)}(f))$]. Thus we identify $C^\infty(\Omega)/_{\text{Ker } p_m^{(k)}}$ with a subspace of $\prod_{|\alpha| \le k} C_\alpha(K_m)$. For each $\alpha \in N^n$ with $|\alpha| \le k$, the α–th projection of $Q_m^{(k)}(V)$:

$$\pi_\alpha(Q_m^{(k)}(V)) = \{D^\alpha f : f \in V\} \quad \text{in } C_\alpha(K_m)$$

(by (3)) is equicontinuous [if $\zeta, t \in K_m$ are such that $\|\zeta - t\|_{\mathscr{P}} \le \delta$ then the line segment joining them lies in C (by (1)), hence $|D^\alpha f(\zeta) - D^\alpha f(t)| = |\Sigma_{i=1}^n \int_t^\zeta \frac{\partial}{\partial z_i} D^\alpha f(z) dz_i| \le n|\zeta - t|$], and bounded in $C_\alpha(K_m)$, hence $\pi_\alpha(Q_m^{(k)}(V))$ is relatively compact in $C_\alpha(K_m)$ [by Arzela–Ascoli's theorem], thus $Q_m^{(k)}(v)$(or exactly, $\Psi(Q_m^{(k)}(V))$) is relatively compact in $\prod_{|\alpha| \le k} C_\alpha(\Omega)$ [by Tychonoff's theorem]; consequently, $Q_m^{(k)}(V)$ is precompact in $C^\infty(\Omega)/_{\text{Ker } p_m^{(k)}}$, which proves our assertion.

(f) Let Ω be a non–empty open subset of $\mathbb{R}^n$ and

$$\mathscr{D}(\Omega) = \{f \in C^\infty(\Omega) : \text{supp } f = \overline{\{x : f(x) \neq 0\}} \text{ is compact}\}.$$

Then $\mathscr{D}(\Omega)$ is a closed subspace of $(C^\infty(\Omega), \mathscr{P}_{(Sc)})$, hence it is a Fréchet, Schwartz space.

Exercises

20–1. Let X and Y be LCS.

(a) Let $T \in L(X,Y)$ and q be a continuous seminorm on Y. Show that if either T or q is precompact then $q \circ T$ is a precompact seminorm on X.

(b) If p and r are precompact seminorms on X, show that $p+r$ and λp (for any $\lambda \geq 0$) are precompact seminorms on (Randtke [1972].)

20–2. Prove (20.a).

20–3. Prove (20.b).

20–4. Prove (20.c).

20–5. Prove (20.d).

20–6. Prove (20.e).

20–7. Show that for any LCS G, the family of all $\tau(G,G')$–precompact seminorms on G determines the finest Schwartz topology $\mathfrak{T}$ on G (i.e. $(G,\mathfrak{T})$ is a Schwartz space) which is consistent with $\langle G,G'\rangle$.

20–8. Show that a LCS X is quasi–normable if and only if for any $V \in \mathcal{U}_X$ there is a $W \in \mathcal{U}_X$ with $W \subset V$ such that

$$\beta(X',X)\Big|_{V^\circ} = r_{W^\circ}(\cdot)\text{–top}\Big|_{V^\circ}.$$

(Grothendieck [1973, p.176])

21

Elementary Riesz-Schauder's Theory

We begin with the following:

(21.1) **Lemma.** Let E be a Banach space, let $T \in \mathscr{L}^c(E)$ $(= \mathscr{L}^c(E,E))$ (i.e., compact operator) and suppose that

$$T_\lambda = \lambda I_E - T \quad (\text{for any } \lambda \neq 0).$$

Then:

(a) dim Ker $T_\lambda < \infty$.

(b) Im $T_\lambda = T_\lambda E$ is closed in E.

(c) codim Im $T_\lambda < \infty$.

Proof. (a) The map $\frac{1}{\lambda}T : \text{Ker } T_\lambda \longrightarrow \text{Ker } T_\lambda$ is an identity map. As $\frac{1}{\lambda}T$ is compact, it follows that the closed unit ball in Ker T_λ is compact, and hence that Ker T_λ is finite.

(b) As E is complete and dim Ker $T_\lambda < \infty$, it follows that Ker T_λ is a complementary subspace of E, and hence that there exists a closed vector subspace M of E such that $E = \text{Ker } T_\lambda \oplus M$; consequently we define $S : M \longrightarrow E$ by setting

$$Sx = \lambda x - Tx \quad (\text{for all } x \in M).$$

It is clear that S is an operator which is one–one (since $M \cap \text{Ker } T_\lambda = \{0\}$) and Im S = Im T_λ. In order to prove the closedness of Im T_λ, it suffices to show (by (6.4)) that S is a topological injection or, equivalently, there is $\gamma > 0$ such that

$$\|x\| \leq \gamma \|Sx\| \quad (\text{for all } x \in M). \tag{1}$$

In fact, suppose, on the contrary, that for any $n \geq 1$ there is an $0 \neq x_n \in M$ such that

$$\|x_n\| > 2^n \|Sx_n\|.$$

Then $u_n = \|x_n\|^{-1}x_n$ is a sequence in M such that

$$\|u_n\| = 1 \text{ and } \|Su_n\| \longrightarrow 0. \tag{2}$$

As T is compact, there exists a subsequence $\{u_{n_k}\}$ of $\{u_n\}$ and $z \in E$ such that $\lim_k \|Tu_{n_k} - z\| = 0$. It then follows from (2) that

$$\begin{aligned} \|\lambda u_{n_k} - z\| &\leq \|\lambda u_{n_k} - Tu_{n_k}\| + \|Tu_{n_k} - z\| \\ &= \|Su_{n_k}\| + \|Tu_{n_k} - z\| \longrightarrow 0. \end{aligned} \tag{3}$$

Therefore $z \in M$ (since $u_{n_k} \in M$) and

$$Sz = \lim_k S(\lambda u_{n_k}) = 0$$

(by (2)), thus $z = 0$ (since S is one–one). But $\|u_{n_k}\| = 1$, it then follows from (3) that

$$\|z\| = \lim_k \|\lambda u_{n_k}\| = |\lambda| \neq 0,$$

which gives a contradiction.

(c) We first observe that

$$\operatorname{codim} \operatorname{Im} T_\lambda = \dim (E/\operatorname{Im} T_\lambda) \quad \text{and} \quad (E/\operatorname{Im} T_\lambda)' \equiv (\operatorname{Im} T_\lambda)^\circ.$$

By (17.12) (a), we have

$$(\operatorname{Im} T_\lambda)^\circ = (T_\lambda')^{-1}(E^\circ) = \operatorname{Ker} T_\lambda' = \operatorname{Ker} (\lambda I_{E'} - T').$$

On the other hand, Schauder's duality theorem (19.3) shows that T' is compact, hence $\dim \operatorname{Ker} T_\lambda' < \infty$ [by part (a)], thus

$$\dim (E/\operatorname{Im} T_\lambda)' = \dim (\operatorname{Im} T_\lambda)^\circ = \dim \operatorname{Ker} T_\lambda' < \infty;$$

consequently, we obtain

$$\operatorname{codim} \operatorname{Im} T_\lambda = \dim (E/\operatorname{Im} T_\lambda) = \dim (E/\operatorname{Im} T_\lambda)' < \infty.$$

(21.a) Let E be a Banach space, let $T \in \mathscr{L}^c(E)$, let $T_\lambda = \lambda I_E - T$ (for all $\lambda \neq 0$) and we write $T'_\lambda = \lambda I_{E'} - T'$. Then

$$\dim \operatorname{Ker} T_\lambda = \dim \operatorname{Ker} T'_\lambda .$$

In other words, the homogeneous linear functional equations

$$(\lambda I_E - T)x = 0 \text{ and } (\lambda I_{E'} - T')x' = 0 \quad (x \in E, x' \in E')$$

have the some number of linearly independent solution.

(21.b) <u>Null spaces and ranges of</u> T_λ : Let E be a Banach space, let $T \in \mathscr{L}^c(E)$, let $\lambda \neq 0$ and $T_\lambda = \lambda I_E - T$.

(i) $\dim \operatorname{Ker} T_\lambda^n < \infty$ (for all $n \geq 0$) and

$$\{0\} = \operatorname{Ker} T_\lambda^o \subset \operatorname{Ker} T_\lambda \subset \operatorname{Ker} T_\lambda^2 \subset \cdots$$

Furthermore, there exists a smallest integer $k \geq 0$ (depending on λ) such that

$$\operatorname{Ker} T_\lambda^k = \operatorname{Ker} T_\lambda^{k+i} \quad (\text{for all } i \geq 0),$$

and that if $k > 0$ then the inclusion

$$\operatorname{Ker} T_\lambda^o \subset \operatorname{Ker} T_\lambda \subset \cdots \subset \operatorname{Ker} T_\lambda^k,$$

are all proper. [Using Riesz's Lemma (2.10) and (21.1) (a).]

(ii) $\operatorname{Im} T_\lambda^n$ is closed in E (for any $n \geq 0$) and

$$E = \operatorname{Im} T_\lambda^o \supset \operatorname{Im} T_\lambda \supset \operatorname{Im} T_\lambda^2 \supset \cdots$$

Furthermore, there exists a smallest integer $k \geq 0$ (depending on λ) such that

$$\operatorname{Im} T_\lambda^k = \operatorname{Im} T_\lambda^{k+i} \quad \text{(for all } i \geq 0\text{)},$$

and that if $k > 0$ then the inclusions

$$\operatorname{Im} T_\lambda^0 \supset \operatorname{Im} T_\lambda \supset \cdots \supset \operatorname{Im} T_\lambda^k$$

are all proper. [Using Riesz's Lemma (2.10) and (21.1) (b).]

(iii) There exists a smallest integer $k \geq 0$ (depending on λ) such that

$$\operatorname{Ker} T_\lambda^k = \operatorname{Ker} T_\lambda^{k+i} \quad \text{and} \quad \operatorname{Im} T_\lambda^k = \operatorname{Im} T_\lambda^{k+i} \quad \text{(for all } i \geq 0\text{))},$$

and that

$$E = \operatorname{Ker} T_\lambda^k \oplus \operatorname{Im} T_\lambda^k.$$

(21.2) **Theorem.** Let E be a Banach space, let $T \in \mathscr{L}^c(E)$ and $\lambda \neq 0$. Then the following statements are equivalent:

(a) $T_\lambda = \lambda I_E - T$ is surjective (i.e. onto)

(b) T_λ is injective (i.e.,one–one)

(c) $T_\lambda' = \lambda I_{E'} - T'$ is surjective.

(d) T_λ' is injective.

Proof. (a) $\Leftrightarrow$ (d): Follows from (6.6),(21.1) and (19.13) [since E is complete].

(b) $\Rightarrow$ (c):By (21.1)(b), T_λ is a topological injection, hence T_λ' is a topological surjection (by (6.6)).

(c) $\Rightarrow$ (b): Follows from (6.6) and Banach's open mapping theorem (or (17.12)(c)).

(a) $\Rightarrow$ (b): For any $m \geq 1$, let

$$E_m = \operatorname{Ker} (\lambda I_E - T)^m.$$

Then E_m are closed vector subspaces of E such that

$$E_1 \subset E_2 \subset E_3 \subset \cdots.$$

Suppose, on the contrary, that T_λ is not injective. Then there exists an $0 \neq x_1 \in E_1$, hence the surjectivity of T_λ ensures that there is an $x_2 \in E$ such that

$$x_1 = T_\lambda x_2 .$$

Continue this process, we obtain a sequence $\{x_n\}$ in E such that

$$x_m = T_\lambda x_{m+1} \quad (\text{for all } m \geq 1).$$

As

$$T_\lambda^m x_{m+1} = x_1 \neq 0 \quad \text{and} \quad T_\lambda^{m+1} x_{m+1} = T_\lambda x_1 = 0,$$

it follows that

$$x_{m+1} \in E_{m+1} \setminus E_m \quad (\text{for all } m \geq 1),$$

and hence from Riesz's Lemma (2.10) that there exists a $u_{m+1} \in E_{m+1}$ such that

(1) $$\|u_{m+1}\| = 1 \quad \text{and} \quad \operatorname{dist}(u_{m+1}, E_m) > \frac{1}{2}.$$

For any positive integers $k < m$, we have $k \leq m-1$, hence

(2) $$E_k \subset E_m \quad \text{and} \quad T_\lambda(E_k) \subset T_\lambda(E_m) \subset E_{m-1} .$$

Observe that

$$Tu_m - Tu_k = \lambda[u_m - (u_k - T_\lambda(\frac{u_k}{\lambda}) + T_\lambda(\frac{u_m}{\lambda}))] .$$

It follows from(2) that

$$u_k \in E_{m-1}, \quad T_\lambda(\frac{u_k}{\lambda}) \in E_{m-1} \quad \text{and} \quad T_\lambda(\frac{u_m}{\lambda}) \in E_{m-1},$$

and hence from (1) that

$$\|Tu_m - Tu_k\| = |\lambda| \, \|u_m - (u_k - T_\lambda(\frac{u_k}{\lambda}) + T_\lambda(\frac{u_m}{\lambda}))\|$$

$$\geq |\lambda| \operatorname{dist}(u_m, E_{m-1}) > \frac{|\lambda|}{2}.$$

Therefore $\{Tu_m\}$ cannot have any convergent subsequence,which contradicts the compactness of T; thus T_λ must be injective.

(c) ⇒ (d): Schauder's duality theorem (19.3) ensures that T' is compact, so that we apply the implication (a) ⇒ (b) of the theorem to conclude that T'_λ is injective.

(21.3) <u>Definition</u>. Let E be a normed space and $T \in L(E)$ (=L(E,E)). A complex number λ is called a <u>regular value of</u> T if it satisfies the following three conditions:

(R1) Ker $(\lambda I_E - T) = \{0\}$ (the inverse of $T_\lambda = \lambda I_E - T$ is denoted by $R_\lambda(T)$ and called the <u>resolvent operator of</u> T or, simply, the <u>resolvent</u> of T);

(R2) Im T_λ is dense in E;

(R3) $R_\lambda(T)$ is bounded.

The set consisting of all regular values of T is called the <u>resolvent set</u> of T, and denoted by $\rho(T)$. The set

$$\sigma(T) = \mathbb{C} \setminus \rho(T)$$

is called the <u>spectrum of</u> T, and members in $\sigma(T)$ are referred to as <u>spectral values</u> of T.

The spectrum $\sigma(T)$ of T can be decomposed into the following three disjoint sets:

(a) The set, defined by

$$\sigma_e(T) = \{ \lambda \in \mathbb{C} : T_\lambda = \lambda I_E - T \text{ is not one–one } \},$$

is called the <u>point spectrum</u> or <u>discrete spectrum</u> of T, and members in $\sigma_e(T)$ are called <u>eigenvalues</u> of T.

(b) The set, defined by

$$\sigma_c(T) = \{ \lambda \in \sigma(T) : \text{Ker } T_\lambda = \{0\} \text{ and } \overline{\text{Im } T_\lambda} = E \},$$

is called the continuous spectrum of T.

(c) The set, defined by

$$\sigma_r(T) = \{ \lambda \in \sigma(T) : \text{Ker } T_\lambda = \{0\} \text{ and } \overline{\text{Im } T_\lambda} \neq E \},$$

is called the residual spectrum of T.

Remarks: (i) If $0 \neq x \in E$ is such that

$$(\lambda I_E - T)x = 0$$

(i.e., λ is an eigenvalue of T), then x is called the eigenvector of T corresponding to λ.

(ii) Each $\lambda \in \sigma_c(T)$ is such that $R_\lambda(T)$ is not bounded.

(iii) If E is complete and $\lambda \in \rho(T)$, then Banach's open mapping theorem shows that $R_\lambda(T)$ is defined on the whole space E.

(21.4) **Theorem (Eigenvalues)** Let E be a Banach space and $T \in \mathscr{L}^c(E)$ (i.e., a compact operator).

(a) $\sigma(T)\backslash\{0\} \subset \sigma_e(T)$ (i.e., any non–zero spectrum of T is an eigenvalue of T).

(b) $\sigma(T) = \sigma(T')$.

(c) For any $k > 0$ the set, defined by

$$\{\lambda \in \sigma_e(T) : |\lambda| \geq k\},$$

is finite, hence the set $\sigma_e(T)$ of all eigenvalues of T is at most countable, and only 0 can be the limit point of $\sigma(T)$.

Proof. (a) If $\lambda \neq 0$ is not an eigenvalue of T, then $\mathrm{Ker}\, T_\lambda = \{0\}$ (where $T_\lambda = \lambda I_E - T$); in other words, T_λ is one–one, hence T_λ is surjective (by (21.2)) and surely, a topological isomorphism (by Banach's open mapping theorem); consequently,

$$R_\lambda(T) = T_\lambda^{-1} : E \longrightarrow E \text{ is bounded,}$$

thus $\lambda \in \rho(T)$.

(b) We first claim that if $\dim E = \infty$ then

$$0 \in \sigma(T) \text{ and } 0 \in \sigma(T').$$

Indeed, since $\dim E = \infty$, every compact operator T on E does not have a bounded inverse (otherwise, $I_E = T^{-1}T$ is compact, hence $\dim E < \infty$), thus $0 \in \sigma(T)$. On the other hand, $\dim E' = \infty$ and T' is compact, a similar argument shows that $0 \in \sigma(T')$.

Suppose now that $0 \neq \lambda \notin \sigma(T)$. Then $\lambda \in \rho(T)$ [since $\sigma(T)=\mathbb{C}\setminus\rho(T)$], hence $T_\lambda = \lambda I_E - T$ is one–one, and thus $T'_\lambda = \lambda I_{E'} - T'$ is surjective (by (21.2)); consequently, T'_λ is a topological isomorphism by (21.2) (since $T' \in \mathscr{L}^c(E')$); in other words, $\lambda \in \rho(T')$, thus $\lambda \notin \sigma(T')$. This proves that $\sigma(T') \subset \sigma(T)$.

Conversely, if $0 \neq \lambda \notin \sigma(T')$, then $\lambda \in \rho(T')$, hence T'_λ is one–one, thus T_λ is one–one (by (21.2)). Consequently, $\lambda \in \rho(T)$ or, equivalently, $\lambda \notin \sigma(T)$. This proves that $\sigma(T) \subseteq \sigma(T')$.

(c) Suppose that there is some $k > 0$ such that

(1) $$\lambda_m \in \{\lambda \in \sigma_e(T) : |\lambda| \geq k\} \text{ with } \lambda_m \neq \lambda_n \ (m \neq n) \quad (\text{for all } m \geq 1).$$

For each λ_m, let $e_m \in E$ be such that

$$(\lambda_m I_E - T)e_m = 0.$$

Then $\{e_1,\cdots,e_m\}$ is linearly independent (for any $m \geq 1$), hence $E_m = <\{e_1,\cdots,e_m\}>$ is an m–dimensional closed subspace of E such that $E_m \subsetneqq E_{m+1}$ (for all $m \geq 1$). By Riesz's Lemma (2.10), there exists an $u_{m+1} \in E_{m+1}$ such that

$$(2) \qquad \|u_{m+1}\| = 1 \quad \text{and} \quad \text{dist}(u_{m+1}, E_m) > \frac{1}{2}.$$

As $|\lambda_m| \geq k$ (by(1)), it follows that $\{\frac{1}{\lambda_n} u_n\}$ is a bounded sequence in E. We claim that

$$(3) \qquad \left\|\frac{1}{\lambda_m} Tu_m - \frac{1}{\lambda_n} Tu_n\right\| > \frac{1}{2} \quad \text{whenever } m > n.$$

To this end, we first observe that

$$(4) \qquad \frac{1}{\lambda_m} Tu_m - \frac{1}{\lambda_n} Tu_n = u_m - [u_m - \frac{1}{\lambda_m} Tu_m + \frac{1}{\lambda_n} Tu_n],$$

and that $TE_n \subset E_n \subset E_{m-1}$ (since $n \leq m-1$). Thus it suffices to show by (2) and (4) that $u_m - \frac{1}{\lambda_m} Tu_m \in E_{m-1}$ or, more generally,

$$(I_E - \frac{1}{\lambda_m} T) E_m \subset E_{m-1},$$

thus, (3) follows. Indeed, any $x \in E_m$ can be uniquely represented as $x = \Sigma_{i=1}^{m} \alpha_i e_i$, thus we have

$$\begin{aligned}(I_E - \frac{1}{\lambda_m} T)x &= \Sigma_{i=1}^{m} \alpha_i (I_E - \frac{1}{\lambda_m} T) e_i \\ &= \Sigma_{i=1}^{m} \alpha_i (e_i - \frac{\lambda_i}{\lambda_m} e_i) \\ &= \Sigma_{i=1}^{m-1} \alpha_i (1 - \frac{\lambda_i}{\lambda_m}) e_i \in E_{m-1},\end{aligned}$$

which obtains our assertion.

Now the formula (3) shows that $\{\frac{1}{\lambda_m} Tu_m\}$ does not contain any convergent subsequence; this contradicts the compactness of T [since $\{\frac{1}{\lambda_m} u_m\}$ is a bounded sequence in E]. Consequently, the set $\{ \lambda \in \sigma_e(T) : |\lambda| \geq k \}$ must be finite, and thus $\sigma_e(T)$ is at most countable.

Finally, if $\{\lambda_n\}$ is a sequence of distinct eigenvalues of T such that $\lim_n \lambda_n = \lambda$, and suppose that $\lambda \neq 0$. Then there is a natural number $N \geq 1$ such that

$$| \, |\lambda_n| - |\lambda| \, | < 1 \quad (\text{for all } n \geq N),$$

hence

$$|\lambda_n| > |\lambda| - 1 \quad (\text{for all } n \geq N);$$

in other words, the set

$$\{ \mu \in \sigma_e(T) : |\mu| > |\lambda| - 1 \}$$

is infinite. This is impossible; thus $\lambda = 0$, that is , only 0 can be the limit point of $\sigma_e(T)$.

It should be noted that the preceding result does not guarantee the existence of eigenvalues of $T \in \mathscr{L}^c(E)$, namely there exist compact operators on Banach spaces which do not have any eigenvalues (for instance, the Voltera operator on C[a,b] is such an example). But self–adjoint compact operators on Hilbert spaces always have eigenvalues as shown by the following result. To do this, we require the following notation: Let H be Hilbert. Recall (see(3.m)) that for any $\psi \in H'$ there is a unique $y \in H$ such that

$$\psi(x) = [x,y] \quad (\text{for all } x \in H).$$

Thus the Hilbert adjoint of any $T \in L(H)$, denoted by T^*, is defined by

$$[x,T^*y] = [Tx,y] \quad (\text{for all } x,y \in H).$$

We say that T is self–adjoint (or Hermitian) if $T = T^*$.

(21.c) Spectral properties of self–adjoint operators: Let H be a Hilbert space and $T \in L(H)$. Then T is self–adjoint if and only if

$$[Tx,x] \in \mathbb{R} \quad (\text{for all } x \in H).$$

Suppose now that T is self–adjoint. Then:

(i) $\sigma_e(T) \subset \mathbb{R}$.

(ii) Let $Tx = \lambda x$ and $Ty = \mu y$ $(x \neq 0, y \neq 0)$. If $\lambda \neq \mu$ then

$$[x,y] = 0.$$

(iii) An $\lambda \in \mathbb{C}$ is a regular value of T if and only if there exists an $\alpha > 0$ such that

$$\|(\lambda I_H - T)x\| \geq \alpha\|x\| \quad (\text{for all } x \in H).$$

(iv) $\sigma(T) \subset \mathbb{R}$.

(v) $\sigma_r(T) = \phi$.

(21.5) **Theorem.** Let H be a Hilbert space, let $0 \neq T \in L(H)$ be self–adjoint and suppose that

(1) $$m = \inf\{ [Tx,x] : \|x\| = 1 \} \quad \text{and} \quad M = \sup\{ [Tx,x] : \|x\| = 1 \}.$$

Then:

(a) $\|T\| = \max\{|m|,|M|\} = \sup\{|[Tx,x]| : \|x\| = 1\}$.

(b) If, in addition, T is assumed to be compact (i.e., compact, self–adjoint), then M (resp. m) is the largest (resp. smallest) eigenvalue of T; consequently,

(2) $$\sigma_e(T) \neq \phi.$$

Proof. (a) For any $x \in H$ with $\|x\| = 1$, we have

$$|[Tx,x]| \leq \|Tx\| \; \|x\| \leq \|T\|,$$

thus

$$-\|T\| \leq m \leq M \leq \|T\| \; ;$$

consequently,

$$\max \{|m|,|M|\} \leq \|T\| \quad \text{and} \quad \sup \{|[Tx,x]| : \|x\| = 1\} \leq \|T\|.$$

Conversely, let $\mu = \max \{ |m|,|M| \}$. Then

$$-\mu \leq m \leq [Tx,x] \leq M \leq \mu \quad (\text{for all } x \text{ with } \|x\| = 1),$$

hence

$$(3) \qquad |[Tx,x]| \leq \mu \|x\|^2 \quad (\text{for all } 0 \neq x \in H).$$

On the other hand, for any $u \in H$ and $\lambda > 0$, since $T = T^*$, it follows that

$$[T(\lambda u + \frac{1}{\lambda}Tu) \, , \lambda u + \frac{1}{\lambda}Tu] - [T(\lambda u - \frac{1}{\lambda}Tu) \, , \lambda u - \frac{1}{\lambda}Tu]$$

$$= \quad 2[Tu,Tu] + 2[T^2u,u] = 4[Tu,Tu],$$

and hence from (3) that

$$\begin{aligned} \|Tu\|^2 \quad &= \frac{1}{4}\{[T(\lambda u + \frac{1}{\lambda}Tu) \, , \lambda u + \frac{1}{\lambda}Tu] \\ &\qquad - [T(\lambda u - \frac{1}{\lambda}Tu),\lambda u - \frac{1}{\lambda}Tu] \} \\ &\leq \frac{\mu}{4}(\|\lambda u + \frac{1}{\lambda}Tu\|^2 + \|\lambda u - \frac{1}{\lambda}Tu\|^2) \\ &\leq \frac{\mu}{4}(2\|\lambda u\|^2 + 2\|\frac{1}{\lambda}Tu\|^2) \end{aligned}$$

$$(4) \qquad = \frac{\mu}{2}(\lambda^2\|u\|^2 + \frac{1}{\lambda^2}\|Tu\|^2) \quad (\text{for all } \lambda > 0).$$

In particular, we take $\lambda^2 = \frac{\|Tu\|}{\|u\|}$ in (4), we obtain

$$\|Tu\|^2 \leq \mu\|Tu\| \; \|u\|,$$

thus

$$\|T\| \leq \mu = \max \{|m|,|M|\}.$$

This proves part (a).

(b) The operator $S = T - mI_H$ is self–adjoint on H with

$$\sup \{[Sx,x] : \|x\| = 1\} = M - m \quad \text{and} \quad \inf \{[Sx,x] : \|x\| = 1\} = 0.$$

By part (a), we have

$$\|S\| = M - m.$$

On the other hand, the definition of M shows that there exists a sequence $\{x_n\}$ in H such that

$$\|x_n\| = 1 \quad \text{and} \quad \lim_n [Tx_n,x_n] = M.$$

As T is compact, there exists a subsequence $\{x_{n_k}\}$ and $\{x_n\}$ and $u \in H$ such that $u = \lim_k Tx_{n_k}$. Observe that

$$\|Tx_n - Mx_n\|^2 = \|(T - mI_H)x_n - (M - m)x_n\|^2$$

$$= \|(T - mI_H)x_n\|^2 + (M - m)^2\|x_n\|^2 - 2(M - m)[(T - mI_H)x_n ,x_n]$$

$$\leq 2(M - m)^2 - 2(M - m)([Tx_n,x_n] - m) \longrightarrow 0 \quad \text{as } n \longrightarrow 0.$$

Since $u = \lim_k Tx_{n_k}$, it follows from (5) that $u = \lim_k M x_{n_k}$, and hence that

$$Tu = \lim_k T(Mx_{n_k}) = M \lim_k Tx_{n_k} = Mu,$$

thus M is an eigenvalue of T (since $M \neq 0$).

To prove that M is the largest one, let $\lambda \in \sigma_e(T)$, and let $0 \neq x \in H$ be such that

$$Tx = \lambda x.$$

Without loss of generality, one can assume that $\|x\| = 1$. Then we have, from the definition of M, that

$$\lambda = [Tx,x] \leq M,$$

which proves that M is the largest eigenvalue of T.

A similar argument can be applied to conclude that $m \neq 0$ is the smallest eigenvalue of T.

Finally, since $T \neq 0$, it follows that $M \neq 0$, and hence that $M \in \sigma_e(T)$; in other words, $\sigma_e(T) \neq \phi$.

Exercises

21–1. Let E be a normed space, let $T \in L^c(E)$, let $\lambda \neq 0$ and $T_\lambda = \lambda I_E - T$. If $\{x_n\}$ is a sequence in E such that $u = \lim_n T_\lambda x_n$, show that there exist a subsequence $\{x_{n_k}\}$ of $\{x_n\}$ and $x \in E$ such that

$$\lim_k x_{n_k} = x \quad \text{and} \quad u = T_\lambda x.$$

21–2. Prove (21.a).

21–3. Prove (21.b).

21–4. Prove (21.c).

21–5. (Fredholm integral equations). Let $K_f(\cdot,\cdot)$ be a continuous function on $[a,b]\times[a,b]$ (where $a < b$), and define

$$(T_f g)(t) = \int_b^a K_f(t,s)g(s)ds \quad \text{for all } g \in C[a,b].$$

(a) Show that $T_f : C[a,b] \to C[a,b]$ is a compact operator (It is called the Fredholm operator).

(b) Suppose that $|K_f(t,s)| \leq \alpha$ (for all $t,s \in [a,b]$) and that

$$0 < |\lambda| \leq \frac{1}{\alpha(b-a)}.$$

Then for any $h \in C[a,b]$ there exists a unique $g \in C[a,b]$ such that

$$g(t) - \lambda(T_f g)(t) \equiv h(t) \text{ (for all } t \in [a,b]).$$

Moreover, this function g is the limit of the iterative sequence $\{g_n, n \geq 0\}$, where g_o is any element in $C[a,b]$ and

$$g_{n+1}(t) = v(t) + \lambda \int_b^a K_f(t,s) g_n(s) ds \quad \text{(for all } n \geq 0).$$

21–6. (Volterra integral equations). Let $K_V(\cdot,\cdot)$ be a continuous function on the triangle in the ts–plane given by

$$\{(t,s) \in \mathbb{R}^2 : a \leq t \leq b,\ a \leq s \leq t\},$$

and define

$$(T_V g)(t) = \int_a^t K_V(t,s) g(s) ds \qquad \text{for all } g \in C[a,b].$$

(a) Show that $T_V : C[a,b] \to C[a,b]$ is a compact operator. (It is called the Volterra operator.)

(b) For any $\lambda \in \mathbb{C}$ and $h \in C[a,b]$, there always exists a unique $g \in C[a,b]$ such that

$$g(t) - \lambda \int_a^t K_V(t,s) g(s) ds = h(t).$$

In other words, the Volterra operator T_V does not have any eigenvalue.

22

An Introduction to Operator Ideals

This section is devoted to a study of elementary theory of operator ideals, together with its application to classify types of locally convex spaces. Unfortunately, other applications, for instance, probability on Banach spaces and the structure theory of Banach spaces, seemed impossible to include in this section, we refer the reader to the excellent books (or monographs) written by Pietsch [1980] and Schwartz [1981].

The oldest operator ideal is the ideal $\sigma_2(H)$ of Hilbert–Schmidt operators, acting on a separable Hilbert space H, introduced by Hilbert in 1905 and Schmidt in 1907. As we shall see, to obtain an ideal theory on Banach spaces (or more generally on locally convex spaces) that is suitable for applications, it is not sufficient to consider only the algebra L(X), we must consider the space L(X, Y) for arbitrary LCS (or B–spaces) X and Y. The first example of operator ideal on Banach spaces is the operator ideal of all compact operators introduced by Riesz in 1918. The general theory of operator ideals, connecting with the fundamental work on tensor products has been initiated by Schatten [1950] and Grothendieck [1955]. However, the nontrivial translation from the language of tensor products to operators on Banach spaces was accomplished by Pietsch [1963–1971, 1980].

In the sequel, $\mathscr{L}$ will always denote the class of all operators between arbitrary locally convex spaces; while L (resp. B, V) the class of all locally convex spaces (resp. all Banach spaces, normed spaces). For any $A \subseteq L$, we write

$$\mathscr{L}_A = \cup \{ L(X, Y) : X, Y \in A \}.$$

Throughout this section, A will be assumed to be a subclass of L containing K.

(22.1) <u>Definition</u>. For a given $A \subset L$, a subclass $\mathfrak{A}$ of $\mathscr{L}_A$ is called an <u>operator ideal</u> on A if the components

$$\mathfrak{A}(X, Y) = \mathfrak{A} \cap L(X, Y) \quad (\text{for all } X, Y \in \mathbf{A})$$

satisfy the following condition:

(OI_0) $\mathfrak{A}(X, Y)$ contains $\mathfrak{F}(X, Y)$ (the set of all finite operators).

(OI_1) $\mathfrak{A}(X, Y)$ is a vector subspace of $L(X, Y)$.

(OI_2) For any $X_0, Y_0 \in \mathbf{A}$, any $R \in L(X_0, X)$, $S \in L(Y, Y_0)$, and any $T \in \mathfrak{A}(X, Y)$, one has $STR \in \mathfrak{A}(X_0, Y_0)$.

Operator ideals on B (resp. V, L) are called <u>operator ideals</u> on <u>B–spaces</u> (resp. <u>normed spaces</u>, <u>LCS</u>).

An operator ideal $\mathfrak{A}$ on B–spaces is said to be <u>closed</u> if all components $\mathfrak{A}(E, F)$ are closed in $L(E, F)$ (w.r.t. operator norm) whenever $E, F \in \mathbf{B}$.

(22.a) <u>Space ideal</u> (Pietsch [1980]) : A subclass D of L is called a <u>space ideal</u> (Stephani [1976], Pietsch [1980, p53]) if it satisfies the following conditions:

(SI_0) $K \in D$.

(SI_1) $X \times Y \in D$ for all $X, Y \in D$.

(SI_2) If $X_0 \prec X$ with $X \in D$ then $X_0 \in D$ (where $X_0 \prec X$ means that X_0 is topologically isomorphic to a subspace of X which is complementary in X).

Let D be a space ideal. A subclass D_0 of D is called a <u>base</u> for D if it satisfies the following conditions:

(BSI_0) D_0 contains an $E \neq \{0\}$.

(BSI_1) For any $X \in D$ there exists an $X_0 \in D_0$ such that

$$X \prec X_0.$$

It is not hard to show (see Stephani [1976]) that a subclass D_0 of B is a base for a space ideal if and only if it satisfies the following two conditions:

(i) D_0 contains an $E \neq \{0\}$.

(ii) For any $E, F \in D_0$ there exists a $G \in D_0$ such that

$$E \times F \prec G.$$

In this case, the class, defined by

$$\tilde{D}_0 = \{ E : E \prec G \text{ for some } G \in D_0 \},$$

is a space ideal with D_0 as a base.

(22.b) <u>Space(·)</u> and <u>Op(·)</u> (Pietsch [1980, p294]) : Let $\mathfrak{A}$ be an operator ideal on $\mathbf{A}$ (where $\mathbf{A}$ is either $\mathbf{L}$ or $\mathbf{B}$). The class of LCS, defined by

$$\text{Space}(\mathfrak{A}) = \{ X \in \mathbf{A} : I_X \in \mathfrak{A}(X, X) \},$$

is a space ideal (called the <u>space ideal associated with</u> $\mathfrak{A}$). Let $D_0 \subseteq \mathbf{L}$ be given. Operator $T \in L(X, Y)$ $(X, Y \in \mathbf{L})$ is said to be D_0–<u>factorable</u> if there exist $Z \in D_0$ and $R \in L(X, Z)$ and $L \in L(Z, Y)$ such that

$$T = LR.$$

We denote by $\text{Op}(D_0)$ the set of all D_0–factorable operators. It can be shown (see Stephani [1976]) that $\text{Op}(D_0)$ is an operator ideal if and only if D_0 is a base for a space ideal. Moreover, we have the following:

(i) If D is a space ideal, then $D = \text{space}[\,\text{Op}(D_0)\,]$.

(ii) Let $\mathfrak{A}$ be an operator ideal on $\mathbf{A}$. Then

$$\text{Op}[\,\text{Space}(\mathfrak{A})\,] \subset \mathfrak{A}.$$

Moreover, they are equal if and only if $\mathfrak{A}$ has the <u>factorization property</u> in the sense that there is a base D_0 for a space ideal such that $\mathfrak{A} = \text{Op}(D_0)$.

(22.2) <u>Examples</u>. (a) <u>Finite operators</u> $\mathfrak{F}_{\mathcal{L}}$(or $\mathfrak{F}$) : Let X and Y be LCS. An operator $T \in L(X, Y)$ is called a <u>finite operator</u> if $\dim T(X)$ is finite.

The class of all finite operators between arbitrary LCS(resp. B–spaces), denoted by $\mathfrak{F}_L$(resp. $\mathfrak{F}$), is the smallest operator ideal on LCS(resp. on B–spaces). $\mathfrak{F}$ is not closed.

(b) Bounded operators $\mathscr{L}_L^b$: Let X and Y be LCS. An operator $T \in L(X, Y)$ is called a bounded operator if there exists a 0–neighbourhood V in X such that T(V) is bounded in Y.

The class of all bounded operators between arbitrary LCS, denoted by $\mathscr{L}_L^b$, is an operator ideal on LCS.

(c) Compact operator $\mathscr{L}_L^c$(or $\mathscr{L}^c$) : Let X and Y be LCS. An operator $T \in L(X, Y)$ is said to be compact (resp. precompact) if there exists a 0–neighbourhood V in X such that T(V) is relatively compact (resp. totally bounded) in Y.

The class of all compact operators between arbitrary LCS(resp. B–spaces), denoted by $\mathscr{L}_L^c$(resp. $\mathscr{L}^c$), is an operator ideal on LCS(resp. on B–spaces). Moreover, $\mathscr{L}^c$ is closed.

The class of all precompact operators between arbitrary LCS(resp. B–spaces), denoted by $\mathscr{L}_L^p$(resp. $\mathscr{L}^p$), is an operator ideal on LCS(resp. B–spaces). Moreover, if $Y \in L$ is complete, then

$$\mathscr{L}_L^c(X, Y) = \mathscr{L}_L^p(X, Y) \text{ (for all } X \in L).$$

(d) Weakly compact operators $\mathscr{L}_L^{wc}$(or $\mathscr{L}^{wc}$) : Let X and Y be LCS. An operator $T \in L(X, Y)$ is said to be weakly compact if there exists a 0–neighbourhood V in X such that T(V) is relatively $\sigma(Y, Y')$–compact in Y.

The class of all weakly compact operators between arbitrary LCS(resp. B–spaces), denoted by $\mathscr{L}_L^{wc}$(resp. $\mathscr{L}^{wc}$), is an operator ideal on LCS(resp. on B–spaces). Moreover, $\mathscr{L}^{wc}$ is closed.

(e) Quasi–Schwartz operators $\mathscr{L}_L^{pb}$: Let X and Y be LCS. An operator $T \in L(X, Y)$ is called a b–precompact operator (or quasi–Schwartz operator (see (20.8))) if there exists a 0–neighbourhood V in X such that T(V) is b–precompact in Y.

The class of all quasi–Schwartz operators between arbitrary LCS, denoted by

$\mathscr{L}_L^{pb}$, is an operator ideal on LCS. If Y is metrizable, then (11.2) and (15.2) (a) show that

$$\mathscr{L}_L^{pb}(X, Y) = \mathscr{L}_L^{p}(X, Y) \text{ (for all } X \in L).$$

(f) <u>Completely continuous operators</u> $\mathscr{L}^{cc}$: Let E and F be B–spaces. An operator $T \in L(E, F)$ is said to be <u>completely continuous</u> if T sends every $\sigma(E, E')$–null sequence in E into a null sequence in F.

The class of all completely continuous operators between arbitrary B–spaces, denoted by $\mathscr{L}^{cc}$, is an operator ideal on B–spaces.

(g) <u>Approximable operators</u> $\mathfrak{S}$: Let E and F be B–spaces. An operator $T \in L(E, F)$ is said to be <u>approximable</u> if there exists a sequence $\{T_n\}$ in $\mathfrak{F}(E, F)$ such that $\lim_{n\to\infty} \| T_n - T \| = 0$.

The class of all approximable operators between arbitrary B–spaces, denoted by $\mathfrak{S}$, is an operator ideal on B–spaces. Moreover, $\mathfrak{S}$ is the <u>smallest closed</u> operator ideal on B–spaces.

(h) <u>Separable operators</u> $\mathfrak{X}$: Let E and F be B–spaces. An operator $T \in L(E, F)$ is said to be <u>separable</u> if T(E) contains a dense, countable subset.

The class of all separable operators between arbitrary B–spaces, denoted by $\mathfrak{X}$, is an operator ideal on B–spaces.

Other examplesof operator ideals can be found in Pietsch [1980].

If $\mathfrak{A}$ and $\mathfrak{B}$ are operator ideals on A, then the following notation will be used in the sequel:

(i) For a given $X_0 \in A$, $\mathfrak{A}(X_0, \cdot) \subset \mathfrak{B}(X_0, \cdot)$ means that $\mathfrak{A}(X_0, Y) \subseteq \mathfrak{B}(X_0, Y)$ for all $Y \in A$.

(ii) For a given $Y_0 \in A$, $\mathfrak{A}(\cdot, Y_0) \subset \mathfrak{B}(\cdot, Y_0)$ means that $\mathfrak{A}(X, Y_0) \subset \mathfrak{B}(X, Y_0)$ for all $X \in A$.

(22·3) **Proposition.** <u>Let</u> $A \subset L$ <u>contain the finite product spaces of members in</u> A, and let $\mathfrak{A}$ and $\mathscr{B}$ be operator ideals on $\mathbf{A}$. We define, for any $X, Y \in \mathbf{A}$,

$$\mathfrak{A} \circ \mathscr{B}(X, Y) = \{ T \in L(X, Y) : T = LS \text{ with } L \in \mathfrak{A}(Z, Y), S \in \mathscr{B}(X, Z) \text{ for some } Z \in \mathbf{A}\},$$

and

$$\mathfrak{A}^{-1} \circ \mathscr{B}(X, Y) = \{ T \in L(X, Y) : LT \in \mathscr{B}(X, \cdot) \text{ (for any } L \in \mathfrak{A}(Y, \cdot)) \}$$

(simply we write $\mathfrak{A}^{-1} \circ \mathscr{B} = \{ T \in \mathscr{L}: LT \in \mathscr{B} \text{ (for all } L \in \mathfrak{A}) \}$,

and similarly

$$\mathfrak{A} \circ \mathscr{B}^{-1} = \{ T \in \mathscr{L}: TR \in \mathfrak{A} \text{ (for all } R \in \mathscr{B}) \}.$$

Then $\mathfrak{A} \circ \mathscr{B}$, $\mathfrak{A}^{-1} \circ \mathscr{B}$ and $\mathfrak{A} \circ \mathscr{B}^{-1}$ are operator ideals on $\mathbf{A}$ (called respectively the product of $\mathfrak{A}$ and $\mathscr{B}$, the left–hand quotient and the right–hand quotient).

Proof. To show that $\mathfrak{A} \circ \mathscr{B}$ is an operator ideal on $\mathbf{A}$, it has to check the condition (OI_1) since the other two conditions are obvious, hence we let $T_i \in \mathfrak{A} \circ \mathscr{B}(X, Y)$. Then there exist $Z_i \in \mathbf{A}$ and operators $L_i \in \mathfrak{A}(Z_i, Y)$ and $S_i \in \mathscr{B}(X, Z_i)$ such that

$$T_i = L_i S_i \ (i = 1, 2).$$

Hence we have the following commutative diagram:

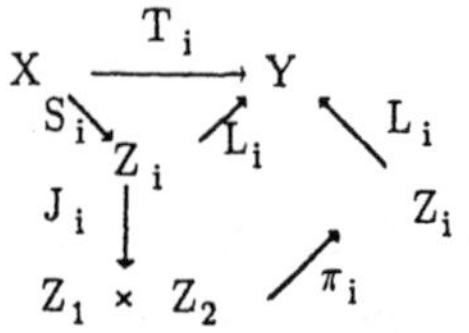

where J_i are canonical embeddings and π_i are projections $(i = 1, 2)$. It is clear that

$$J_1S_1 + J_2S_2 \in \mathscr{B}(X, Z_1 \times Z_2) \text{ and } L_1\pi_1 + L_2\pi_2 \in \mathfrak{A}(Z_1 \times Z_2, Y);$$

it then follows that

$$(L_1\pi_1 + L_2\pi_2)(J_1S_1 + J_2S_2) \in \mathfrak{A}\circ\mathscr{B}(X, Y),$$

and hence from

$$(L_1\pi_1 + L_2\pi_2)(J_1S_1 + J_2S_2) = L_1I_{Z_1}S_1 + L_2I_{Z_2}S_2 = T_1 + T_2$$

that $T_1 + T_2 \in \mathfrak{A}\circ\mathscr{B}(X, Y)$. Thus $\mathfrak{A}\circ\mathscr{B}$ is an operator ideal on **A** .

Finally, it is easy to check that $\mathfrak{A}^{-1}\circ\mathscr{B}$ and $\mathfrak{A}\circ\mathscr{B}^{-1}$ are operator ideals on **A**.

<u>Remark.</u> Let $\mathfrak{A}$, $\mathscr{B}$ and $\mathfrak{B}$ be operator ideals on **A**. Then

(i) $\mathfrak{A} \circ \mathscr{B} \subset \mathfrak{A} \cap \mathscr{B} \subset \mathfrak{A}$.

(ii) $\mathscr{B} \subset \mathfrak{A}^{-1} \circ \mathscr{B}$ and $\mathfrak{A} \subset \mathfrak{A} \circ \mathscr{B}^{-1}$.

(iii) $\mathfrak{A}^{-1} \circ \mathfrak{A} = \mathfrak{A} \circ \mathfrak{A}^{-1} = \mathscr{L}$.

(iv) If $\mathfrak{A} \subset \mathscr{B}$ then

$$\mathscr{B}^{-1} \circ \mathfrak{B} \subset \mathfrak{A}^{-1} \circ \mathfrak{B} \text{ and } \mathfrak{B} \circ \mathscr{B}^{-1} \subset \mathfrak{B} \circ \mathfrak{A}^{-1}.$$

(v) $(\mathfrak{A}^{-1} \circ \mathscr{B}) \circ \mathfrak{B}^{-1} = \mathfrak{A}^{-1} \circ (\mathscr{B} \circ \mathfrak{B}^{-1})$, hence we write

$$\mathfrak{A}^{-1} \circ \mathscr{B} \circ \mathfrak{B}^{-1} = (\mathfrak{A}^{-1} \circ \mathscr{B}) \circ \mathfrak{B}^{-1}.$$

(vi) The following facts should be compared with Remark (v):

$$(\mathfrak{A}^{-1} \circ \mathscr{B}) \circ \mathfrak{B} \subset \mathfrak{A}^{-1} \circ (\mathscr{B} \circ \mathfrak{B}) \text{ and}$$
$$\mathfrak{A} \circ (\mathscr{B} \circ \mathfrak{B}^{-1}) \subset (\mathfrak{A} \circ \mathscr{B}) \circ \mathfrak{B}^{-1}.$$

(22·c) <u>Notation</u>. Space $(\mathfrak{A}^{-1} \circ \mathscr{B})$ (Pietsch [1980]) : Let $\mathfrak{A}$ and $\mathscr{B}$ be operator ideals on **A**. Then

$$\text{Space}(\mathfrak{A}^{-1} \circ \mathscr{B}) = \{X \in \mathbf{A}: \mathfrak{A}(X,\cdot) \subset \mathscr{B}(X,\cdot)\} \text{ and}$$
$$\text{Space}(\mathfrak{A} \circ \mathscr{B}^{-1}) = \{Y \in \mathbf{A}: \mathscr{B}(\cdot, Y) \subset \mathfrak{A}(\cdot, Y)\}.$$

Moreover, we have (see Pietsch [1980]):

(i) $\mathscr{L}^{cc} \circ \mathscr{L}^{wc} = \mathscr{L}^{c}$ but $\mathscr{L}^{wc} \circ \mathscr{L}^{cc} \neq \mathscr{L}^{c}$;

(ii) $\mathscr{L}^{cc} = \mathscr{L}^{c} \circ (\mathscr{L}^{wc})^{-1}$ but $\mathscr{L}^{wc} \neq (\mathscr{L}^{cc})^{-1} \circ \mathscr{L}^{c}$.

Because of (i) and (ii) of (22.c), it is natural to consider the following interesting facts:

(22.d) Special properties of operators (resp. B–spaces) (Pietsch [1980]): Operators in $(\mathscr{L}^{cc})^{-1} \circ \mathscr{L}^{c}$(resp. $(\mathscr{L}^{wc})^{-1} \circ \mathscr{L}^{cc}$, $\mathfrak{X}^{-1} \circ \mathscr{L}^{wc}$) are called Rosenthal (resp. Dunford–Pettis, Grothendieck) operators. While B–spaces E belonging to Space($(\mathscr{L}^{cc})^{-1} \circ \mathscr{L}^{c}$) (resp. Space($(\mathscr{L}^{wc})^{-1} \circ \mathscr{L}^{cc}$), Space($\mathfrak{X}^{-1} \circ \mathscr{L}^{wc}$)) are said to have the Rosenthal (resp. Dunford–Pettis, Grothendieck) property. In view of (22.c), we see that a B–space E has the Rosenthal (resp. Dunford–Pettis, Grothendieck) property if and only if

$$\mathscr{L}^{cc}(E, \cdot) \subset \mathscr{L}^{c}(E, \cdot) \text{ (resp. } \mathscr{L}^{wc}(E, \cdot) \subset \mathscr{L}^{cc}(E, \cdot)\text{, } \mathfrak{X}(E, \cdot) \subset \mathscr{L}^{wc}(E, \cdot)\text{) (on } \mathfrak{B}\text{).}$$

(22.4) Definition. Let $A \subsetneq L$ and $\mathfrak{A}$ an operator ideal on A.

(a) $\mathfrak{A}$ is said to be injective if the following is true : For any $X, Y \in A$ and $T \in L(X, Y)$,

$T \in \mathfrak{A}(X, Y)$ if and only if there exists $Y_0 \in A$ and a topological injection $J : Y \longrightarrow Y_0$ such that $JT \in \mathfrak{A}(X, Y_0)$.

(b) $\mathfrak{A}$ is said to be surjective if the following is true: For any $X, Y \in A$ and $T \in L(X, Y)$,

$T \in \mathfrak{A}(X, Y)$ if and only if there exists $X_0 \in A$ and an open operator $Q : X_0 \longrightarrow X$ such that $TQ \in \mathfrak{A}(X_0, Y)$.

(c) The injective hull (resp. surjective hull) of $\mathfrak{A}$, denoted by $\mathfrak{A}^{inj}$(resp. $\mathfrak{A}^{sur}$), is defined to be the intersection of all injective (resp. surjective) operator ideals on A containing $\mathfrak{A}$.

(d) Suppose that A is closed w.r.t. strong duals(i.e., $X \in A$ implies $X'_\beta \in A$). Then the dual operator ideal of $\mathfrak{A}$, denoted by $\mathfrak{A}^{dual}$, is defined by

$$\mathfrak{A}^{dual}(X, Y) = \{ T \in L(X, Y) : T' \in \mathfrak{A}(Y'_\beta, X'_\beta) \} \ (X, Y \in \mathbb{A}).$$

Moreover, we say that $\mathfrak{A}$ is symmetric (resp. completely symmetric) if

$$\mathfrak{A} \subset \mathfrak{A}^{dual} (\text{resp. } \mathfrak{A} = \mathfrak{A}^{dual}).$$

(e) Suppose that $A = B$, i.e., $\mathfrak{A}$ is an operator ideal on B–spaces. Then the regular hull of $\mathfrak{A}$, denoted by $\mathfrak{A}^{reg}$, is defined by

$$\mathfrak{A}^{reg}(E, F) = \{T \in L(E, F) : K_F T \in \mathfrak{A}(E, F'')\} \quad (E, F \in \mathbb{B}).$$

Moreover, we say that $\mathfrak{A}$ is regular if $\mathfrak{A} = \mathfrak{A}^{reg}$.

(22.2) Examples: (j) The operator ideals $\mathfrak{F}$, $\mathscr{L}^c$ and $\mathscr{L}^{wc}$ on B–spaces are completely symmetric, injective and surjective.

(k) The operator ideal $\mathscr{L}^{cc}$ on B–spaces is regular, but not symmetric; also $\mathscr{L}^{cc}$ is injective but not surjective; moreover, one can show that $(\mathscr{L}^{cc})^{sur} = \mathscr{L}_B$.

(ℓ) The operator ideal $\mathfrak{S}$ on B–spaces is completely symmetric, but neither injective nor surjective; moreover, one can show that $\mathfrak{S}^{inj} = \mathscr{L}^c$.

(m) The operator ideal $\mathfrak{X}$ on B–spaces is regular, injective and surjective, but not symmectric.

(22·e) A characterization of completely symmetric operator ideals(Pietsch [1980, p.70]) : It is not hard to show that an operator ideal $\mathfrak{A}$ on B–spaces is completely symmetric if and only if it is symmetric and regular.

The injectivity and surjectivity of operator ideals on B–spaces can be characterized, in terms of some sort of domination by the following result, due to Stephani [1970] (see Pietsch [1980, pp.109 and 112]).

(22·5) Theorem. *Let* $\mathfrak{A}$ *be an operator ideal on* B–*spaces(or normed spaces).*

(a) $\mathfrak{A}$ *is injective if and only if for any* B–*spaces(resp. normed spaces)* E, F *and* F_0, *it follows from*

(1) $$T \in L(E, F).\ S \in \mathfrak{A}(E, F_0)\ \text{and}\ \|Tx\| \leq \|Sx\|\ \ (x \in E)$$

that $T \in \mathfrak{A}(E, F)$.

(b) $\mathfrak{A}$ *is surjective if and only if for any* B–*spaces(resp. normed spaces)* E_0, E *and* F, *it follows from*

(2) $$T \in L(E, F),\ S \in \mathfrak{A}(E_0, F)\ \text{and}\ T(U_E) \subset S(U_{E_0})$$

that $T \in \mathfrak{A}(E, F)$.

Proof. We verify the theorem for $\mathfrak{A}$ on B–spaces.

(a) Necessity: Let E, F and F_0 be B–spaces, and let $T \in L(E, F)$ and $S \in \mathfrak{A}(E, F_0)$ be such that $\|Tx\| \leq \|Sx\|$ $(x \in E)$. Let us define $T_1 : E \longrightarrow \overline{T(E)}$ and $S_1 : E \longrightarrow \overline{S(E)}$ by setting

$$T_1x = Tx \ \text{and}\ S_1x = Sx \quad (\text{for all}\ x \in E).$$

Then the surjectivity of $S_1 : E \longrightarrow S(E)$ together with $\|Tx\| \leq \|Sx\|$ ensures that there exists an operator $R \in L(\overline{S(E)}, \overline{T(E)})$ (since $\|R\| \leq 1$) such that $T_1 = RS_1$(by the continuity of R and usual extension), hence we obtain the following commutative diagram

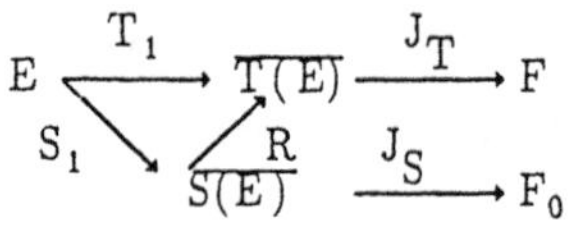

where J_T and J_S are canonical embeddings.

As J_S is a metric injection and $\mathfrak{A}$ is injective, it follows from $S = J_SS_1 \in \mathfrak{A}(E, F_0)$ that $S_1 \in \mathfrak{A}(E, \overline{S(E)})$, and hence that $T_1 = RS_1 \in \mathfrak{A}(E, \overline{T(E)})$, consequently

$$T = J_T T_1 \in \mathfrak{A}(E, F).$$

<u>Sufficiency</u>: We first observe that for any B–space G, the space $(\ell^\infty(U_{G'}), \|\cdot\|_\infty)$ is a B–space, which is denoted by G^{inj}, and the map $J_G : G \longrightarrow G^{inj}$, defined by

$$J_G(x) = [\langle x, f\rangle, f \in U_{G'}] \quad (x \in G),$$

is a metric injection (since $\|J_G(x)\|_\infty = \|x\|$).

Now let E and F be B–spaces, let $T \in L(E, F)$ and suppose that $J_F T \in \mathfrak{A}(E, F^{inj})$. It is required to show that $T \in \mathfrak{A}(E, F)$, so that $\mathfrak{A}$ is injective. Indeed, since

$$\|Tx\| = \|J_F Tx\|_\infty \leq \|J_F T(x)\|_\infty \quad (x \in E)$$

and $J_F T \in \mathfrak{A}(E, F^{inj})$, it follows from the assumption of the sufficiency that $T \in \mathfrak{A}(E, F)$, which obtains our assertion.

(b) <u>Necessity</u>: Let E_0, E and F be B–spaces, let $T \in L(E, F)$ and $S \in \mathfrak{A}(E_0, F)$ be such that $T(U_E) \subset S(U_{E_0})$. Suppose that

$$\hat{T} : E/\text{Ker } T \longrightarrow F \text{ and } \hat{S} : E_0/\text{Ker } S \longrightarrow F$$

are the injections associated with T and S respectively. As $T(U_E) \subset S(U_{E_0})$, it follows that $\text{Im } \hat{T} \subset \text{Im } \hat{S}$, and hence from the injectivity of $\hat{S}$ that there is a linear map $R: E/\text{Ker } T \longrightarrow E_0/\text{Ker } S$ such that $\hat{T} = \hat{S}R$; hence we obtain the following commutative diagram

$$\begin{array}{ccccc} E & \xrightarrow{Q_T} & E/\text{Ker } T & \xrightarrow{\hat{T}} & F \\ & & \downarrow R & \nearrow \hat{S} & \\ E_0 & \xrightarrow{Q_S} & E_0/\text{Ker } S & & \end{array}$$

We claim that R is continuous. Indeed, since $R = (\hat{S})^{-1} \circ \hat{T}$ and $T(U_E) \subset S(U_{E_o})$, it follows that

$$R(Q_T(O_E)) \subset R(Q_T(U_E)) = (\hat{S})^{-1} \circ \hat{T}(Q_T(U_E)) = (\hat{S})^{-1}(TU_E)$$
$$\subset (\hat{S})^{-1}(SU_{E_o}) = Q_S(U_{E_o}),$$

and hence that R is continuous.

Now the openness of Q_S and the surjectivity of $\mathfrak{A}$, together with $S = (\hat{S})Q_S \in \mathfrak{A}(E_0, F)$ imply that $\hat{S} \in \mathfrak{A}(E_0/\text{Ker } S, F)$, so that $\hat{T} = \hat{S}R \in \mathfrak{A}(E/\text{Ker } T, F)$; consequently,

$$T = \hat{T}Q_T \in \mathfrak{A}(E, F).$$

<u>Sufficiency</u>. We first observe that for any B–space G, the space $(\ell^1(U_G), \|\cdot\|_1)$ is a B–space, which is denoted by G^{sur}, and that the map $Q_G : G^{sur} \longrightarrow G$, defined by

$$Q_G([\xi_x, x \in U_G]) = \sum_{U_G} \xi_x \cdot x \quad ([\xi_x, x \in U_G] \in \ell^1(U_G)),$$

is a continuous, onto linear map, hence open (by Banach's open mapping theorem).

Now let E and F be B–spaces, let $T \in L(E, F)$ and suppose that $TQ_E \in \mathfrak{A}(E^{sur}, F)$. It is required to show that $T \in \mathfrak{A}(E, F)$, so that $\mathfrak{A}$ is surjective. Indeed, by the openness of Q_E

$$T(U_E) \subset (T \circ Q_E)(U_{E^{sur}}),$$

it follows from the assumption of the sufficiency that $T \in \mathfrak{A}(E, F)$, which obtains our assertion.

(22·f) <u>Some characterizations of injective hulls, surjective hulls and regular hulls</u>

<u>of operator ideals on</u> B–<u>space</u>(Pietsch [1980] and Wong [1982b]): We say that a Banach space F has(or is):

(i) <u>the extension property</u> if for any B–spaces E_0 and E, any topological injection $J : E_0 \rightarrowtail E$ and any $S_0 \in L(E_0, F)$, there exists an extension $S \in L(E, F)$ such that

$$S_0 = SJ;$$

We denoted by F_∞ the class of all B–spaces with the extension property and define

$$\mathfrak{F}_\infty = Op(F_\infty) \text{ (for definition see } (22\cdot b));$$

(ii) <u>the lifting property</u> if for any B–spaces G_0 and G, any open operator $Q: G \twoheadrightarrow G_0$ and any $T_0 \in L(F, G_0)$ there exists a lifting $T \in L(F, G)$ such that

$$T_0 = QT;$$

we denote by F_1 the class of all B–spaces with the lifting property and define

$$\mathfrak{F}_1 = Op(F_1);$$

(iii) <u>quasi–dual</u> if there exists a projection (operator) from F'' into F. We denote by D the class of all quasi–dual B–spaces and define

$$\mathscr{D} = Op(D).$$

Suppose now that $\mathfrak{A}$ is an operator ideal on B–spaces. Then the following assertions hold:

(iv) (Pietsch [1980, pp.70, 73]). We have

$$\mathfrak{A}^{inj} = (\mathfrak{F}_\infty)^{-1} \circ \mathfrak{A} \text{ and } \mathfrak{A}^{sur} = \mathfrak{A} \circ (\mathfrak{F}_1)^{-1}.$$

Moreover, for any B–spaces E and F,

$$T \in \mathfrak{A}^{inj}(E, F) \text{ if and only if } J_F T \in \mathfrak{A}(E, F^{inj})$$

and

$$T \in \mathfrak{A}^{sur}(E, F) \text{ if and only if } TQ_E \in \mathfrak{A}(E^{sur}, F).$$

(v) (Wong [1982]) $\mathfrak{A}^{reg} = \mathscr{D}^1 \circ \mathfrak{A}$.

It is of great importance that almost all operator ideals on B–spaces appearing in the applications permit a complete metrizable vector topology which is determined by quasi–norms as given by the following:

(22·6) Definition. Let $\mathfrak{A}$ be an operator ideal on B–spaces. A map $A : \mathfrak{A} \longrightarrow \mathbb{R}_+$ is called an ideal–quasi–norm (or simply, quasi–norm) if it satisfies the following conditions:

(QOI_0) $A(I_K) = 1$.

(QOI_1) There exists a constant $\lambda \geq 1$ such that

$$A(T_1+T_2) \leq \lambda[A(T_1)+A(T_2)] \quad (T_1, T_2 \in \mathfrak{A}(E, F)).$$

(QOI_2) For any B–spaces E_0, E, F, F_0, and any $R \in L(E_0, E)$, $T \in \mathfrak{A}(E, F)$ and $S \in L(F, F_0)$, we have $A(STR) \leq \|S\|A(T)\|R\|$.

Moreover, if (QOI_1) is replaced by the following stronger condition (called the p–triangle inequality):

$(QOI_1)^*$ There exists a constant $p \in (0,1)$ such that

$$[A(T_1+T_2)]^p \leq [A(T_1)]^p+[A(T_2)]^p \quad (T_1, T_2 \in \mathfrak{A}(E, F)),$$

then $A(\cdot)$ is called an ideal p–norm(or simply, p–norm). In case $p = 1$, it is called an ideal–norm (or simply, norm).

An ideal–quasi–norm (resp. ideal p–norm, ideal–norm) $A(\cdot)$ on $\mathfrak{A}$ is said to be complete if each component $\mathfrak{A}(E, F)$, where $E, F \in B$, is complete under the metrizable vector topology induced by $A(\cdot)$.

By a quasi–normed operator ideal (resp. p–normed operator ideal, normed operator ideal) on B–spaces is meant an operator ideal $\mathfrak{A}$ on B–spaces equipped with a complete ideal–quasi–norm (resp. ideal p–norm, ideal–norm).

Remark. Every ideal p–norm must be an ideal–quasi–norm; conversely, if $[\mathfrak{A}, A]$ is a quasi–normed operator ideal on B–space, then there exists a complete p–norm A_p on $\mathfrak{A}$ which is equivalent to $A(\cdot)$(see Pietsch [1980, p.92]).

The topology induced by an ideal quasi–norm $A(\cdot)$ is a metrizable vector topology as shown by the following:

(22.7) **Lemma.** Let $\mathfrak{A}$ be an operator ideal on B–spaces, let A be an ideal–quasi–norm on $\mathfrak{A}$ and $E, F \in B$.

(a) $A(\lambda S) = |\lambda| A(S)$ (for all $S \in \mathfrak{A}(E, F)$ and $\lambda \in K$).

(b) $\|S\| \leq A(S)$ (for all $S \in \mathfrak{A}(E, F)$).

(c) $A(\cdot)$ has the cross– property:

$$A(x' \otimes y) = \|x'\| \, \|y\| \quad (\text{for all } x' \in E' \text{ and } y \in F).$$

Proof. (a) By (QOI_2), we have

$$A(\lambda S) = A[(\lambda I_F) S I_E] \leq \|\lambda I_F\| A(S) \|I_E\| = |\lambda| A(S);$$

it then follows that

$$A(S) = A(\tfrac{1}{\lambda}(\lambda S)) \leq \tfrac{1}{|\lambda|} A(\lambda S) \quad (\text{for } \lambda \neq 0).$$

Therefore $A(\lambda S) = |\lambda| A(S)$ (for all $\lambda \neq 0$). Clearly $A(0) = 0$, thus (a) holds.

(b) For any $x \in U_E$ and $h \in U_{F'}$,

$$1 \otimes x : \lambda \longrightarrow \lambda x : K \longrightarrow E \text{ and}$$
$$h \otimes 1 : y \longrightarrow \langle y, h \rangle : F \longrightarrow K$$

are operators for which

$$\langle Sx, h \rangle I_K = (h \otimes 1)S(1 \otimes x) : K \longrightarrow K.$$

By (QOI_2), (QOI_0) and part (a)

$$|\langle Sx,h \rangle| = A(\langle Sx,h \rangle I_K) \leq \|h \otimes 1\| \, \|A(S)\| \, \|1 \otimes x\| \leq A(S),$$

hence

$$\|S\| = \sup\{ \, |\langle Sx, h \rangle| : x \in U_E \, , \; h \in U_{F'} \} \leq A(S).$$

(c) It is known from part (b) that

$$\|x'\| \|y\| = \|x' \otimes y\| \leq A(x' \otimes y).$$

On the other hand,

$$x' \otimes 1 : u \longrightarrow \langle u, x' \rangle : E \longrightarrow K \text{ and}$$
$$1 \otimes y : \lambda \longrightarrow \lambda y : K \longrightarrow F.$$

are operators for which

$$x' \otimes y = (1 \otimes y)I_K(x' \otimes 1) : E \longrightarrow F,$$

it then follows from (QOI_0) and (QOI_2) that

$$A(x' \otimes y) = A((1 \otimes y)I_K(x' \otimes 1)) \leq \|1 \otimes y\| A(I_K) \|x' \otimes 1\| = \|y\| \, \|x'\|,$$

and hence that (c) is true.

On every operator ideal on B–spaces, there exist a lot of different ideal–quasi–norms. However, the "nice" ideal–quasi–norms are selected by the completeness as shown by the following fundamental result, due to Pietsch [1980, p.90].

(22.8) **Theorem** (Pietsch). Let $\mathfrak{A}$ and $\mathscr{C}$ be operator ideals on B–spaces, let $A(\cdot)$ and $C(\cdot)$ be ideal–quasi–norms on $\mathfrak{A}$ and $\mathscr{C}$ resp., and suppose that

$$\mathfrak{A} \subset \mathscr{C}.$$

If $A(\cdot)$ and $C(\cdot)$ are complete, then for each pair of B–spaces E and F, there exists a constant $\mu > 0$ such that

(1) $$C(\cdot) \leq \mu A(\cdot) \text{ on } \mathfrak{A}(E, F);$$

in other words, the canonical embedding map

(2) $$J : (\mathfrak{A}(E, F), A(\cdot)) \longrightarrow (\mathscr{C}(E, F), C(\cdot))$$

is continuous. Consequently, there exists at most one complete, ideal–quasi–norm (up to equivalence) on any given operator ideal on B–spaces.

Proof. We employ the closed graph theorem to verify the continuity of the embedding map. Indeed, let $\overline{G}(J)$ be the $A\times C$–closure of the graph $G(J)$ in the product space $\mathfrak{A}(E, F) \times \mathscr{C}(E, F)$, and let $(S, T) \in \overline{G(J)}$. Then there exists an $S_n \in \mathfrak{A}(E, F)$ such that

$$(S, T) = \lim_n (S_n, S_n) \quad \text{in} \quad \mathfrak{A}(E, F) \times \mathscr{C}(E, F).$$

As $\|S_n\| \leq A(S_n)$ and $\|S_n\| \leq C(S_n)$, it follows that

$$S = \|\cdot\|\text{–}\lim_n S_n \text{ and } T = \|\cdot\|\text{–}\lim_n S_n,$$

and hence that $S = T$; in other words $(S, T) \in G(J)$.

Remark. Because of the preceding result, we agree to use the following notation:

For two quasi–normed operator ideals $[\mathfrak{A}, A]$ and $[\mathscr{C}, C]$ on B–spaces,

$$[\mathfrak{A}, A] \subseteq [\mathscr{C}, C] \text{ if and only if } \mathfrak{A} \subseteq \mathscr{C} \text{ and } C(\cdot) \leq A(\cdot) \text{ on } \mathfrak{A}.$$

(22·g) A characterization of complete ideal p–norms on operator ideal on B–spaces (Pietsch [1980, p.91]): Let $\mathfrak{A}$ be an operator ideal on B–spaces, let $A : \mathfrak{A} \longrightarrow \mathbb{R}_+$ be a map and $p \in (0,1]$. Then A is a complete ideal p–norm on $\mathfrak{A}$ (i.e., $[\mathfrak{A}, A]$ is a p–normed ideal) if and only if it satisfies the following conditions (see (7·13)):

(i) $A(I_{\mathbb{K}}) = 1$.

(ii) If $\{S_n\}$ is a sequence in $\mathfrak{A}(E, F)$(where $E, F \in \mathfrak{B}$) such that $\Sigma_{n=1}^{\infty} A(S_n)^p < \infty$, then $\Sigma_{n=1}^{\infty} S_n = S \in \mathfrak{A}(E, F)$ and $A(S)^p \leq \Sigma_{n=1}^{\infty} A(S_n)^p$.

(iii) For any B–spaces E_0, E, F, F_0 and any

$$R \in L(E_0, E), \ T \in \mathfrak{A}(E, F) \text{ and } S \in L(F, F_0),$$

it follows that $A(STR) \leq \|S\|A(T)\|R\|$.

(22·2) Examples: (n) Nuclear operators: Let E, F be B–spaces. An operator $T \in L(E, F)$ is called a nuclear operator if there exists a sequence $\{f_n\}$ in E' and a sequence $\{y_n\}$ in F with $\Sigma_{n=1}^{\infty} \|f_n\|\|y_n\| < \infty$ such that

$$Tx = \Sigma_{n=1}^{\infty} \langle x, f_n\rangle y_n = (\Sigma_{n=1}^{\infty} f_n \otimes y_n)\, x \ (x \in E).$$

We denote by $\mathfrak{N}$ the class of all nuclear operators between arbitrary B–spaces, and define for any $T \in \mathfrak{N}(E, F)$ $(E, F \in \mathfrak{B})$ that

$$N_{\ell}(T) = \inf \{\Sigma_{n=1}^{\infty} \|f_n\| \, \|y_n\| : T = \Sigma_n f_n \otimes y_n \text{ with } f_n \in E' \text{ and } y_n \in F\}.$$

Then $[\mathfrak{N}, N_{\ell}]$ is a normed operator ideal, it is the smallest normed operator ideal. Moreover, $\mathfrak{N}$ is symmetric but not regular; also $\mathfrak{N}$ is neither injective nor surjective (see Pietsch [1980, pp.110–112]).

(p) Integral operators: Let E and F be B–spaces. It is known (see(19.a)) that every finite operator $T \in \mathfrak{F}(E, F)$ admits a finite representation

(p.1) $$T = \Sigma_{i=1}^{n} f_i \otimes y_i \text{ with } f_i \in E' \text{ and } y_0 \in F \ (i = 1,...,n).$$

Of course, the above finite representation of T is not unique. In particular, if $F = E$, one can define, in terms of (p·1), the trace of T (or the trace functional on $\mathfrak{F}(E, F)$) by

(p·2) $$tr(T) = \Sigma_{i=1}^{n} f_i(y_i) \quad (\text{whenever } T = \Sigma_{i=1}^{n} f_i \otimes y_i \in \mathfrak{F}(E; E)).$$

It is not hard to show that the trace of T (i.e., (p·2)) does not depend on the special choice of the finite representation of T.

An operator $T \in L(E, F)$ is called an integral operator if there exists a constant $\mu \geq 0$ such that

(p·3) $$|tr(TR)| \leq \mu\|R\| \quad (\text{for all } R \in \mathfrak{F}(F, E)).$$

We denote by $\mathfrak{I}$ the class of all integral operators between arbitrary B–spaces, and define, for any $T \in \mathfrak{I}(E, F)$, that

(p·4) $$\rceil(T) = \sup\{|tr(TR)| : \|R\| \leq 1, \ R \in \mathfrak{F}(F, E)\}.$$

Then $[\mathfrak{I}, \rceil]$ is a normed operator ideal on B–spaces such that

$$[\mathfrak{N}, N_{\mu}] \subset [\mathfrak{I}, \rceil].$$

Moreover, $\mathfrak{I}$ is completely regular, but neither injective nor surjective (Pietsch [1980, pp.110–112]).

(q) Absolutely summing operators: Let E and F be B–spaces. An operator $T \in L(E, F)$ is said to be absolutely summing if there exists an $\mu \geq 0$ such that

$$(q\cdot 1)\qquad \Sigma_{i=1}^{n}\|Tx_i\| \leq \mu \sup\{ \Sigma_{i=1}^{n} |\langle x_i, f\rangle| : f \in U_{E'} \}$$
for any finite subset $\{x_1,\cdots,x_n\}$ of E.

We denote by $\mathscr{A}$ the class of all absolutely summing operators between arbitrary B–spaces, and define for any $T \in \mathscr{A}(E, F)$ that

$$\pi_{\ell 1}(T) = \inf\{ \mu \geq 0 : (q\cdot 1) \text{ holds } \}.$$

Then $[\mathscr{A}, \pi_{\ell 1}]$ is a normed operator ideal on B–spaces such that

$$[\mathfrak{N}, N_{\ell 1}] \subsetneq [\mathfrak{I},]] \subset [\mathscr{A}, \pi_{\ell 1}].$$

Moreover, $\mathscr{A}$ is regular but not symmetric; also $\mathscr{A}$ is injective but not surjective (see Pietsch [1980, pp.110–112]).

For any normed operator ideal $[\mathfrak{A}, A]$, there are three natural and important procedures associated with $[\mathfrak{A}, A]$. One of them is the dual procedure which has been studied in (22·4). Another one is a natural extension of the definition of integral operators (see (22·2) (p)), which we call the conjugate operator. The final one is a natural extension of the conjugate operation, which we call the adjoint operation. Before doing these, we recall that an operator $T \in L(E, F)$(where $E, F \in \mathbb{B}$) is an integral operator if there exists a $\mu \geq 0$ such that

$$|tr(TR)| \leq \mu\|R\| \quad (\text{for all } R \in \mathfrak{F}(F, E));$$

namely, a $T \in L(E, F)$ is integral if and only if the composite of the trace functional and T is a continuous linear functional on $\mathfrak{F}(F, E)$ w.r.t. operator norm $\|\cdot\|$; in other words

$$(tr) \circ T \in (\mathfrak{F}(F, E), \|\cdot\|)'.$$

As any operator ideal contains the finite operator ideal $\mathfrak{F}$, we extend naturally this notion to the general case as defined by the following:

(22·9) **Definition.** Let $[\mathfrak{A}, A]$ be a normed operator ideal on B–spaces and $T \in L(E, F)$(where $E, F \in \mathbf{B}$). We say that:

(i) T belongs to the conjugate operator ideal $\mathfrak{A}^{\Delta}$ if there exists a $\mu \geq 0$ such that

(22·9·1) $$|\mathrm{tr}(TR)| \leq \mu A(R) \quad (\text{for all } R \in \mathfrak{F}(F, E)).$$

In this case, we define

(22·9·2) $$A^{\Delta}(T) = \inf\{\mu \geq 0 : (22\cdot 9\cdot 1) \text{ holds}\},$$

and also write

(22·9·3) $$[\mathfrak{A}, A]^{\Delta} = [\mathfrak{A}^{\Delta}, A^{\Delta}].$$

(ii) T belongs to the adjoint operator ideal $\mathfrak{A}^{*}$ if there exists a $\mu \geq 0$ such that

(22·9·4) $$|\mathrm{tr}(TSL_0R)| \leq \mu \|S\| A(L_0) \|R\|,$$

whenever $R \in \mathfrak{F}(F, F_0)$, $L_0 \in \mathfrak{A}(F_0, E_0)$, $S \in \mathfrak{F}(E_0, E)$ for arbitrary B–spaces E_0 and F_0; it is represented in the following non–commutative diagram:

$$\begin{array}{ccc} E & \xrightarrow{\;T\;} & F \\ S\big\uparrow & L_0 & \big\downarrow R \\ E_0 & \longleftarrow & F_0 \end{array}$$

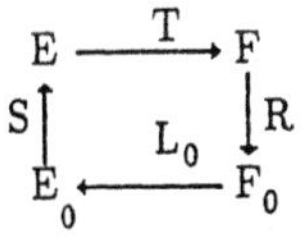

In this case, we define

(22·9·5) $$A^{*}(T) = \inf\{\mu \geq 0 : (22\cdot 9\cdot 4) \text{ holds for any } E_0, F_0 \in B\},$$

and also write

$$[\mathfrak{A},A]^* = [\mathfrak{A}^*,A^*].$$

Remarks : (i) As the trace is a linear functional on $\mathfrak{F}(F, F)$, it follows from (22·9·1) that an $T \in L(E,F)$ belongs to $\mathfrak{A}^\Delta(E,F)$ if and only if

(22·9·6) $$(tr)\circ T \in (\mathfrak{F}(F,E),A)',$$

thus one can show that

(22·9·7) $$A^\Delta(T) = \sup\{|tr(TR)| : A(R) \leq 1, R \in \mathfrak{F}(F,E)\}.$$

Consequently, if $\mathfrak{F}(F,E)$ is dense in $(\mathfrak{A}(F,E),A)$, then

$$(\mathfrak{A}^\Delta(E,F),A^\Delta) \rightarrowtail (\mathfrak{A}(F,E),A)' \quad \text{(metrically injective)}.$$

In particular, we have

$$[\mathfrak{I}, \,]] = [\mathfrak{C},\|\cdot\|]^\Delta$$

(by (22.2)(p) and (22.9)(i)) and one can show that

$$(\mathfrak{C}(F,E),\|\cdot\|)' \equiv (\mathfrak{I}(E,F''),])$$

(since $\mathfrak{F}(F,E)$ is dense in $(\mathfrak{C}(F,E),\|\cdot\|)$ (see (22.2)(g)).

(ii) By a similar argument given in the proof that $[\mathfrak{I},]]$ be a normed operator ideal, one can show that $[\mathfrak{A}^\Delta,A^\Delta]$ and $[\mathfrak{A}^*,A^*]$ are normed operator ideals on B–spaces.

(22·10) Theorem (Pietsch). <u>For any normed operator ideal</u> $[\mathfrak{A},A]$ <u>on</u> B–<u>spaces,</u> <u>one has</u>

$$[\mathfrak{A}^\Delta,A^\Delta] \subset [\mathfrak{A}^*,A^*].$$

<u>Note</u> : If the B–spaces E and F have the metric approximation property, then

$$\mathfrak{A}^{\Delta}(E, F) = \mathfrak{A}^{*}(E, F) \text{ and } A^{\Delta} = A^{*} \text{ on } \mathfrak{A}^{\Delta}(E, F).$$

<u>Proof of</u> (22·10): Let E and F be B–spaces and $T \in \mathfrak{A}^{\Delta}(E, F)$. For any B–spaces E_0 and F_0, given the following operators

$$\begin{array}{ccc} E & \xrightarrow{T} & F \\ S \uparrow & & \downarrow B \\ E_0 & \xleftarrow{L_0} & F_0 \end{array} \quad \text{with } B \in \mathfrak{F}(F, F_0),\ L_0 \in \mathfrak{A}(F_0, E_0) \text{ and } S \in \mathfrak{F}(E_0, E).$$

Then $R = SL_0B \in \mathfrak{F}(F, E)$, hence

$$|tr(TR)| \le A^{\Delta}(T)A(R) \le A^{\Delta}(T)\|S\|A(L_0)\|B\|;$$

it then follows that

$$T \in \mathfrak{A}^{*}(E, F) \text{ and } A^{*}(T) \le A^{\Delta}(T).$$

(22·h) <u>Conjugate operator ideals and adjoint operator ideals</u> (Pietsch [1980]): Let $[\mathfrak{A}, A]$ and $[\mathscr{C}, C]$ be normed operator ideals on B–spaces.

(i) $[\mathfrak{A}, A] \subset [\mathfrak{A}^{**}, A^{**}] = [\mathfrak{A}, A]^{**}$

(ii) If $[\mathfrak{A}, A] \subset [\mathscr{C}, C]$, then

$$[\mathscr{C}^{\Delta}, C^{\Delta}] \subset [\mathfrak{A}^{\Delta}, A^{\Delta}] \text{ and } [\mathscr{C}^{*}, C^{*}] \subset [\mathfrak{A}^{*}, A^{*}].$$

(iii) $[\mathfrak{I}, \mathrm{I}]^{*} = [\mathscr{L}\ \|\cdot\|]$ and $[\mathscr{L}\ \|\cdot\|]^{*} = [\mathfrak{I}, \mathrm{I}]$.

(iv) For any finite–dimensional B–spaces E and F, we have

$$(\mathfrak{A}^{*}(F, E), A^{*}) \equiv (\mathfrak{A}(E, F), A)' \text{(metrically isomorphic)}.$$

It is trivial that $\mathbf{B} \subset \mathbf{L}$, and also known (see (22.2)) that for any $E, F \in \mathbf{B}$, we have

$$\mathscr{L}_L^{c}(E, F) = \mathscr{L}^{c}(E, F);\ \ \mathscr{L}_L^{wc}(E, F) = \mathscr{L}^{wc}(E, F)$$

$$\text{and} \quad \mathscr{L}_L^{pb}(E, F) = \mathscr{L}^c(E, F).$$

This observation leads to the following notion: Let

$$A_0 \subset A \subset L$$

and let $\mathfrak{A}_0$ and $\mathfrak{A}$ be operator ideals on A_0 and A respectively. We say that $\mathfrak{A}$ is an <u>extension</u> of $\mathfrak{A}_0$ if

$$\mathfrak{A}(X, Y) = \mathfrak{A}_0(X, Y) \quad (\text{for all } X, Y \in A_0).$$

For a given operator ideal $\mathfrak{A}_0$ on A_0, Pietsch [1980, pp.393–405] has demonstrated six types of extension procedures. Here we only give three of them as follows:

(22·11) <u>Definition</u> (Pietsch [1980]). Let $\mathfrak{A}_0$ be an operator ideal on B–spaces. For any LCS X and Y and $S \in L(X, Y)$, we say that S belongs to:

(i) The <u>superior extension</u> of $\mathfrak{A}_0$, denoted by $\mathfrak{A}_0^{sup}$, if

$$LSR \in \mathfrak{A}_0(E, F) \text{ for any } R \in L(E, X) \text{ and } L \in L(Y, F),$$

where E and F are arbitrary B–spaces;

(ii) the <u>inferior extension</u> of $\mathfrak{A}_0$, denoted by $\mathfrak{A}_0^{inf}$, if S admits a factorization:

$$\begin{array}{ccc} X & \xrightarrow{\quad S \quad} & Y \\ {\scriptstyle L_0}\downarrow & S_0 & \uparrow{\scriptstyle R_0} \\ E_0 & \xrightarrow{\qquad} & F_0 \end{array}$$

for <u>some</u> B–spaces E_0, F_0 and <u>some</u> operators $L_0 \in L(X, E_0)$, $S_0 \in \mathfrak{A}_0(E_0, F_0)$ and $R_0 \in L(F_0, Y)$;

(iii) The <u>right–superior extension</u> of $\mathfrak{A}_0$, denoted by $\mathfrak{A}_0^{rup}$, if for any B–space F and any $T \in L(Y, F)$, the operator $TS \in L(X, F)$ admits a factorization:

$$\begin{array}{ccc} X & \xrightarrow{S} & Y \\ {\scriptstyle R_0}\downarrow & S_0 & \downarrow{\scriptstyle T} \\ E_0 & \longrightarrow & F \end{array}$$

for some B–space E_0 and some operators $R_0 \in L(X, E_0)$ and $S_0 \in \mathfrak{A}_0(E_0, F)$.

In order to give other criteria for operators belonging to either $\mathfrak{A}_0^{sur}$ or $\mathfrak{A}_0^{inf}$ or $\mathfrak{A}_0^{rup}$, we require the following notation: Let X and Y be LCS. Recall that (see (22·b)) an operator $T \in (X, Y)$ is B-<u>factorable</u>(in this case, it is also called a <u>Banach operator</u> or abbreviated B–<u>operator</u>) if T admits a factorization

$$\begin{array}{ccc} X & \xrightarrow{T} & Y \\ & {\scriptstyle R_0}\searrow \quad \nearrow{\scriptstyle L_0} & \\ & G_0 & \end{array}$$

for some B–space G_0 and some operators $R_0 \in L(X, G_0)$ and $L_0 \in L(G_0, Y)$. The class of all B–operators, denoted by $\mathfrak{B}_0$, is an operator ideal on LCS. In terms of $\mathfrak{B}_0$, we are able to give the following result, due to the author.

(22·12) **Theorem.** <u>Let $\mathfrak{A}_0$ be an operator ideal on B–spaces, let $\mathfrak{A}$ be an operator ideal on LCS which is an extension of $\mathfrak{A}_0$ and $\mathfrak{B}_0$ the class of all B–operators. Then</u>

$$\mathfrak{A}_0^{sup} = \mathfrak{B}_0^{-1} \circ \mathfrak{A} \circ \mathfrak{B}_0^{-1};\ \mathfrak{A}_0^{inf} = \mathfrak{B}_0 \circ \mathfrak{A} \circ \mathfrak{B}_0 \ \underline{\text{and}}$$

$$\mathfrak{A}_0^{rup} = \mathfrak{B}_0^{-1} \circ (\mathfrak{A} \circ \mathfrak{B}_0).$$

<u>**Proof.**</u> For the proof, we assume that X and Y are arbitrary LCS and $S \in L(X, Y)$.

(i) $\mathfrak{A}_0^{sup} = \mathfrak{B}_0^{-1} \circ \mathfrak{A} \circ \mathfrak{B}_0^{-1}$: The inclusion $\mathfrak{B}_0^{-1} \circ \mathfrak{A} \circ \mathfrak{B}_0^{-1} \subset \mathfrak{A}_0^{sup}$ is obvious since every operator with either domain or range in a B–space must be a B–operator.

Conversely, let $S \in \mathfrak{A}_0^{\sup}$ and suppose that

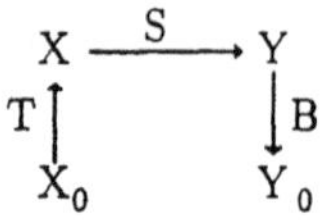

where X_0, Y_0 are arbitrary LCS and T and B are arbitrary B–operators with

$$T \in \mathfrak{B}_0(X_0, X) \text{ and } B \in \mathfrak{B}_0(Y, Y_0).$$

Then there exist B–spaces E_0 and F_0 and operators T_i and B_i $(i = 1, 2)$ such that the following diagram commutes:

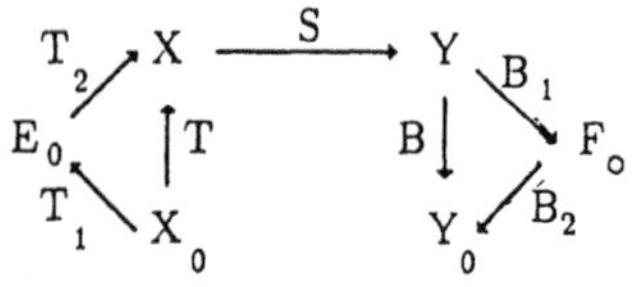

As $S \in \mathfrak{A}_0^{\sup}$ and $\mathfrak{A}$ is an extension of $\mathfrak{A}_0$, it follows that

$$B_1ST_2 \in \mathfrak{A}_0(E_0, F_0) = \mathfrak{A}(E_0, F_0),$$

and hence that

$$BST = B_2B_1ST_2T_1 \in \mathfrak{A}(X_0, Y_0);$$

Thus $S \in \mathfrak{B}_0^{-1} \circ \mathfrak{A} \circ \mathfrak{B}_0^{-1}$. This proves that $\mathfrak{A}_0^{\sup} \subset \mathfrak{B}_0^{-1} \circ \mathfrak{A} \circ \mathfrak{B}_0^{-1}$.

(ii) $\mathfrak{A}_0^{\inf} = \mathfrak{B}_0 \circ \mathfrak{A} \circ \mathfrak{B}_0$: The inclusion $\mathfrak{A}_0^{\inf} \subset \mathfrak{B}_0 \circ \mathfrak{A} \circ \mathfrak{B}_0$ is obvious (by the same reason as in (i)). Conversely, let $S \in \mathfrak{B}_0 \circ \mathfrak{A} \circ \mathfrak{B}_0$. Then S admits a factorization

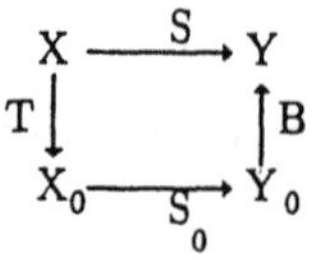

for <u>some</u> LCS X_0 and Y_0 and some operators T, S_0 and B with

$$T \in \mathfrak{B}_0(X, X_0), \; S_0 \in \mathfrak{A}(X_0, Y_0) \text{ and } B \in \mathfrak{B}_0(Y_0, Y).$$

It then follows from the definition of B–operators that there exist <u>some</u> B–spaces E_0 and F_0 and operators T_i and $B_i (i = 1, 2)$ such that the following diagram commutes:

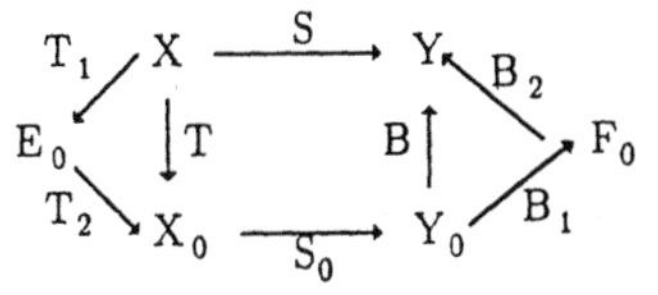

Since $S_0 \in \mathfrak{A}(X_0, Y_0)$ and $\mathfrak{A}$ is an extension of $\mathfrak{A}_0$, it follows that

$$B_1 S_0 T_2 \in \mathfrak{A}(E_0, F_0) = \mathfrak{A}_0(E_0, F_0),$$

and hence from $S = BS_0T = B_2(B_1S_0T_2)T_1$ that $S \in \mathfrak{A}_0^{\text{inf}}$. This proves that $\mathfrak{B}_0 \circ \mathfrak{A} \circ \mathfrak{B}_0 \subset \mathfrak{A}_0^{\text{inf}}$.

(iii) $\mathfrak{A}_0^{\text{rup}} = \mathfrak{B}_0^{-1} \circ (\mathfrak{A} \circ \mathfrak{B}_0)$.

Suppose that $S \in \mathfrak{A}_0^{\text{rup}}$. For any LCS Z and any $L \in \mathfrak{B}_0(Y, Z)$, it is required to show that

$$LS \in \mathfrak{A} \circ \mathfrak{B}_0(X, Z).$$

Indeed, as $L \in \mathfrak{B}_0(Y, Z)$, there exists a B–space F_0 and some operators L_i such that the following diagram commutes:

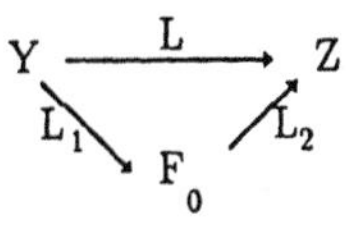

Since $S \in \mathfrak{A}_0^{\text{rup}}$, the operator $L_1 S \in L(X, F_0)$ admits a factorization

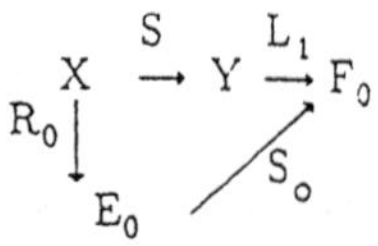

$$\begin{array}{ccccc} X & \xrightarrow{S} & Y & \xrightarrow{L_1} & F_0 \\ {\scriptstyle R_0}\downarrow & & & \nearrow{\scriptstyle S_0} & \\ E_0 & & & & \end{array}$$

for some B–space E_0 and some $S_0 \in \mathfrak{A}_0(E_0, F_0)(=\mathfrak{A}(E_0, F_0))$, hence $L_2S_0 \in \mathfrak{A}(E_0, Z)$, and thus we obtain

$$LS = L_2L_1S = L_2S_0R_0 = (L_2S_0)R_0 \in \mathfrak{A} \circ \mathfrak{B}_0(X, Z)$$

(on account of $R_0 \in \mathfrak{B}_0(X, E_0)$ since $E_0 \in \mathbb{B}$), which proves our assertion.

Conversely, let $S \in \mathfrak{B}_0^{-1} \circ (\mathfrak{A} \circ \mathfrak{B}_0)$. For any B–space F and any $L \in L(Y, F)$, it is required to show that $LS \in L(X, F)$ admits a factorization:

$$\begin{array}{ccccc} X & \xrightarrow{S} & Y & \xrightarrow{L} & F \\ {\scriptstyle R_0}\downarrow & & & \nearrow{\scriptstyle S_0} & \\ E_0 & & & & \end{array}$$

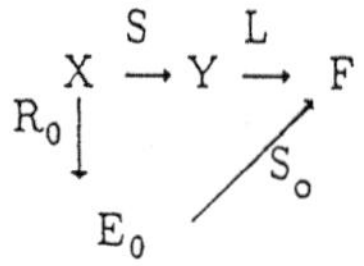

for some B–space E_0 and some operators R_0 and S_0 with

$$R_0 \in L(X, E_0) \text{ and } S_0 \in \mathfrak{A}_0(E_0, F).$$

Indeed, as F is a B–space, it follows that L is a B–operator, and hence from $S \in \mathfrak{B}_0^{-1} \circ (\mathfrak{A} \circ \mathfrak{B}_0)$ that $LS \in \mathfrak{A} \circ \mathfrak{B}_0(X, F)$; thus LS admits a factorization

$$\begin{array}{ccccc} X & \xrightarrow{S} & Y & \xrightarrow{L} & F \\ {\scriptstyle B}\downarrow & & & \nearrow{\scriptstyle T} & \\ X_0 & & & & \end{array}$$

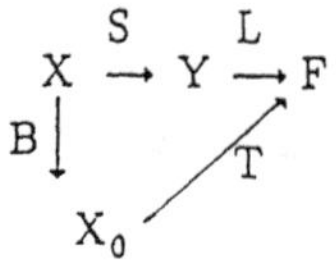

for some LCS X_0 and some operators B and T with

$$T \in \mathfrak{A}(X_0, F) \text{ and } B \in \mathfrak{B}_0(X, X_0).$$

It then follows from $B \in \mathfrak{B}_0(X, X_0)$ that there exists a B–space E_0 and operators $B_1 \in L(X, E_0)$ and $B_2 \in L(E_0, X_0)$ such that the following diagram is commutative:

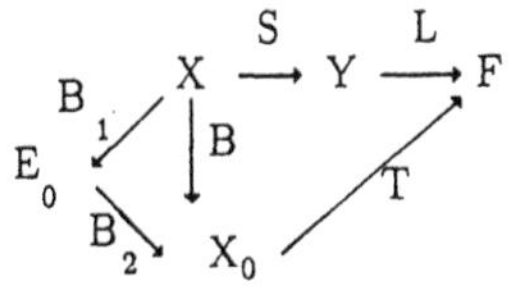

Let

$$R_0 = B_1 \text{ and } S_0 = TB_2.$$

Then $S_0 \in \mathfrak{A}(E_0, F) = \mathfrak{A}_0(E_0, F)$(since $T \in \mathfrak{A}(X_0, F)$) and

$$LS = TB = TB_2B_1 = S_0R_0,$$

which obtains our assertion.

<u>Remark.</u> From the preceding result, we see that

$$\mathfrak{A}_0^{\text{inf}} \subset \mathfrak{A}_0^{\text{rup}} \subset u_0^{\text{sup}}.$$

Moreover, $\mathfrak{A}_0^{\text{sup}}$ is the largest extension of $\mathfrak{A}_0$, while $\mathfrak{A}_0^{\text{inf}}$ is the smallest extension of $\mathfrak{A}_0$ on LCS.

(22·j) <u>Characterizations of</u> $\mathfrak{A}_0^{\text{sup}}$, $\mathfrak{A}_0^{\text{inf}}$ <u>and</u> $\mathfrak{A}_0^{\text{rup}}$(Junek [1983] and Pietsch [1980]): Let $\mathfrak{A}_0$ be an operator ideal on B–spaces. The class of $\tilde{\mathfrak{A}}_0$, defined by $\tilde{\mathfrak{A}}_0(X, Y) = \{T \in L(X, Y): \tilde{T} \in \mathfrak{A}_0(\tilde{X}, \tilde{Y})\}$(for any normed spaces X, Y), (where $\tilde{X}$ denotes the completion of X and $\tilde{T} : \tilde{X} \longrightarrow \tilde{Y}$ is the canonical extension of T) is clearly an operator ideal on V which is an extension of $\mathfrak{A}_0$(called the <u>extension</u> of $\mathfrak{A}_0$ on V). Suppose that X and Y are LCS and that $T \in L(X, Y)$.

(i) (Junek [1983, p.107]) $T \in \mathfrak{A}_0^{\text{sup}}(X, Y)$ if and only if for any bounded disk B in X and any disked 0–neighbourhood U in Y ,

$$Q_U^Y \, T \, J_B^X \in \tilde{\mathfrak{A}}_0(X(B), Y/U),$$

where $J_B^X : X(B) \longrightarrow X$ is the canonical embedding and $Q_U^Y : Y \longrightarrow Y/U$ is the quotient map.

(ii) (Junek [1983, p.107]) $T \in \mathfrak{A}_0^{\text{inf}}(X, Y)$ if and only if T admits the following factorization

$$\begin{array}{ccc} X & \xrightarrow{\ T\ } & Y \\ {\scriptstyle Q_V^X}\downarrow & & \uparrow{\scriptstyle J_A^Y} \\ X_V & \xrightarrow[T_0]{} & Y(A) \end{array}$$

for some disked 0–neighbourhood V in X, some bounded disk A in Y and $T_0 \in \bar{\mathfrak{A}}_0(X_V, Y(A))$.

(iii) (Pietsch [1980, p.397]). $T \in \mathfrak{A}_0^{\text{sup}}(X, Y)$ if and only if for any continuous seminorm q on Y there exists a continuous seminorm p on X with $q(Tx) \leq p(x)$ $(x \in X)$ such that the induced map $T(p,q) : X_p \longrightarrow Y_q$, defined by

$$T(p, q)Q_p^X(x) = Q_q^Y(Tx) \qquad (\text{for any } x \in X),$$

belongs to $\bar{\mathfrak{A}}_0(X_p, Y_q)$.

(22·2) Examples (r) $\mathfrak{F}^{\text{inf}} = \mathfrak{F}^{\text{sup}}$; up to now, no interior characterization of the operators $T \in \mathfrak{F}^{\text{sup}}$ (see Junek [1983, p.107]).

(s) $\mathscr{L}_{\mathrm{B}}^{\text{inf}} = \mathscr{L}_L^{\mathrm{b}}$ and $\mathscr{L}_{\mathrm{B}}^{\text{sup}} = \mathscr{L}$.

(t) $(\mathscr{L}^c)^{\text{inf}} = \mathscr{L}_L^{\text{pb}}$ and $(\mathscr{L}^c)^{\text{sup}} = \mathscr{L}_L^{\text{lp}}$ (locally precompact operators).

(22·k) Factorization problem (Junek [1983, p.111]): Let $\mathfrak{A}_0$ be an operator ideal on B–spaces. Find $\mathbf{A}_1, \mathbf{A}_2 \subset \mathbf{L}$ such that

$$\mathfrak{A}_0^{\text{inf}}(X, Y) = \mathfrak{A}_0^{\text{sup}}(X, Y) \quad (\text{for all } X \in \mathbf{A}_1 \text{ and } Y \in \mathbf{A}_2).$$

Some sufficient conditions are found by Junek [1983,pp.111–128]; for instance, he shows that if $[\mathfrak{A}, A]$ is an injective, surjective quasi–normed operator ideal on B–spaces then

$$\mathfrak{A}^{inf}(X, Y) = \mathfrak{A}^{sup}(X, Y),$$

whenever X is any bornological (DF)–space(for definition, see Junek [1983]) and Y is any metrizable LCS.

(22·13) Definition. (Pietsch [1980, p.400]). Let $\mathfrak{A}_0$ be an operator ideal on B–spaces. We say that a LCS X is:

(i) an $\mathfrak{A}_0$–space if $I_X \in \mathfrak{A}_0^{rup}(X, X)$;

(ii) a co–$\mathfrak{A}_0$–space if $L(G, X) \subset \mathfrak{A}_0^{inf}(G, X)$ (for any normed space G);

(iii) a mix–$\mathfrak{A}_0$–space if $I_X \in \mathfrak{A}_0^{sup}(X, X)$.

We denote by sp $\mathfrak{A}_0$(resp. co $\mathfrak{A}_0$ and mix $\mathfrak{A}_0$) the class of all $\mathfrak{A}_0$–spaces(resp. co–$\mathfrak{A}_0$–spaces, mix–$\mathfrak{A}_0$–spaces); also we write

$$\text{sp}\,\mathfrak{A}_0 = \text{Groth}(\mathfrak{A}_0)$$

and call the Grothendieck space ideal.

In particular, the $\mathscr{L}^c$–spaces(resp. co–$\mathscr{L}^c$–spaces) are called Schwartz spaces (resp. co–Schwartz spaces), the $\mathscr{L}^{wc}$–spaces are called infra–Schwartz spaces, and the $\mathfrak{N}$–spaces(resp. co–$\mathfrak{N}$–spaces) are called nuclear spaces (resp. co–nuclear spaces).

As a consequence of (22·j) and (22·13), we obtain the following :

(22·14) **Theorem** (Pietsch [1980]). Let $\mathfrak{A}_0$ be an operator ideal on B–spaces and X a LCS.

(a) X is an $\mathfrak{A}_0$–space if and only if

$$L(X, F) \subseteq \mathfrak{A}_0^{inf}(X, F) \quad \text{(for any B–space F)},$$

and this is the case if and only if for any continuous seminorm q on X there exists a continuous seminorm p on X with $q \le p$ such that the canonical map $Q_{q,p} : \tilde{X}_p \longrightarrow \tilde{X}_q$ (for definition, see §10) belongs to $\mathfrak{A}_0(\tilde{X}_p, \tilde{X}_q)$.

(b) X is a co–$\mathfrak{A}_0$–space if and only if for any bounded disk A in X there exists a bounded disk B in X with $A \subset B$ such that the canonical embedding $J_{A,B} : \tilde{X}(A) \longrightarrow \tilde{X}(B)$ (for definition, see §10) belongs to $\mathfrak{A}_0(\tilde{X}(A), \tilde{X}(B))$.

(c) X is a mix–$\mathfrak{A}_0$–space if and only if for any bounded disk A in X and any continuous seminorm p on X, the unique extension $\widetilde{Q_p J_A} : \tilde{X}(A) \longrightarrow \tilde{X}_p$ of the map $Q_p J_A : X(A) \longrightarrow X_p$ belongs to $\mathfrak{A}_0(\tilde{X}(A), \tilde{X}_p)$.

Different operator ideals can generate the same class of operators. In order to study this, we require the following notion which is important for the theory of $\mathfrak{A}_0$–spaces. Let $\mathfrak{A}_1$ and $\mathfrak{A}_2$ be operator ideals on B–spaces. We say that $\mathfrak{A}_1$ and $\mathfrak{A}_2$ are equivalent, denoted by $\mathfrak{A}_1 \sim \mathfrak{A}_2$, if it satisfies the following conditions:

(i) there is some natural number $n \ge 1$ such that $\mathfrak{A}_1^n \subset \mathfrak{A}_2$ (denoted by $\mathfrak{A}_1 \prec \mathfrak{A}_2$);

(ii) $\mathfrak{A}_2 \prec \mathfrak{A}_1$.

(22·15) **Proposition** (Pietsch [1980, p.401]) Let $\mathfrak{A}_0$ and $\mathscr{B}_0$ be two operator ideals on B–spaces.

(a) If $\mathfrak{A}_0 \subset \mathscr{B}_0$, then

$$\mathrm{sp}\, \mathfrak{A}_0 \subset \mathrm{sp}\, \mathscr{B}_0 \ \text{ and } \ \mathrm{co}\, \mathfrak{A}_0 \subset \mathrm{co}\, \mathscr{B}_0.$$

(b) For any natural number $n \ge 1$, we have

$$\mathrm{sp}\, \mathfrak{A}_0^n = \mathrm{sp}\, \mathfrak{A}_0 \ \text{ and } \ \mathrm{co}\, \mathfrak{A}_0^n = \mathrm{co}\, \mathfrak{A}_0.$$

(c) If $\mathfrak{A}_0 \sim \mathscr{B}_0$ then

$$\mathrm{sp}\, \mathfrak{A}_0 = \mathrm{sp}\, \mathscr{B}_0 \ \text{ and } \ \mathrm{co}\, \mathfrak{A}_0 = \mathrm{co}\, \mathscr{B}_0.$$

Proof. (a) Follows from (22·14) (a) and (b).

(b) As $\mathfrak{A}_0^n \subset \mathfrak{A}_0$, it follows from (a) that

$$\mathrm{sp}\,\mathfrak{A}_0^n \subset \mathrm{sp}\,\mathfrak{A}_0 \quad \text{and} \quad \mathrm{co}\,\mathfrak{A}_0^n \subset \mathrm{co}\,\mathfrak{A}_0.$$

Hence, it has to show that

$$\mathrm{sp}\,\mathfrak{A}_0 \subset \mathrm{sp}\,\mathfrak{A}_0^2 \quad \text{and} \quad \mathrm{co}\,\mathfrak{A}_0 \subset \mathrm{co}\,\mathfrak{A}_0^2.$$

In fact, let $X \in \mathrm{sp}\,\mathfrak{A}_0$. For any continuous seminorm q on X there exists a continuous seminorm p on X with $q \leq p$ such that $Q_{q,p} \in \mathfrak{A}_0(\tilde{X}_p, \tilde{X}_q)$. For this p, there exists a continuous seminorm r on X with $p \leq r$ such that $Q_{p,r} \in \mathfrak{A}_0(\tilde{X}_r, \tilde{X}_p)$, thus

$$Q_{q,r} = Q_{q,p} \circ Q_{p,r} \in \mathfrak{A}_0^2(\tilde{X}_r, \tilde{X}_q).$$

Consequently, $X \in \mathrm{sp}\,\mathfrak{A}_0^2$ (by (22·15) (a)).

The proof of $\mathrm{co}\,\mathfrak{A}_0 \subset \mathrm{co}\,\mathfrak{A}_0^2$ is similar to that of $\mathrm{sp}\,\mathfrak{A}_0 \subset \mathrm{sp}\,\mathfrak{A}_0^2$, and hence will be omitted.

(c) Follows from (a) and (b) of this result.

Permanence properties and dual spaces of $\mathfrak{A}_0$–spaces can be found in the book of Junek [1983, pp.136–143].

Exercises

22–1. Prove (22.a).

22–2. Prove (22.b).

22–3. Prove (22.c).

22–4. Prove (22.d).

22–5. Prove (22.e).

22–6. Prove (22.f)(iv) and (v).

22–7. Prove (22.g).

22–8. Prove (22.h).

22–9. Prove (22.j).

22–10. Prove (22.k).

22–11. Let $\mathcal{U}$ be an operator ideal on B–spaces.

(a) $S_0 \in \mathcal{U}^{inj}(E,F_0)$ if and only if there exists a B–space F and $S \in \mathcal{U}(E,F)$ such that

$$\|S_0x\| \leq \|Sx\| \quad \text{(for all } x \in E) .$$

(b) Dually $T_0 \in \mathcal{U}^{sur}(E_0,F)$ if and only if there exists a B–space E and $T \in \mathcal{U}(E,F)$ such that

$$T_0(U_{E_0}) \subset T(U_E) \quad \text{(or equivalently } \operatorname{Im} T_0 \subset \operatorname{Im} T).$$

(Hint : Use Ex. 6–18.)

23

An Aspect of Fixed Point Theory

(by **Shih Mau-Hsiang**, Chung-yuan University)

This section is devoted to an exposition of an aspect of fixed point theory that provides a significant part of linear and nonlinear functional analysis. The main contribution of this part of fixed point theory is due to Ky Fan.

(23.1) Definiton. Let $v_0, v_1, \cdots, v_n$ be $n+1$ points in a Euclidean space which are not contained in any linear manifold (i.e., a translation of a vector subspace) of dimension less than n. The convex hull of these $n+1$ points is called an n–simplex and is denoted by $v_0v_1\cdots v_n$. The points $v_0, v_1, \cdots, v_n$ are called vertices of the simplex.

For $0 \leq k \leq n$ and $0 \leq i_0 < i_1 < \cdots < i_k \leq n$, the k–simplex $v_{i_0}v_{i_1}\cdots v_{i_k}$ is a subset of the n–simplex $v_0v_1\cdots v_n$; it is called a k–dimensional face (or simply k–face) of $v_0v_1\cdots v_n$.

Suppose that $T = \{ \tau_0, \tau_1, \cdots, \tau_m \}$ is a finite collection of n– simplexes in $v_0v_1\cdots v_n$ such that

(i) $\bigcup_{i=1}^{m} \tau_i = v_0v_1\cdots v_n$, and

(ii) $\tau_i \cap \tau_j$ is either empty or a common face (of any dimension) of τ_i and τ_j $(i, j = 1, \cdots, m)$.

Then T is called a triangulation of $v_0v_1\cdots v_n$.

Let T be a triangulation. A vertex of a simplex in T is called a vertex of T. An $(n-1)$–face of a simplex from T is called a boundary (n–1)–simplex of T if it is a face of exactly one n–simplex of T.

We begin with the celebrated Sperner's combinatorial lemma (Sperner [1928]).

(23.2) **Sperner's Lemma.** Let T be a triangulation of an n–simplex $\sigma = v_0v_1\cdots v_n$. To each vertex v of T, let $\varphi(v) \in \{ 0, 1, \cdots, n \}$ be such that whenever

$v \in v_{i_0}v_{i_1}\cdots v_{i_k}$, $\varphi(v) \in \{ i_0, i_1, \cdots, i_k \}$ (for $0 \leq k \leq n$ and $0 \leq i_0 < i_1 < \ldots < i_k \leq n$) . Then the number of n–simplexes in T whose vertices receive $n+1$ different values is odd.

Proof. We may think of T as a house; the n–simplexes of T are the rooms of the house. A room will be considered a good room if its $n+1$ vertices are labelled with $n+1$ diffferent integers, otherwise it will be called a bad room. If the vertices of an $(n-1)$–face of a simplex τ of T are labelled with the integers $0, 1, \cdots, n-1$, then that face is considered to be a door to the room τ. A door which is a boundary $(n-1)$–simplex of T is called an external door. Clearly a good room has exactly one door and a bad room has either no doors or two doors. (The accompaning figure shows a triangulation of a 2–simplex in $\mathbb{R}^2$; the external doors are marked with ——•——•——

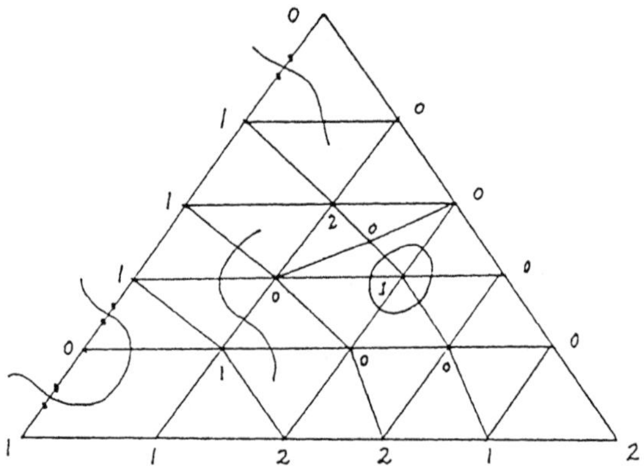

Let us trace paths through the house, going through each door exactly one. To start a path we either enter a room through an external door or we leave a good room through its unique door. Every time we enter a room through a door and see a second door in the room, we go out of the room through the second door. This will lead us either to the outside of the house or to a new room. According to this rule, our path will be stopped either when we get to the outside of the house through an external door, or when we enter a good room. After going through each door exactly once by tracing various paths,we pair off the two ends of each path. In this way each good room is paired either with another good room or with an external door. Also each external door is paired either with another external door or with a good room. This pairing implies that the number of good rooms and the number of external doors are congruent modulo 2, i.e., these numbers have the same parity.

To complete the proof of Sperner's lemma, it suffices to show that the number of boundary $(n-1)$–simplexes in T with vertices labelled $0, 1, \cdots, n-1$ is odd. This is immediate by induction on n.

Remark. (i) The above proof is due to Cohen [1967], but the illustration with doors and rooms of a house is due to Ky Fan [1967], who uses this argument to prove a combinatorial result more general than Sperner's lemma.

(ii) Sperner's lemma with the consideration of orientation is due to Brown/Cairns [1961].

(iii) Ky Fan's conjecture in 1978, related to Sperner's lemma, was proved by Shih/Lee [1992].

(23.3) **Lemma.** (Knaster–Kuratowski–Mazurkiewicz [1929]). Let $F_0, F_1, \cdots, F_n$ be $n+1$ closed subsets of an n–simplex $v_0v_1\cdots v_n$. Suppose that for each $0 \leq k \leq n$ and $0 \leq i_0 < i_1 < \ldots < i_k \leq n$, we have

$$(1) \qquad v_{i_0}v_{i_1}\cdots v_{i_k} \subset F_{i_0} \cup F_{i_1} \cup \cdots \cup F_{i_k} .$$

Then $\bigcap_{i=0}^{n} F_i \neq \phi$.

Proof. Let T be an arbitrary triangulation of the simplex $v_0v_1\cdots v_n$. For each vertex v of T, we assign an integer $\varphi(v) \in \{0, 1, \cdots, n\}$ such that $v \in F_{\varphi(v)}$ and $\varphi(v) \in \{i_0, i_1, \cdots, i_k\}$ whenever $v \in \{v_{i_0}, v_{i_1}, \cdots, v_{i_k}\}$, this is possible by virtue of (1). Thus, by Sperner's lemma, there exists a simplex $w_0w_1 \ldots w_n \in T$ such that

$$\varphi(w_i) = i \quad (i = 0, 1, \cdots, n).$$

It follows that

$$w_i \in F_{\varphi(w_i)} = F_i \qquad (i = 0, 1, \cdots, n).$$

Consider now a sequence of triangulations $\{T^{(m)}\}$ of $v_0v_1...v_n$ such that the mesh of $T^{(m)}$ tends to zero (as $m \longrightarrow \infty$). From the above argument, one can find, for each m, that an n–simplex $w_0^{(m)}w_1^{(m)}...w_n^{(m)}$ in T such that $w_i^{(m)} \in F_i$ $(i = 0,1, ... ,n)$. By the compactness of $v_0v_1...v_n$, we can choose a subsequence $\{m_j\}$ of $\{m\}$ with $m_1 < m_2 < ...$ such that $\lim_j w_i^{(m_j)} = q_i$ exists $(i = 0,1,..,n)$. As the mesh of $T^{(m_j)}$ tends to zero (as $j \to \infty$), it follows that $q_0=q_1=..=q_n$, and hence that $\lim_j w_i^{(m_j)} = q$ (for all $i = 0,1,..,n$). Since F_i $(i=0,1,..,n)$ are closed, we conclude that $q \in F_i$ (for all $i = 0,1,..,n$), and hence that $\bigcap_{i=0}^{n} F_i \neq \emptyset$. This completes the proof.

(23.4) **Definition.** Let B be a convex set in a vector space X, and let A be a non–empty subset of B. The family $\{ F(x) : x \in A \}$ of subsets of B is said to be a KKM covering for B if

$$co\{ x : x \in \alpha \} \subset \cup \{ F(x) : x \in \alpha \}$$

for any finite set $\alpha \subset A$.

The following infinite dimensional generalization of Knaster–Kuratowski–Mazurkiewicz's lemma (or simply KKM lemma) is due to Ky Fan [1961].

(23.5) **Theorem** (KKM coverings). Let B be a convex set in a Hausdorff TVS X, let $\emptyset \neq A \subset B$ and $\{F(x) : x \in A\}$ a KKM covering for B. If there exists an $a \in A$ such that $\overline{F(a)}$ is compact, then $\bigcap_{x \in A} \overline{F(x)} \neq \phi$.

Proof. By compactness of $\overline{F(a)}$, it suffices to show that the family $\{ \overline{F(x)}: x \in A \}$ has the finite intersection prpoerty. Let $\{ x_0, x_1, \cdots, x_n \} \subset A$ and let $\sigma = v_0 v_1 \cdots v_n$ be an n–simplex. Define $\psi : \sigma \longrightarrow B$ by setting

(1) $$\psi(\Sigma_{i=0}^{n} \mu_i v_i) = \Sigma_{i=0}^{n} \mu_i x_i$$

for each $\mu_0, \mu_1, \cdots, \mu_n \geq 0$ with $\Sigma_{i=0}^{n} \mu_i = 1$. Then ψ is continuous, hence the set $G_i = \psi^{-1}(\overline{F(x_i)})$ is closed in σ (for each $i = 0, 1, 2, \cdots, n$). If $0 \leq k \leq n$ and $0 \leq i_0 < i_1 < \cdots < i_k \leq n$, we claim that

$$v_{i_0} v_{i_1} \cdots v_{i_k} = \subset \bigcup_{j=0}^{k} G_{i_j} .$$

Indeed, since $co\{x_{i_0}, x_{i_1}, \cdots, x_{i_k}\} \subset \bigcup_{j=0}^{k} F(x_{i_j}) \subset \bigcup_{j=0}^{k} \overline{F(x_{i_j})}$, it follows from (1) that

$$\begin{aligned} v_{i_0} v_{i_1} \cdots v_{i_k} &= co\{ v_{i_0}, v_{i_1}, \cdots, v_{i_k} \} \\ &= \psi^{-1}(co\{ x_{i_0}, x_{i_1}, \cdots, x_{i_k} \}) \\ &\subset \psi^{-1}(\bigcup_{j=0}^{k} \overline{F(x_{i_j})}) = \bigcup_{j=0}^{k} \psi^{-1}(\overline{F(x_{i_j})}) = \bigcup_{j=0}^{k} G_{i_j} . \end{aligned}$$

By (23.3), $\bigcap_{i=0}^{n} G_i \neq \emptyset$, it then follows from $\bigcap_{j=0}^{n} G_j = \bigcap_{i=0}^{n} \psi^{-1}(\overline{F(x_i)}) = \psi^{-1}(\bigcap_{i=0}^{n} \overline{F(x_i)})$ that $\bigcap_{i=0}^{n} \overline{F(x_i)} \neq \emptyset$, which obtains our assertion.

Finite dimensional generalizations of KKM lemma may be found in Ky Fan [1970], Shapley [1973] and Gale [1984]. Selection theorem of KKM lemma may be found in Shih [1986 a] while infinite dimensional generalizations of KKM lemma can be found in Brézis–Nirenberg– Stampacchia [1972], Ky Fan [1981] and [1984], and Shih [1986 b].

In order to give some important applications of (23.5), we require the following terminology. Recall that a real–valued function φ on a topological space Ω is <u>lower</u> (respectively <u>upper</u>) <u>semi–continuous</u> if the set $\{ x \in \Omega : \varphi(x) \leq \lambda \}$ (respectively $\{ x \in \Omega : \varphi(x) \geq \lambda \}$) is closed in Ω for each $\lambda \in \mathbb{R}$; if B is a convex set in a vector space, then a real–valued function f on B is said to be <u>quasi–concave</u> (respectively <u>quasi–convex</u>) if $\{ x \in B : f(x) > \lambda \}$ (respectively $\{ x \in B : f(x) < \lambda \}$ is convex for each

$\lambda \in \mathbb{R}$.

(23.6) **Theorem** (Ky Fan's minimax inequality [1972]). Let K be a non-empty compact convex subset of a Hausdorff TVS X, and let φ be a real-valued function on the product space $K \times K$ satisfying the following conditions:

(i) for each fixed $x \in K$, the function $\varphi(x, \cdot)$ is lower semi-continuous on K;

(ii) for each fixed $y \in K$, the function $\varphi(\cdot, y)$ is quasi-concave on K.

Then there exists $\hat{y} \in K$ such that $\varphi(x, \hat{y}) \leq \sup_{x \in K} \varphi(x,x)$; consequently,

$$\min_{y \in K} \sup_{x \in K} \varphi(x,y) \leq \sup_{x \in K} \varphi(x,x).$$

Proof. Let $\rho = \sup_{x \in K} \varphi(x,x)$. We may assume that $\rho \neq +\infty$. Suppose, for any $x \in K$, that

$$F(x) = \{ y \in K : \varphi(x, y) \leq \rho \}.$$

By (i), each $F(x)$ is closed in K, and hence compact in K (by the compactness of K). We claim that $\{ F(x): x \in K \}$ is a KKM covering for K. Indeed, suppose, on the contrary, that there exists $\{x_0, x_1, \ldots, x_n\} \subset K$ and $\alpha_i \geq 0$ $(i = 0, 1, \cdots, n)$ with $\Sigma_{i=0}^{n} \alpha_i = 1$ such that $w = \Sigma_{i=0}^{n} \alpha_i x_i \notin \bigcup_{i=0}^{n} F(x_i)$. Then $\varphi(x_i, \Sigma_{i=0}^{n} x_i) > \rho$ (for $i = 0, 1, \cdots, n$), hence the quasi-concavity of $\varphi(\cdot, w)$ ensures that $\varphi(w, w) > \rho$, which gives a contradiction. By (23.5), $\bigcap_{x \in K} F(x) \neq \emptyset$. Take $\hat{y} \in \bigcap_{x \in K} F(x)$, then this $\hat{y}$ has the required property.

(23.7) **Theorem** (Ky Fan [1959, p.118]). Let K be a non-empty compact convex subset of a Hausdorff TVS X, and let ψ be a real-valued continuous function defined on $K \times K$. If for each fixed $y \in K$, the function $\psi(\cdot, y)$ is quasi-convex on K, then there exists a point $\hat{y} \in K$ such that

$$\psi(x, \hat{y}) \geq \psi(\hat{y}, \hat{y}) \qquad (\text{for all } x \in K).$$

Proof. Define φ on $K \times K$ by

$$\varphi(x, y) = \psi(y, y) - \psi(x, y) .$$

Then the assertion follows from (23.6).

(23.a) A geometric formulation of Ky Fan's minimax inequality (see Ky Fan [1972,pp.104–106]) : Let K be a non–empty compact convex set in a Hausdorff TVS, let B be a subset of $K \times K$ such that :

(i) For each $x \in K$, the set $\{y \in K : (x,y) \in B\}$ is open in K.

(ii) For each $y \in K$, the set $\{x \in K : (x,y) \in B\}$ is non–empty and convex.

Then there exists a point $\hat{x} \in K$ such that $(\hat{x}, \hat{x}) \in B$.

The following generalization of Ky Fan's minimax inequality, which is due to Shih/Tan, is more useful for the study of variational inequalities.

(23.8) Theorem. Let B be a non–empty convex subset of a Hausdorff TVS X, and φ, ψ two real–valued functions on the product space $B \times B$. Suppose that

(i) $\varphi(x,y) \leq \psi(x,y)$ (for all $(x,y) \in B \times B$) and $\psi(x,x) \leq 0$ (for all $x \in B$).

(ii) For each fixed $x \in B$, the function $\varphi(x,\cdot)$ is lower semi–continuous on B.

(iii) For each fixed $y \in B$, the function $\psi(x,y)$ is quasi–concave on B.

(iv) There exists a non–empty compact subset K of B and $x_0 \in B$ such that $\psi(x_0,y) > 0$ for all $y \in B \setminus K$.

Then there exists $\hat{y} \in B$ such that $\varphi(x,\hat{y}) \leq 0$ for all $x \in B$.

Proof. For each $x \in B$, let

$$F(x) = \{y \in B : \psi(x,y) \leq 0\},$$

$$G(x) = \{ y \in B : \varphi(x, y) \leq 0 \}.$$

Then the following assertions hold :

(a) by (i), $F(x) \subset G(x)$ (for each $x \in B$);

(b) by (ii), $G(x)$ is closed in B (for each $x \in B$);

(c) by (iv), $F(x_0) \subset K$ so that $\overline{F(x_0)}$ is compact;

(d) by (i) and (iii), $\{ F(x) : x \in B \}$ is a KKM covering for B.

Thus by (23.5), we have $\bigcap_{x \in B} \overline{F(x)} \neq \emptyset$. By (a) and (b), $\bigcap_{x \in B} G(x) \neq \emptyset$. Take $\hat{y} \in \bigcap_{x \in B} G(x) \neq \emptyset$; then $\varphi(x, \hat{y}) \leq 0$ (for all $x \in B$).

As an immediate application of (23.7), we obtain Tychonoff's fixed point theorem (Tychonoff [1935]) as follows :

(23.9) **Theorem** (Tychonoff). Let K be a non–empty compact convex set in a LCS X. Then every continuous map $f : K \longrightarrow K$ has a fixed point, i.e., there exists $\hat{x} \in K$ such that $f(\hat{x}) = \hat{x}$.

Proof. As X is locally convex, there exists a family $\{ p_\nu \}_{\nu \in I}$ of continous semi–norms on X determining the topology. For each $\nu \in I$, let

$$F_\nu = \{ x \in K : p_\nu(x - f(x)) = 0 \}.$$

Then each F_ν is a closed subset of K. A point $\hat{x} \in K$ is a fixed point of f if and only if $\hat{x} \in \bigcap_{\nu \in I} F_\nu$. By the compactness of K, it suffices to show that $\bigcap_{i=1}^{n} F_{\nu_i} \neq \emptyset$ (for any finite set $\{\nu_1, \nu_2, \cdots, \nu_n\} \subset I$). Given $\{ \nu_1, \nu_2, \cdots, \nu_n \} \subset I$, let us define $\psi : K \times K \longrightarrow \mathbb{R}$ by

$$\psi(x, y) = \Sigma_{i=1}^{n} p_{\nu_i}(x - f(y)).$$

Then ψ is continuous, and for each $y \in K$, the function $\psi(\cdot, y)$ is quasi–convex on K, hence by (23.7), there exists $\hat{y} \in K$ such that $\psi(x, \hat{y}) \geq \psi(\hat{y}, \hat{y})$ (for all $x \in K$), i.e.,

$$(1) \qquad \Sigma_{i=1}^{n} p_{\nu_i}(x - f(\hat{y})) \geq \Sigma_{i=1}^{n} p_{\nu_i}(\hat{y} - f(\hat{y})) \quad \text{(for all } x \in K).$$

It then follows from $f(\hat{y}) \in K$ that $\inf_{x \in K} \Sigma_{i=1}^{n} p_{\nu_i}(x - f(\hat{y})) = 0$, and hence from (1) that

$p_{\nu_i}(x - f(\hat{y})) = 0$ $(i=1,2, \cdots ,n)$; in other words, $\hat{y} \in \bigcap_{i=1}^{n} F_{\nu_i}$, which proves our assertion.

Remark. When $X = \mathbb{R}^n$, Tychonoff's theorem reduces to the celebrated Brouwer's fixed point theorem. Brouwer's theorem was extended to Banach spaces by Schauder [1930].

The following version of Schauder's fixed point theorem is more suitable for applications.

(23.10) Theorem. (Schauder). Let C be a non–empty closed convex set in a B–space E and $f: C \to C$ a mapping. If f is continuous and $f(C)$ is relatively compact (in E), then f has a fixed point.

Proof. The completeness of E ensures that $K = \overline{co}\, f(C)$ is a compact (convex) set in E. As $f(C) \subset C$, it follows from the closedness and convexity of C that $\overline{co}\, f(C) \subset C$, and hence that

$$f(K) = f(\overline{co}\, f(C)) \subset f(C) \subset \overline{co}\, f(C) = K.$$

Therefore, (23.9) implies that f has a fixed point in K (since K is compact convex) and *a fortiori* in C (since $K = \overline{co}\, f(C) \subset C$).

Schauder's fixed point theorem has found many interesting applications, one of them which deals with linear problem is the following Lomonosov's partial solution of invariant subspace problem.

(23.b) (Lomonosov). Let T be a bounded linear operator on a normed space E. If there exists a non–zero compact operator S such that $TS = ST$, then T has a non–trivial closed invariant subspace.

(23.11) **Theorem** (Markoff [1936] and Kakutani [1938]). Let K be a non–empty compact convex set in a LCS X, and let $\mathfrak{F}$ be a commuting family of continuous affine maps of K into itself. Then $\mathfrak{F}$ has a common fixed point, i.e., there exists $\hat{x} \in K$ such that $f(\hat{x}) = \hat{x}$ for all $f \in \mathfrak{F}$.

Proof. For $f \in \mathfrak{F}$, let $F(f) = \{ x \in K : f(x) = x \}$. Then $F(f) \neq \emptyset$ (for each $f \in \mathfrak{F}$) by Tychonoff's fixed point theorem. As K is compact and each $f \in \mathfrak{F}$ is continuous and affine, $F(f)$ is compact convex. A point $\hat{x} \in K$ is a common fixed point for $\mathfrak{F}$ if and only if $\cap \{ F(f) : f \in \mathfrak{F} \} \neq \emptyset$. By the compactness of K, it suffices to show that $\{ F(f) : f \in \mathfrak{F} \}$ has the finite intersection property. Let $f_1, f_2, \cdots, f_n \in \mathfrak{F}$. We claim that $\bigcap_{i=1}^{n} F(f_i) \neq \emptyset$. The result is true for $n = 1$. Assume that $\bigcap_{i=1}^{n-1} F(f_i) \neq \emptyset$. If $x \in \bigcap_{i=1}^{n-1} F(f_i)$, then $f_i(f_n(x)) = f_n f_i(x) = f_n(x)$ (for each $i < n$), so that $f_n(x) \in \bigcap_{i=1}^{n-1} F(f_i)$. Thus

$$f_n\Big(\bigcap_{i=1}^{n-1} F(f_i)\Big) \subset \bigcap_{i=1}^{n} F(f_i) .$$

As $\bigcap_{i=1}^{n-1} F(f_i)$ is non–empty compact convex, by Tychonoff's fixed point theorem, there exists $u \in \bigcap_{i=1}^{n-1} F(f_i)$ such that $f_n(u) = u$. Thus $\bigcap_{i=1}^{n} F(f_i) \neq \emptyset$. This completes the proof.

Remark. Markoff–Kakutani's fixed point theorem can be used to give a rather simple proof for the existence of Haar measures on compact Abelian groups (see, e.g., Larsen [1973, pp.362–365]). Kakutani showed that Markoff–Kakutani's fixed point theorem implies the Hahn–Banach's extension theorem (see Kakutani [1938]).

Given a Cartesian product $K = \prod_{i=1}^{n} K_i$, let

$$\hat{K}^i = \prod_{j \neq i} K_j \text{ and } \hat{x}^i = (x_1, \cdots, x_{i-1}, x_{i+1}, \cdots, x_n) \in \hat{K}_i.$$

For $x = (x_1, \cdots, x_n)$, $y = (y_1, \cdots, y_n)$, let

$$(x_i, \hat{y}^i) = (y_1, \cdots, y_{i-1}, x_i, y_{i+1}, \cdots, y_n).$$

(23.12) **Theorem** (Ky Fan [1964]). Let $n \geq 2$, let $K_1, K_2, \cdots, K_n$ be non-empty compact convex sets in a Hausdorff TVS X, and let $f_1, f_2, \cdots, f_n$ be real-valued functions on the product space $K = \prod_{i=1} K_i$ satisfying the following conditions:

(i) for each $i = 1, 2, \cdots, n$ and for each fixed $x_i \in K_i$, $f_i(x_i, \hat{x}^i)$ is a lower semi-continuous function of $\hat{x}^i$ on $\hat{K}^i$;

(ii) for each $i = 1, 2, \ldots, n$ and for each fixed $\hat{x}^i \in \hat{K}^i$, $f_i(x_i, \hat{x}^i)$ is a quasi-concave function of x_i on K_i.

Suppose that t_i $(1 \leq i \leq n)$ are real numbers, and that for every i and for each point $\hat{y}^i \in \hat{K}^i$, there exists a point $x_i \in K_i$ such that $f_i(x_i, \hat{y}^i) > t_i$. Then there is a point

$$\tilde{x} = (\tilde{x}_1, \tilde{x}_2, \cdots, \tilde{x}_n) \in K$$

such that $f_i(\tilde{x}_1, \tilde{x}_2, \cdots, \tilde{x}_n) > t_i$ $(i = 1, 2, \cdots, n)$.

Proof. Define $\varphi : K \times K \longrightarrow \mathbb{R}$ by setting

$$\varphi(x, y) = \min_{1 \leq i \leq n} \{ f_i(x_i, \hat{y}^i) - t_i \},$$

for $x = (x_1, x_2, \cdots, x_n)$, $y = (y_1, y_2, \cdots, y_n) \in K$. Then condition (i) implies that for each $x \in K$, $\varphi(x, y)$ is a lower semi-continuous function of y on K. Condition (ii) implies that for each $y \in K$, $\varphi(x, y)$ is a quasi-concave function of x on K. On the other hand, the last part of the hypothesis is equivalent to saying that for every $y \in K$, there is an $x \in K$ satisfying $\varphi(x, y) > 0$. Thus by Ky Fan's minimax inequality (23.6), there exists $\tilde{x} \in K$ such that $\varphi(\tilde{x}, \tilde{x}) > 0$, which means the existence of a point $(\tilde{x}_1, \tilde{x}_2, \cdots, \tilde{x}_n) \in K$ satisfying $f_i(\tilde{x}_1, \tilde{x}_2, \cdots, \tilde{x}_n) > t_i$ $(i = 1, 2, \ldots, n)$. This proves the theorem.

The following is a geometric formulation of (23.11) (see Ky Fan [1964]).

(23.c) (Ky Fan). Let $n \geq 2$, let $K_1, K_2, \cdots, K_n$ be non-empty compact convex subsets of a Hausdorff TVS X, and let $A_1, A_2, \cdots, A_n$ be subsets of $\prod_{i=1}^{n} K_i$ having the following two properties:

(i) for each i and for every $x_i \in K_i$, the section

$$A_i(x_i) = \{ \hat{x}^i : x_i \in A_i \} \text{ is open in } \hat{K}^i ;$$

(ii) for each i and for every $\hat{x}^i \in \hat{K}^i$, the section

$$A_i(\hat{x}^i) = \{ x_i : \hat{x}^i \in A_i \} \text{ is non-empty and convex.}$$

Then $\bigcap_{i=1}^{n} A_i \neq \emptyset$.

(23.13) Theorem (Nash's equilibrium point theorem (Nash [1950])). Let $n \geq 2$, let $K_1, K_2, \cdots, K_n$ be non-empty compact convex sets in a Hausdorff TVS X, and let $f_1, f_2, \cdots, f_n$ be continuous real-valued functions defined on the product space $K = \prod_{i=1}^{n} K_i$. Suppose that for each $i = 1, 2, \cdots, n$ and for each fixed $\hat{x}^i \in \hat{K}^i$, $f_i(x_i, \hat{x}^i)$ is a quasi-concave function of x_i on K_i. Then there exists a point $\overline{x} = (\overline{x}_1, \overline{x}_2, \cdots, \overline{x}_n) \in K$ (called an equilibrium point) such that

$$f_i(\overline{x}_1, \overline{x}_2, \cdots, \overline{x}_n) = \max \{ f_i(y_i, \hat{x}^i) : y_i \in K_i \} \quad (i = 1,2,\cdots,n).$$

Proof. For each $i = 1,2,\cdots,n$ and $x = (x_1,x_2,\cdots,x_n) \in K$, let

$$g_i(x) = f_i(x) - \max \{ f_i(y_i, \hat{x}^i) : y_i \in K_i \} .$$

Suppose that $\epsilon > 0$ and $t_i = -\epsilon$ $(i = 1, 2, \cdots, n)$. Then $g_1, g_2, \cdots, g_n$ satisfy the hypotheses of (23.12), so by (23.12) we have

$$A(\epsilon) = \{ x \in K : f_i(x) \geq \max \{ f_i(y_i, \hat{x}^i) : y_i \in K_i \} - \epsilon \} \neq \emptyset.$$

As K is compact, it follows that $\bigcap_{\epsilon>0} A(\epsilon) \neq \emptyset$. Let $\bar{x} \in \bigcap_{\epsilon>0} A(\epsilon)$. Then $\bar{x}$ is the required equilibrium point.

(23.14) Von Neumann's minimax principle (Sion [1958]). Let K_1 and K_2 be non–empty compact convex sets in a Hausdorff TVS X, and let φ be a real–valued function on $K_1 \times K_2$ satisfying the following two conditions:

(i) for each fixed $x \in K_1$, the function $y \longrightarrow \varphi(x, y)$ is lower semi–continuous and quasi–convex on K_2;

(ii) for each fixed $y \in K_2$, the function $x \longrightarrow \varphi(x, y)$ is upper semi–continuous and quasi–concave on K_1.

Then

$$\min_{y \in K_2} \max_{x \in K_1} \varphi(x, y) = \max_{x \in K_1} \min_{y \in K_2} \varphi(x, y) .$$

Proof. We always have $\min_{y \in K_2} \varphi(t,y) \leq \min_{y \in K_2} \max_{x \in K_1} \varphi(x,y)$ (for all $t \in K_1$), hence

$$\max_{x \in K_1} \min_{y \in K_2} \varphi(x, y) \leq \min_{y \in K_2} \max_{x \in K_1} \varphi(x, y) .$$

To prove the reverse inequality, let $f_1 = \varphi$, $f_2 = -\varphi$, $\epsilon > 0$ and define

$$t_1 = \min_{y \in K_2} \max_{x \in K_1} \varphi(x, y) - \epsilon ,$$

$$t_2 = -\max_{x \in K_1} \min_{y \in K_2} \varphi(x, y) - \epsilon.$$

Then for each $y \in K_2$, there exists $\tilde{x} \in K_1$ such that $f_1(\tilde{x}, y) > t_1$; and for each $x \in K_1$, there exists $\tilde{y} \in K_2$ such that $f_2(x, \tilde{y}) > t_2$. Applying (23.12), there exists $(x_\epsilon, y_\epsilon) \in K_1 \times K_2$ such that $f_1(x_\epsilon, y_\epsilon) > t_1$ and $f_2(x_\epsilon, y_\epsilon) > t_2$, hence

$$\min_{y \in K_2} \max_{x \in K_1} \varphi(x, y) - \epsilon < \varphi(x_\epsilon, y_\epsilon) < \max_{x \in K_1} \min_{y \in K_2} \varphi(x, y) + \epsilon ,$$

and thus

$$\min_{y\in K_2}\max_{x\in K_1}\varphi(x,y)-\epsilon\le\max_{x\in K_1}\min_{y\in K_2}\varphi(x,y)+\epsilon .$$

As $\epsilon > 0$ was arbitrary, we conclude that

$$\min_{y\in K_2}\max_{x\in K_1}\varphi(x,y)\le\max_{x\in K_1}\min_{y\in K_2}\varphi(x,y),$$

thus the proof is complete.

(23.15) Definition (Pietsch [1980, p.40]). A collection $\mathscr{M}$ of real–valued functions, defined on a set X, is called concave if, given any finite subset $\{ f_1, f_2, \cdots ,f_n \}$ of $\mathscr{M}$ and $\alpha_1, \alpha_2,\cdots,\alpha_n \ge 0$ with $\Sigma_{i=1}^{n} \alpha_i = 1$, there exists $f \in \mathscr{M}$ such that

$$f(x) \ge \Sigma_{i=1}^{n} \alpha_i f_i(x) \quad \text{for all } x \in X .$$

The following theorem is a formulation given by Pietsch [1980, p.40] of Ky Fan's [1953] generalization of a result of Kneser [1952]. This generalization of Kneser's result is stated below as (23.18).

(23.16) Theorem. Let K be a non–empty compact convex subset of a Hausdorff TVS X, and let $\mathscr{M}$ be a concave collection of lower semi–continuous convex real–valued functions f on K. Then for any given $\rho \in \mathbb{R}$, one of the following properties holds:

(a) There is an $f \in \mathscr{M}$ such that $\inf_{x\in K} f(x) > \rho$.

(b) There exists a point $\dot{y} \in K$ such that $f(\dot{y}) \le \rho$ for all $f \in \mathscr{M}$ simultaneously.

Proof. Without loss of generality we may assume that $\rho = 0$. For each $f \in \mathscr{M}$, let

$$Q(f) = \{ x \in K : f(x) \le 0 \} .$$

Then $Q(f)$ is closed in K by lower semi–continuity of f. If the family $\{ Q(f) : f \in \mathcal{M} \}$ has the finite intersection property, then by compactness of K, we obtain the alternative (b). If $\{ Q(f) : f \in \mathcal{M} \}$ has no finite intersection property then we show that the alternative (a) holds.

In fact, there are $f_1, f_2, \cdots, f_n \in \mathcal{M}$ such that $\bigcap_{i=1}^{n} Q(f_i) = \phi$. For each $i = 1, 2, \cdots, n$, let $V_i = K \setminus Q(f_i)$. Then each V_i is open in K for which $\{ V_1, V_2, \cdots, V_n \}$ is an open covering of the compact space K. Let $\{ \beta_1, \beta_2, \cdots, \beta_n \}$ be a continuous partition of unity subordinate to this open covering. Then $\beta_1, \beta_2, \cdots, \beta_n$ are continuous non–negative real–valued functions on K such that for each $i = 1, 2, \cdots, n$, $\beta_i(x) = 0$ if $x \in K \setminus V_i$ and

$$\Sigma_{i=1}^{n} \beta_i(x) = 1 \quad (\text{for all } x \in K).$$

Define $\varphi : K \times K \longrightarrow \mathbb{R}$ by setting

$$\varphi(x, y) = \Sigma_{i=1}^{n} \beta_i(y) f_i(y) - \Sigma_{i=1}^{n} \beta_i(y) f_i(x) \quad ((x, y) \in K \times K).$$

Then $\varphi(x,x) = 0$ (for all $x \in K$). As each β_i is continuous with $\beta_i \geq 0$, and each f_i is lower semi–continuous, it follows that the function $y \longrightarrow \varphi(x, y)$ is lower semi–continuous on K. On the other hand, the convexity of each $f_i \in \mathcal{M}$ ensures that the function $x \longrightarrow \varphi(x,y)$ is quasi–concave. It then follows from Ky Fan's minimax inequality (23.6) that there exists $\hat{y} \in K$ such that $\varphi(x,\hat{y}) \leq 0$ (for all $x \in K$), i.e.,

$$\Sigma_{i=1}^{n} \beta_i(\hat{y}) f_i(\hat{y}) \leq \Sigma_{i=1}^{n} \beta_i(\hat{y}) g_i(x) \quad (\text{for all } x \in K),$$

and hence from the concavity of $\mathcal{M}$ that there is an $f \in \mathcal{M}$ such that

$$f(x) \geq \Sigma_{i=1}^{n} \beta_i(\hat{y}) f_i(x) \quad (\text{for all } x \in K).$$

Consequently,

$$0 < \Sigma_{i=1}^{n} \beta_i(\hat{y}) f_i(\hat{y}) \leq \Sigma_{i=1}^{n} \beta_i(\hat{y}) f_i(x) \leq f(x) \quad (\text{for all } x \in K).$$

This proves the alternative (a), and the proof is complete.

(23.17) Definition (Ky Fan [1953]). Let φ be a real–valued function defined on the product set $X \times Y$ of two arbitrary sets X, Y. Then φ is said to be concave on X, if for any two elements $x_1, x_2 \in X$ and two numbers $\alpha_1 \geq 0$, $\alpha_2 \geq 0$ with $\alpha_1 + \alpha_2 = 1$, there exists $x_0 \in X$ such that

$$\varphi(x_0,y) \geq \alpha_1\varphi(x_1,y) + \alpha_2\varphi(x_2,y) \quad (\text{for all } y \in Y).$$

(23.18) Theorem (Kneser [1952] and Ky Fan [1953]). Let B be the arbitrary non–empty set and K a non–empty compact convex subset of a Hausdorff TVS X, and let $\varphi : B \times K \longrightarrow \mathbb{R}$ be a function satisfying :

(i) for each fixed $x \in B$, the function $\varphi(x,\cdot)$ is a lower semi–continuous convex function on K;

(ii) φ is concave on B.

Then

$$\min_{y \in K} \sup_{x \in B} \varphi(x,y) = \sup_{x \in B} \min_{y \in K} \varphi(x,y).$$

Proof. Let $\rho = \sup_{x \in B} \min_{y \in K} \varphi(x,y)$. Applying (23.16) with B being the index set, there is a $\hat{y} \in K$ such that $\varphi(x,y) \leq \rho$ (for all $x \in B$). The conclusion follows.

The following existence theorem of variational inequalities is one of the basic results in nonlinear functional analysis.

(23.19) Theorem (Browder [1965] and Hartman–Stampacchia [1966]). *Let* $(E, \|\cdot\|)$ *be a reflexive Banach space and* B *a non–empty closed convex subset of* E. *Suppose that* $A : B \longrightarrow E'$ *is monotone on* B (i.e., $\langle y - x, A(y) - A(x)\rangle \geq 0$, *for all* $x, y \in B$) *which is continuous along the line segments in* B *for the norm topology on* B *and the topology* $\sigma(E', E)$ *on* E'. *Assume that there exists* $x_0 \in B$ *such that*

(1) $$\lim_{\substack{\|y\| \to \infty \\ y \in B}} \mathrm{Re}\langle y - x_0, A(y)\rangle > 0 \quad \text{(called the coercive condition)}.$$

Then there exists $\dot{y} \in B$ *such that*

(2) $$\mathrm{Re}\langle \dot{y} - x, A(\dot{y})\rangle \leq 0 \quad \text{for all } x \in B.$$

Proof. Define $\varphi, \psi : B \times B \longrightarrow \mathbb{R}$ by

$$\varphi(x, y) = \mathrm{Re}\langle y - x, A(x)\rangle \quad (\text{for all } x, y \in B)$$

and

$$\psi(x, y) = \mathrm{Re}\langle y - x, A(y)\rangle \quad (\text{for all } x, y \in B).$$

As A is monotone, it follows that $\varphi(x, y) \leq \psi(x, y)$ (for all $x, y \in B$). By coercive condition (1), there exists a sufficiently large $R > 0$ such that

$$\mathrm{Re}\langle y - x_0, A(y)\rangle > 0 \quad \text{for all } y \in B \text{ with } \|y\| > R.$$

Let

$$K = \{ y \in B : \|y\| \leq R \}.$$

Then K is a $\sigma(E,E')$–compact convex subset of B (by the reflexivity of E) and $\psi(x_0, y) > 0$ (for all $x \in B \setminus K$). It is clear that $\psi(x, x) \leq 0$ (for all $x \in B$) and that $\psi(x, y)$ is a quasi–concave function of x on B. Thus when E is equipped with $\sigma(E, E')$, all conditions in (23.8) are satisfied, so that by (23.8), there exists $\dot{y} \in B$ such that

$\varphi(x, \dot{y}) \leq 0$ for all $x \in B$, i.e.,

$$\mathrm{Re}\langle \dot{y} - x, A(x)\rangle \leq 0 \quad (\text{for all } x \in B) .$$

The remaining part of argument is due to Minty [1962]. Let $x \in B$ and $w_t = tx + (1-t)\dot{y}$ $(0 \leq t \leq 1)$. Then $w_t \in B$ for all $t \in [0, 1]$ [by the convexity of B], hence

$$\mathrm{Re}\langle \dot{y} - w_t, A(w_t)\rangle \leq 0 ,$$

and thus

$$\mathrm{Re}\langle \dot{y} - x, A(w_t)\rangle \leq 0 \quad (\text{for all } 0 < t \leq 1) .$$

Since A is continuous along the line segments in B for the norm topology on B and the topology $\sigma(E', E)$ on E' ,it follows that $A(w_t) \longrightarrow A(\dot{y})$ in $\sigma(E', E)$ as $t \longrightarrow 0^+$, and hence that $\mathrm{Re}\langle \dot{y} - x, A(\dot{y})) \leq 0$. As $x \in B$ was arbitrary, we conclude that (2) holds.

(23.20) Corollary (Browder [1965b]). Let B be a non–empty bounded closed convex set in a Hilbert space $(H,\langle\cdot,\cdot\rangle)$. Then each non–expansive map $f : B \longrightarrow B$ (i.e., $\|f(x) - f(y)\| \leq \|x - y\|$ for all $x, y \in B$) has a fixed point.

Proof. Let I be the identity map on B. We claim that $I - f$ is monotone. Indeed, for each $x, y \in B$, the non–expansiveness of f ensures that

$$\mathrm{Re}\langle y - x, (I - f)(y) - (I - f)(x)\rangle$$

$$= \langle y - x, y - x\rangle - \mathrm{Re}\langle y - x, f(y) - f(x)\rangle$$

$$\geq \|y - x\|^2 - \|y - x\| \, \|f(y) - f(x)\|$$

$$= \|y - x\| \left(\|y - x\| - \|f(y) - f(x)\|\right) \geq 0 ,$$

which obtains our assertion.

By (23.19),there exists $\dot{y} \in B$ such that

$$\mathrm{Re}<\hat{y}-x,\,(I-f)(\hat{y})>\ \leq 0 \quad \text{for all} \quad x \in B\,.$$

This implies that $f(\hat{y}) = \hat{y}$, and the proof is complete.

Remark. A convex set B in a Banach space is said to have normal structure (Brodskii–Milman [1948]) if for each bounded convex subset C of B which contains more than one point, there exists a point x in C such that $\sup\{\|x-y\| : y \in C\} \neq \operatorname{diam}(C)$.

The following result is more general than (23.19), because every convex set in a Hilbert space has normal structure.

(23.d) (Kirk [1965]). Let K be a weakly compact convex set in a B–space E which has the normal structure. Then every nonexpansive map $f : K \longrightarrow K$ has a fixed point.

(23.21) Definition. Let Ω_1, Ω_2 be topological spaces, and let $f : \Omega_1 \longrightarrow 2^{\Omega_2}$ be a set–valued map (i.e., for each $x \in \Omega_1$, $f(x) \in 2^{\Omega_2}$; here 2^{Ω_2} denotes the power set of Ω_2). We say that f is upper semi–continuous at $x_0 \in \Omega_1$ if for each set G with $f(x_0) \subset G$ there exists an open neighbourhood $N(x_0)$ of x_0 such that $f(x) \subset G$ for all $x \in N(x_0)$; f is upper semi–continuous on Ω_1 if it is upper semi–continuous at every point of Ω_1.

(23.22) Lemma. Let X be a Hausdorff TVS, let $D \subset X$ be non–empty, and let $S : D \longrightarrow 2^D$ be upper semi–continuous such that $S(x)$ is non–empty and bounded (for any $x \in D$). Then for any $g \in E'$ the map $\varphi_g : D \longrightarrow \mathbb{R}$, defined by

$$\varphi_g(y) = \sup_{x \in S(y)} \mathrm{Re}<x,\,g>$$

is upper semi–continuous (in the sense of real–valued functions).

<u>Proof</u>. Let $y_0 \in D$, let $\epsilon > 0$ be given and

$$U_\epsilon = \{ x \in D : |\langle x,g\rangle| < \epsilon/2 \}.$$

Then U_ϵ is an open neighbourhood of 0. As $S(y_0) + U_\epsilon$ is an open set containing $S(y_0)$, it follows from the upper semi–continuity of S at y_0 that there exists a neighbourhood $N(y_0)$ of y_0 in D such that $S(y) \subset S(y_0) + U_\epsilon$ (for all $y \in N(y_0)$). Thus, for each $y \in N(y_0)$,

$$\varphi_g(y) = \sup_{x \in S(y)} \operatorname{Re}\langle x, g\rangle \leq \sup_{x \in S(y_0)+U_\epsilon} \operatorname{Re}\langle x, g\rangle$$

$$\leq \sup_{x \in S(y_0)} \operatorname{Re}\langle x, g\rangle + \sup_{x \in U_\epsilon} \operatorname{Re}\langle x, g\rangle < \varphi_g(y_0) + \epsilon.$$

Hence φ_g is upper semi–continuous and the proof is complete.

(23.23) Theorem (Ky Fan [1952] and Glicksberg [1952]). <u>Let</u> K <u>be a non–empty compact convex subset of a LCS</u> X, <u>and let</u> $S : K \longrightarrow 2^K$ <u>be upper semi–continuous such that for each</u> $x \in K$, $S(x)$ <u>is a non–empty closed convex subset of</u> K. <u>Then there exists</u> $\hat{y} \in K$ <u>such that</u> $\hat{y} \in S(\hat{y})$.

<u>Proof</u>. Suppose that the assertion is false, i.e., $y \notin S(y)$ for all $y \in K$. By the strong separation theorem (8.3), there exists $h \in X'$ such that

$$\operatorname{Re}\langle y, h\rangle - \sup_{x \in S(y)} \operatorname{Re}\langle x, h\rangle > 0.$$

For each $g \in X'$, let us define

$$V(g) = \{y \in K : \operatorname{Re}\langle y,g\rangle - \sup_{x \in S(y)} \operatorname{Re}\langle x,g\rangle > 0\}.$$

Then $K = \bigcup_{g \in X'} V(g)$. By (23.22), each $V(g)$ is open in K. By the compactness of K, there exist $g_1, g_2, \cdots, g_n \in X'$ such that $K = \bigcup_{i=1}^{n} V(g_i)$. Let $\{\beta_1, \beta_2, \cdots, \beta_n\}$ be a continuous partition of unity on K subordinate to the covering $\{V_1, V_2, \cdots, V_n\}$. Then $\beta_1, \beta_2, \cdots, \beta_n$ are continuous non-negative real-valued functions on K such that β_i vanishes on $K \backslash V_i$ (for each $i = 1, 2, \cdots, n$) and $\Sigma_{i=1}^{n} \beta_i(x) = 1$ (for all $x \in K$). Define $\varphi : K \times K \longrightarrow \mathbb{R}$ by setting

$$\varphi(x,y) = \Sigma_{i=1}^{n} \beta_i(y) \mathrm{Re}\langle y-x, g_i \rangle, \quad ((x,y) \in K \times K).$$

Then all conditions of Ky Fan's minimax inequality are satisfied, applying Ky Fan's minimax inequality to get an $\hat{y} \in K$, such that $\varphi(x,\hat{y}) \leq 0$ for all $x \in K$, i.e.,

$$(*) \qquad \Sigma_{i=1}^{n} \beta_i(\hat{y}) \mathrm{Re}\langle \hat{y} - x, g_i \rangle \leq 0 \quad (\text{for all } x \in K).$$

If $i \in \{1,2,\cdots,n\}$ is such that $\beta_i(\hat{y}) > 0$, then $\hat{y} \in V(g_i)$ and hence

$$\mathrm{Re}\langle \hat{y}, g_i \rangle > \sup_{x \in S(\hat{y})} \mathrm{Re}\langle x, g_i \rangle \geq \mathrm{Re}\langle \hat{x}, g_i \rangle \quad (\text{for all } \hat{x} \in S(\hat{y})).$$

It follows that $\mathrm{Re}\langle \hat{y} - \hat{x}, g_i \rangle > 0$. Note that

$$\beta_i(\hat{y}) \mathrm{Re}\langle \hat{y} - \hat{x}, g_i \rangle > 0 \quad \text{whenever } \beta_i(\hat{y}) > 0 \quad (\text{for } i = 1,\cdots,n).$$

Since $\beta_i(\hat{y}) > 0$ for at least one $i \in \{1,2,\cdots,n\}$, it follows that

$$\Sigma_{i=1}^{n} \beta_i(\hat{y}) \mathrm{Re}\langle \hat{y} - \hat{x}, g_i \rangle > 0,$$

which contradicts (*). This proves the theorem.

<u>Remark</u>. When $X = \mathbb{R}^n$, Fan–Glicksberg's theorem reduces to Kakutani's fixed point theorem (see Kakutani [1941]). Thus, Fan–Glicksberg's theorem is a common generalization of Kakutani's fixed point theorem and Tychonoff's fixed point theorem (23.9).

Exercises

23–1. Prove (23.a).

23–2. Let K be a compact convex set in a normed space E, and let $f : K \longrightarrow E$ be a continuous mapping. Show that there exists at least one $u_0 \in K$ such that

$$\|u_0 - f(u_0)\| = \inf_{x \in K} \|x - f(u_0)\| \quad \text{(Ky Fan).}$$

23–3. Prove (23.b).

23–4. Prove (23.c).

23–5. Prove (23.d).

23–6. Let X be a convex set in a TVS and let $A \subset X$ be a non–empty finite set. If $\{ G(x) : x \in A \}$ is a KKM covering for X such that each G(x) is open in X, then show that there exists a KKM covering $\{ F(x) : x \in A \}$ for X such that each F(x) is closed in X and $F(x) \subset G(x)$.

23–7. Let X be a non–empty convex set in a TVS, let $\{ A_1, \cdots, A_n \}$ be a finite family of n closed subsets of X such that $\bigcup_{i=1}^{n} A_i = X$. Then for any n points $x_1, x_2, \cdots, x_n$ of X, there exists k indices $i_1 < i_2 < \cdots < i_k$ between 1 and n such that $co\{x_{i_1}, x_{i_2}, \cdots, x_{i_n}\}$ contains a point of the intersection $\bigcap_{j=1}^{k} A_{i_j}$. (Ky Fan)

24

An Introduction to Ordered Convex Spaces

Throughout this section the scalar field for vector spaces is assumed to be the real field $\mathbb{R}$.

(24.1) Definition. Let X be a vector space. By a (positive) cone in X is meant a non–empty subset X_+ of X satisfying

$$\lambda X_+ + \mu X_+ \subset X_+ \qquad \text{(for all } \lambda, \mu \geq 0).$$

By a vector ordering in X is meant a relation " $\leq$ " in X which is transitive, reflexive and compatible with the vector space operations, namely

(VO1) if $x \leq y$ then $x + u \leq y + u$ (for all $u \in X$);

(VO2) if $x \leq y$ then $\lambda x \leq \lambda y$ (for all $\lambda \geq 0$).

The pair (X,X_+) or $(X,\leq)$ is referred to as an ordered vector space (abbreviated OVS).

We also say that a cone X_+ in X is:

(a) proper if $X_+ \cap (-X_+) = \{ 0 \}$;

(b) generating (or positively generating) if $X = X_+ - X_+$.

Remarks: (i) Clearly a cone X_+ in X associates naturally a vector ordering " $\leq$ " (called the vector ordering associates with X_+) by setting

(1) $$x \leq y \text{ if and only if } y - x \in X_+ .$$

Conversely, any vector ordering " $\leq$ " defines naturally a cone in X (called the associated cone) by setting

$$X_+ = \{ x \in X : x \geq 0 \}$$

such that " $\leq$ " is exactly the vector ordering associated with X_+.

(ii) A cone X_+ in X is proper (resp. generating) if and only if the vector ordering " $\leq$ " associated with X_+ is antisymmetric (resp. directed upwards), that is

$$x = 0 \quad \text{if and only if} \quad x \geq 0 \text{ and } x \leq 0.$$

(resp. for any $x, y \in X$ there is a $u \in X$ such that $x, y \leq u$).

(24.2) Definitions. Let (X, X_+) be an OVS, let $x, y \in X$ and $\{ x_\tau : \tau \in D \} \subset X$.

(a) If $x \leq y$, the set, defined by

$$[x, y] = \{ z \in X : x \leq z \leq y \} = (x + X_+) \cap (y - X_+),$$

is called an order–interval in X.

(b) A subset B of X is said to be order–bounded if it is contained in some order–interval in X.

(c) $\{ x_\tau : \tau \in D \}$ is directed upwards (in symbols $x_\tau \uparrow$) if for any $\tau_1, \tau_2 \in D$ there is an $\tau \in D$ such that

$$x_{\tau_i} \leq x_\tau \quad (i = 1, 2).$$

Similarly one can define directed downwards (in symbols $x_\tau \downarrow$).

Remarks: (i) If $\{ x_\tau : \tau \in D \}$ is directed upwards in X, then D equipped with the ordering, defined by

$$\tau_1 \leq \tau_2 \quad \text{if} \quad x_{\tau_1} \leq x_{\tau_2},$$

is a directed set, hence $\{ x_\tau : \tau \in D \}$ is an increasing net. Decreasing nets are defined similarly.

(ii) A sequence $\{ x_n \}$ in (X, X_+) is said to be increasing (in symbols $x_n \uparrow$) if $x_n \leq x_{n+1}$ (for all $n \geq 1$). The symbol $x_n \downarrow$ denotes a decreasing sequence and its definition is analogous.

(iii) Let $(X,\mathscr{P})$ be a Hausdorff TVS and X_+ a $\mathscr{P}$-closed cone in X. If $x_\tau \uparrow$ (resp $x_\tau \downarrow$) and $x = \mathscr{P}\text{-}\lim_\tau x_\tau$, it is not hard to show that $x = \sup_\tau x_\tau$ (resp. $x = \inf_\tau x_\tau$).

(iv) A subset B of (X,X_+) is said to be order-complete (resp. σ-order-complete) if every increasing net (resp. increasing sequence) in B, that has an upper bound in X, has a supremum belonging to B. In particular, if X itself is order-complete (resp. σ-order- complete), then we say that X is an order-complete OVS (resp. σ-order-complete OVS).

(v) Suppose that the cone X_+ is generating. Then a subset B of X is order-bounded if and only if there exists $u \in X_+$ such that $B \subset [-u, u]$. Thus the family of all order-bounded subsets of X, denoted by $\mathscr{M}_{ob}$, is a convex bornology with $\{ [-u ,u] : u \in X_+ \}$ as a basis; $\mathscr{M}_{ob}$ is called the order-bornology on X.

(24.3) Definition Let (X,X_+) and (Y,Y_+) be OVS and $T \in L^*(X,Y)$. We say that T is:

(a) positive if $T(X_+) \subset Y_+$;

(b) locally order-bounded if it sends order-bounded subsets of X into order-bounded sets in Y.

We denote by $L^*_+(X,Y)$ (resp. $L^{ob}(X,Y)$) the set of all positive (resp. locally order-bounded) linear maps from X into Y, and write

$$L^r(X,Y) = L^*_+(X,Y) - L^*_+(X,Y);$$

elements in $L^r(X,Y)$ are said to be regular. In particular, we write

$$X^*_+ = L^*_+(X,\mathbb{R}), \quad X^\# = X^*_+ - X^*_+ \text{ and } X^b = L^{ob}(X,\mathbb{R}).$$

X^*_+ is called the dual cone of X_+ (it is actually a cone in X^*); $(X^\#,X^*_+)$ is called the order dual of X; while X^b is called the order-bounded dual (or order-bornological dual) of X.

Remarks: (i) $L^{ob}(X,Y)$ is always a vector subspace of $L^*(X,Y)$ and $L^*_+(X,Y)$

is a cone in $L^{ob}(X,Y)$.

(ii) An $f \in X_+^*$ is <u>strictly positive</u> if $f(u) > 0$ whenever $0 \neq u \in X_+$.

(iii) Let $(X,\mathscr{P})$ be a LCS and X_+ a cone in X. Then the <u>dual cone</u> in X' is defined by

$$X'_+ = X' \cap X_+^*.$$

(iv) Using Hahn–Banach's extension theorem, one can show that X_+^* is proper if and only if X_+ is generating. If X_+^* separates points of X then X_+ is proper.

(v) Let $(X,\mathscr{P})$ be a LCS and X_+ a cone in X. Then order–bounded subsets of X are $\mathscr{P}$-bounded if and only if $(X, \mathscr{M}_{von}(\mathscr{P}))^X \subset X^b$.

If $\{ (X_\alpha,(X_\alpha)_+) : \alpha \in \Lambda \}$ is a family of OVS then $\prod_{\alpha \in \Lambda} (X_\alpha)_+$ is a cone in the product space $\prod_{\alpha \in \Lambda} X_\alpha$. The vector ordering induced by $\prod_{\alpha \in \Lambda} (X_\alpha)_+$ is called the <u>product ordering</u>. [Clearly $\prod_\alpha (X_\alpha)_+$ is proper (resp. generating) if and only if each $(X_\alpha)_+$ is proper (resp. generating).] The direct sum $\oplus_\alpha X_\alpha$, equipped with the relative ordering induced by $\oplus_\alpha (X_\alpha)_+ = (\oplus_\alpha X_\alpha) \cap (\prod_\alpha (X_\alpha)_+)$, is called the <u>ordered direct sum of</u> $\{ (X_\alpha,(X_\alpha)_+) : \alpha \in \Lambda \}$.

For the vector ordering defined on a quotient space, we have the following result whose proof is straightforward and hence is left to the reader.

(24.4) **Lemma.** <u>Let</u> M <u>be a vector subspace of</u> (X,X_+), <u>let</u> $Q_M : X \longrightarrow X/M$ <u>be the quotient map and</u> $\hat{X}_+ = Q_M(X_+)$. <u>Then</u> $\hat{X}_+$ <u>is a cone in</u> X/M. <u>Moreover, we have:</u>

(a) $Q_M(x) \leq Q_M(y)$ <u>if and only if there is an</u> $z \in M$ <u>such that</u> $x \leq y + z$.

(b) $\hat{X}_+$ <u>is generating in</u> X/M <u>if and only if</u> $M + (X_+ - X_+) = X$.

(c) $\hat{X}_+$ <u>is proper if and only if</u> $M = (M + X_+) \cap (M - X_+)$.

Let (X,X_+) be an OVS. A functional q, defined on X_+, is said to be superlinear if $-q$ is sublinear on X_+.

The following Hahn–Banach's extension theorem of ordered type is a very useful result for studying positive extensions and duality theorems in OVS.

(24.5) **Theorem.**(Bonsall [1957]). Let (X,X_+) be an OVS and p a sublinear functional on X. Suppose that q is a superlinear functional on X_+ such that

$$q(u) \leq p(u) \quad (\text{for all } u \in X_+).$$

Then there exists an $f \in X^*$ such that

$$q(u) \leq f(u) \quad (u \in X_+) \quad \text{and} \quad f(x) \leq p(x) \quad (x \in X).$$

Proof. We define, for any $x \in X$,

$$\gamma(x) = \inf \{ p(x + u) - q(u) : u \in X_+ \}$$

Then γ is well–defined since

$$p(x + u) - q(u) \geq p(u) - p(-x) - q(u) \geq -p(-x) > -\infty \quad (u \in X_+).$$

Clearly, γ is a sublinear functional on X such that

(1) $$\gamma(x) \leq p(x) \quad (x \in X) \quad \text{and} \quad \gamma(-u) \leq -q(u) \quad (u \in X_+).$$

Applying Hahn–Banach's extension theorem to get an $f \in X^*$ with $f(x) \leq \gamma(x)$ $(x \in X)$, hence f has the required properties by making use of (1).

Remark. The preceding result is equivalent to the Namioka and Bauer theorem concerning with the positive extension (see Wong [1976 a, p.5]). By Bonsall's argument, the preceding result can be extended to a more general setting (see Wong [1980, p.49]).

Let (X,X_+) and (Y,Y_+) be OVS, and let $\langle X, Y\rangle$ be a dual pair. We say that $\langle X, Y\rangle$ is an

(a) order–duality on the right if $Y_+ = -X_+^o$;

(b) order–duality on the left if $X_+ = -Y_+^o$;

(c) order–duality if $Y_+ = -X_+^o$ and $X_+ = -Y_+^o$.

In order to establish duality theorems for OVS, we present the following result (taken from Wong [1976a, p.7]) which contains the crux of the duality problem.

(24.6) **Lemma.** Let (X,X_+) and (Y,Y_+) be OVS which form an order–duality on the right, and $V \subset X$.

(a) If $0 \in V$ then

(1) $$(V + X_+)^o = V^o \cap X_+^o \ (= V^o \cap (-Y_+))\ ;$$

(2) $$(V - X_+)^o = V^o \cap (-X_+^o) \ (= V^o \cap Y_+);$$

(3) $$(\overline{V} \cap \overline{X}_+)^o = \overline{V^o + X_+^o} \ (= \overline{V^o - Y_+}).$$

If, in addition, V is symmetric, then

(4) $$(V + X_+)^o = -(V^o \cap Y_+).$$

(b) Suppose that V is a convex $\tau(X,Y)$–neighbourhood of 0. Then

(5) $$(V \cap X_+)^o = V^o + X_+^o = V^o - Y_+.$$

If, in addition, V is symmetric, then

(6) $$(-(V \cap X_+))^o = V^o + Y_+.$$

Proof. (a) Clearly $(V + X_+)^o \subset V^o \cap X_+^o$. Conversely, if $f \in V^o \cap X_+^o$, then f is negative on X_+ [since $\lambda X_+ \subset X_+$ (for all $\lambda \geq 0$)], hence $f \in (V + X_+)^o$. This proves (1).

Notice that $0 \in V^o$ and that $\lambda X_+ \subset X_+$ (for all $\lambda \geq 0$). As $V^o + X_+^o$ is convex and contains 0, it follows from the bipolar theorem that

$$\overline{V^o + X_+^o} = \overline{co}\,(V^o + X_+^o) = (V^o + X_+^o)^{oo} = (V^{oo} \cap X_+^{oo})^o = (\overline{V} \cap \overline{X}_+)^o$$

and hence that (3) holds.

The equalities (2) and (4) can be verified in the same way.

(b) Clearly, $V^o + X_+^o \subset (V \cap X_+)^o$ [since members in X_+^o are negative on X_+]. Conversely, let $g \in (V \cap X_+)^o$ and suppose that p_V is the gauge of V. Then

$$g(u) \leq p_V(u) \quad \text{(for all } u \in X_+),$$

and the restriction of g to X_+ is superlinear. By (24.5), there exists an $f \in X_+^*$ such that

$$g(u) \leq f(u) \quad (u \in X_+) \quad \text{and} \quad f(x) \leq p_V(x) \quad (x \in X).$$

Therefore $f \in V^o$ and $f - g = Y_+ = -X_+^o$, thus

$$g = f - (f - g) \in V^o - Y_+ = V^o + X_+^o,$$

this proves (5).

Finally, if V is symmetric then so is V^o, thus

$$(-(V \cap X_+))^o = -(V \cap X_+)^o = -(V^o - Y_+) = V^o + Y_+.$$

(24.7) <u>Definition.</u> Let (X,X_+) be an OVS and $V \subset X$. We write:

$F(V) = (V + X_+) \cap (V - X_+)$;

$D(V) = \{x \in V : x = \lambda x_1 - (1 - \lambda)x_2,\ \lambda \in [0,1],\quad x_1, x_2 \in V \cap X_+\}$;

$F_P(V) = (V - X_+) \cap X_+ \ (= \cup \{ [0,u] : u \in V \cap X_+ \})$;

$D_P(V) = V \cap X_+ - X_+$;

$S(V) = \cup \{ [-u,u] : u \in V \cap X_+ \}$.

Then F(V) is called the order–convex hull (or full hull) of V; while D(V) is referred to as the decomposable kernel of V. Furthermore, we say that V is:

(a) order–convex (or full) if $F(V) \subset V$;

(b) 0–convex if it is both order–convex and convex;

(c) decomposable if $V \subset D(V)$;

(d) positively order–convex if $F_P(V) \subseteq V$;

(e) <u>positively dominated</u> if $V \subset D_P(V)$;

(f) <u>absolutely order–convex</u> if $S(V) \subset V$;

(g) <u>absolutely dominated</u> if $V \subset S(V)$;

(h) <u>solid</u> if $V = S(V)$.

<u>Remark</u>: (i) If V is positively order–convex, then

$$V \cap X_+ = D(V) \cap X_+ = S(V) \cap X_+.$$

If V is convex then so is $F(V)$ and $D(V) = \Gamma(V \cap X_+)$. If V is <u>order–convex and symmetric</u>, then $S(V) \subset V$ and $S(V)$ is the largest solid set contained in V; in this case, $S(V)$ is referred to as the <u>solid kernel</u> of V.

(ii) Let (X,X_+) and (Y,Y_+) be OVS which form an order–duality on the right and $N \subset Y$. Since Y is regarded as a vector subspace of X^*, it follows from $Y_+ = -X_+^\circ$ that $Y_+ = X_+^* \cap Y$, and hence that we have to distinguish the order–convex hull of N w.r.t. Y_+ as well as w.r.t. X_+^*; consequently, we define

$$F(N) = (N + Y_+) \cap (N - Y_+); \qquad F^*(N) = (N + X_+^*) \cap (N - X_+^*);$$

$$F_P(N) = (N - Y_+) \cap Y_+; \qquad F_P^*(N) = (N - X_+^*) \cap X_+^*;$$

$$D_P(N) = N \cap Y_+ - Y_+; \qquad D_P^*(N) = N \cap X_+^* - X_+^*;$$

$$S(N) = \{\, y \in Y : w \pm y \in Y_+ \text{ for some } w \in N \cap Y_+ \,\};$$

$$S^*(N) = \{\, f \in X^* : g \pm f \in X_+^* \text{ for some } g \in N \cap X_+^* \,\}.$$

Clearly, we have

$$F(N) = F^*(N) \cap Y; \qquad F_P(N) = F_P^*(N) \cap Y;$$

$$D_P(N) = D_P^*(N) \cap Y; \qquad S(N) = S^*(N) \cap Y.$$

If Y is order–convex (or positively order–convex) in (X^*,X_+^*), then

$$F(N) = F^*(N);\ F_P(N) = F_P^*(N) \text{ and } S(N) = S^*(N).$$

(iii) Let $(X, \mathscr{P})$ be a LCS and X_+ a cone in X. If there exists a $\sigma(X',X)$–compact convex subset N of X' which is positively order–convex in (X^*, X_+^*) and contains 0, then

$$X = X_+ - X_+.$$

In fact, if $X \neq X_+ - X_+$, there would exist $0 \neq f \in X^*$ such that $f = 0$ on $X_+ - X_+$, hence $0 \leq \lambda f \leq 0$ (for all $\lambda \in \mathbb{R}$), thus $\lambda f \in N$ (for all $\lambda \in \mathbb{R}$) by the positive order–convexity of N in (X^*, X_+^*) and $0 \in N$. As N° is a $\tau(X,X')$–neighbourhood of 0, it follows that $f = 0$, which gives a contradiction.

(24.8) Theorem. (The first duality). <u>Let</u> (X,X_+) <u>and</u> (Y,Y_+) <u>be</u> OVS <u>which form an order–duality on the right</u> (that is, $Y_+ = -X_+^\circ$), <u>and let</u> V <u>be a disked</u> $\tau(X,Y)$–<u>neighbourhood of</u> 0, <u>whose</u> $\tau(X,Y)$–<u>closure is denoted by</u> $\overline{V}$.

(a) (Upward form) <u>We have</u>

(24.8.1) $$(F(V))^\circ = D(V^\circ) \quad \text{and} \quad (D(V))^\circ = F(V^\circ);$$

(24.8.2) $$(F_P(V))^\circ = D_P(V^\circ) \quad \text{and} \quad (D_P(V))^\circ = F_P(V^\circ).$$

(b) (Downward form) <u>We have</u>

(24.8.3) $$(F(V^\circ))^\circ = \overline{D(\overline{V})} \quad \text{and} \quad (D(V^\circ))^\circ = \overline{F(\overline{V})};$$

(24.8.4) $$(F_P(V^\circ))^\circ = \overline{D_P(\overline{V})} \quad \text{and} \quad (D_P(V^\circ))^\circ = \overline{F_P(\overline{V})}.$$

<u>Proof</u>. (a) Since $V + X_+$ and $V - X_+$ are convex $\tau(X,Y)$–neighbourhoods of 0, it follows from Alaoglu–Bourbaki's theorem that $(V + X_+)^\circ$ and $(V - X_+)^\circ$ are convex $\sigma(Y,X)$–compact, and hence from the bipolar theorem and (24.6)(a) that

$$\begin{aligned}(F(V))^\circ &= \overline{\text{co}}\,\{(V + X_+)^\circ \cup (V - X_+)^\circ\} \\ &= \text{co}\,\{(V + X_+)^\circ \cup (V - X_+)^\circ\}\end{aligned}$$

$$= \text{co}\,\{-(V^{\circ} + Y_{+}) \cup (V^{\circ} - Y_{+})\} = D(V^{\circ}).$$

On the other hand, the bipolar theorem and (24.6)(b) show that

$$(D(V))^{\circ} = (-(V \cap X_{+}))^{\circ} \cap (V \cap X_{+})^{\circ}$$

$$= (V^{\circ} + Y_{+}) \cap (V^{\circ} - Y_{+}) = F(V^{\circ});$$

so that (24.8.1) holds.

Since $V - X_{+}$ and V are convex $\tau(X,Y)$–neighbourhoods of 0, it follows from (24.6)(b) that

$$(F_P(V))^{\circ} = ((V - X_{+}) \cap X_{+})^{\circ} = (V - X_{+})^{\circ} + X_{+}^{\circ}$$

$$= V^{\circ} \cap (-X_{+}^{\circ}) + X_{+}^{\circ} \quad \text{(by (2) of (24.6))}$$

$$= V^{\circ} \cap Y_{+} - Y_{+} = D_P(V^{\circ}).$$

As $0 \in V \cap X_{+}$, it follows from (2) of (24.6) that

$$(D_P(V))^{\circ} = (V \cap X_{+} - X_{+})^{\circ} = (V \cap X_{+})^{\circ} \cap (-X_{+})^{\circ}$$

$$= (V^{\circ} + X_{+}^{\circ}) \cap (-X_{+}^{\circ}) \quad \text{(by (5) of (24.6))}$$

$$= (V^{\circ} - Y_{+}) \cap Y_{+} = F_P(V^{\circ}).$$

Therefore, (24.8.2) holds.

(b) As $D(\bar{V})$ and $F(\bar{V})$ are convex and contain 0, $\bar{V}$ is a disked $\tau(X,Y)$–neighbourhood of 0, it follows from the bipolar theorem and part (a) of this result that

$$\overline{D(\bar{V})} = (D(\bar{V}))^{\circ\circ} = (F(\bar{V}^{\circ}))^{\circ} = (F(V^{\circ}))^{\circ} \quad \text{and}$$

$$\overline{F(\bar{V})} = (F(\bar{V}))^{\circ\circ} = (D(\bar{V}^{\circ}))^{\circ} = (D(V^{\circ}))^{\circ},$$

so that (24.8.3) holds.

A similar argument given in the proof of (24.8.3) yields (24.8.4).

Remark. The assumption that V be symmetric can be dropped for the formula (24.8.2) and (24.8.4).

The present form of (24.8.1) is taken from Ng and Duhoux [1973], in fact, it was implicitly given in Ng [1970] and Wong [1970,1973b].

(24.9) **Corollary.** *Let* (X,X_+) *and* (Y,Y_+) *be OVS which form an order–duality on the right, and* V *a disked* $\tau(X,Y)$*–neighbourhood of* 0. *Then the following assertions hold*:

(a) *If* V *is order–convex* (*resp. positively order–conex*) *then* $\overline{V}$ *is order–convex* (*resp. positively order–convex*).

(b) $\overline{F(V^o)} = F(V^o)$ *and* $\overline{D(V^o)} = D(V^o)$.

(c) $\overline{F_p(V^o)} = F_p(V^o)$ *and* $\overline{D_p(V^o)} = D_p(V^o)$.

Proof. (a) Suppose that $F(V) \subset V$. Then

$$\overline{V} = V^{oo} = (F(V))^{oo} = (D(V^o))^o \quad \text{(by (24.8.1))}$$
$$= \overline{F(\overline{V})} \supset F(\overline{V}).$$

Suppose that $F_p(V) \subset V$. Then

$$\overline{V} = V^{oo} \supset (F_p(V))^{oo} = (D_p(V^o))^o \quad \text{(by (24.8.2))}$$
$$= \overline{F_p(\overline{V})} \supset F_p(\overline{V}),$$

thus $\overline{V}$ is positively order–convex.

(b) We have:

$$\overline{F(V^o)} = (F(V^o))^{oo} = (\,\overline{D(\overline{V})}\,)^o \quad \text{(by (24.8.3))}$$
$$= (D(\overline{V}))^o = F(\overline{V}^o) = F(V^o)$$

and

$$\overline{D(V^o)} = (D(V^o))^{oo} = (\overline{F(\overline{V})})^o \quad \text{(by (24.8.3))}$$
$$= (F(\overline{V}))^o = D(\overline{V}^o) = D(V^o).$$

(c) A similar argument given in the proof of part (b) shows that (c) holds.

Let (X,X_+) be an OVS and p a seminorm on X. We say that p is:

(a) <u>strongly monotone</u> if

$$a \leq x \leq b \Rightarrow p(x) \leq \max \{ p(a),p(b) \};$$

(b) <u>monotone</u> if

$$0 \leq u \leq w \Rightarrow p(u) \leq p(w);$$

(c) <u>decomposable</u> if

$$p(x) = \inf \{ p(u) + p(w) : u, w \in X_+, x = u - w \};$$

(d) <u>semi–decomposable</u> if it satisfies the following condition: For any $x \in X$ and $\epsilon > 0$, there exist $u, w \in X_+$ such that

$$-w \leq x \leq u \quad \text{and} \quad \max \{ p(u),p(w) \} < p(x) + \epsilon.$$

It is clear that the following implications hold for a seminorm p:

strongly monotone $\Rightarrow$ monotone and

decomposable $\Rightarrow$ semi–decomposable.

(24.10) **Corollary.** <u>Let</u> (X,X_+) <u>be an OVS</u>, <u>let</u> p <u>be a seminorm on</u> X, <u>let</u>

$$V_o = \{ x \in X : p(x) < 1 \} \quad \underline{\text{and}} \quad V_c = \{ x \in X : p(x) \leq 1 \},$$

<u>and let</u> V_c^π <u>be the polar of</u> V_c <u>taken in</u> X^* (that is, with respect to $<X,X^*>$).

(a) p <u>is strongly monotone</u> $\Leftrightarrow$ V_o <u>is order–convex</u>

$\Leftrightarrow$ V_c <u>is order–convex</u> $\Leftrightarrow$

(1) $$p(x) = \sup \{ \, |f(x)| : f \in V_c^\pi \cap X_+^* \} \quad (x \in X).$$

(b) p <u>is monotone</u> $\Leftrightarrow$ V_o <u>is positively order–convex</u>

$\Leftrightarrow$ V_c <u>is positively order–convex</u> $\Leftrightarrow$

(2) $$p(u) = \sup \{ \, f(u) : f \in V_c^\pi \cap X_+^* \} \quad (u \in X_+).$$

(c) p <u>is decomposable</u> $\Leftrightarrow$ V_o <u>is decomposable</u>.

(d) p <u>is semi–decomposable</u> $\Leftrightarrow$ V_o <u>is positively dominated</u>.

<u>Proof.</u> (a) If p is strongly monotone, then V_o is obviously order–convex. On the other hand, it is easily seen that V_c is the $\sigma(X,X^*)$–closure of V_o, hence the order–convexity of V_c follows from the order–convexity of V_o (by(24.9) (a)).

Suppose now that V_c is order–convex. We are going to show that (1) holds. Indeed, we first notice that

$$\sup \{ \, |f(x)| : f \in V_c^\pi \cap X_+^* \} \leq p(x)$$

[since $p(x) = \sup \{ \, |h(x)| : h \in V_c^\pi \}$]. As V_c is order–convex, it follows from (24.8.1) that V_c^π is decomposable in (X^*, X_+^*), and hence that any $h \in V_c^\pi$ can be expressed as the form

$$h = \lambda f - (1 - \lambda) g \quad \text{for some } \lambda \in [0,1] \text{ and } f, g \in V_c^\pi \cap X_+^*;$$

consequently,

$$|h(x)| \leq \lambda |f(x)| + (1 - \lambda) |g(x)| \leq \sup \{ \, |\Psi(x)| : \Psi \in V_c^\pi \cap X_+^* \};$$

thus (1) holds.

Finally, suppose that (1) is true, and that $a \leq x \leq b$. As $V_c^\pi \cap X_+^*$ is $\sigma(X^*,X)$–compact, there exists an $h \in V_c^\pi \cap X_+^*$ such that $p(x) = |h(x)|$. Now the positivity of h implies that

$$h(a) \leq h(x) \leq h(b) \leq p(b),$$

so that

$$-h(x) \leq -h(a) \leq p(a).$$

Therefore

$$p(x) = |h(x)| \leq \max\{p(a),p(b)\}.$$

(b) By making use of (24.8.2), the proof of this part is similar to that given in part (a) and therefore will be omitted.

(c) Suppose that p is decomposable and that $x \in V_0$. Then $p(x) < 1$, hence there exist $u, w \in X_+$ such that

$$x = u - w \quad \text{and} \quad p(u) + p(w) < 1.$$

Choose two positive number λ_1 and λ_2 satisfying

$$p(u) < \lambda_1, \ p(w) < \lambda_2 \quad \text{and} \quad \lambda_1 + \lambda_2 = 1.$$

[Indeed, choose $\delta > 0$ with $p(u) + p(w) + \delta < 1$, and then there is a $\lambda_1 > 0$ with $p(u) < \lambda_1 < p(u) + \delta$; let $\lambda_2 = 1 - \lambda_1$; then λ_1 and λ_2 have the required properties.] Clearly, u/λ_1, w/λ_2 belong to $V_o \cap X_+$ and

$$x = \lambda_1\left(\frac{u}{\lambda_1}\right) - \lambda_2\left(\frac{w}{\lambda_2}\right) \in D(V_o).$$

Conversely, suppose that V_o is decomposable. For any $x \in X$ and $\epsilon > 0$, $(p(x) + \epsilon)^{-1}x \in V_0$, hence there exist $\lambda \in [0,1]$ and $x_1, x_2 \in V_0 \cap X_+$ such that $(p(x) + \epsilon)^{-1}x = \lambda x_1 - (1 - \lambda)x_2$. Now the elememts, defined by

$$u = \lambda(p(x) + \epsilon))x_1 \quad \text{and} \quad w = (1 - \lambda)(p(x) + \epsilon))x_2,$$

are positive elements with $x = u - w$ and

$$p(u) + p(w) = (p(x) + \epsilon)(\lambda p(x_1) + (1 - \lambda)p(x_2)) < p(x) + \epsilon.$$

This proves (1) since we always have

$$p(x) \leq \inf\{\, p(u) + p(w) \; ; \; u, w \in X_+, u - w = x \,\}.$$

(d) A similar argument given in the proof of part (c) yields this part.

Remark. It can be shown that V_0 is decomposable (resp. positively dominated) if and only if for any $\delta > 0$.

$$V \subset (1+\delta)D(V) \quad \text{(resp. } V \subset (1+\delta)D_P(V)\text{)}.$$

Roughly speaking, the first duality theorem, concerned with the order–convexity and decomposability (resp. the positive order–convexity and positively dominated property), is satisfactory for the study of ordered LCS. We state the second duality theorem (without proof), concerned with order–intervals of the form $[-u,u]$ and better suited than the first duality theorem for applications to locally convex Riesz spaces (in particular, normed vector lattices) .

(24.11) **Theorem.** (The second duality). Let (X,X_+) and (Y,Y_+) be OVS which form an order–duality on the right (that is, $Y_+ = -X_+^{\circ}$), let V be a disked $\tau(X,Y)$–neighbourhood of 0 and V_0 the $\tau(X,Y)$–interior of V.

(a) (Upward form) $(S(V))^{\circ} = S(V^{\circ})$.

(b) (Downward form) Assume that V is symmetric. Then the following assertions hold:

(i) $(S(V^{\circ}))^{\circ} = \overline{S(V)} = \overline{S(V_0)} = \overline{S(\overline{V})}$.

(ii) If $V^{\circ} \subset S(V^{\circ})$ then $S(V_0) \subset V_0$.

(iii) Suppose that $D(V_0)$ is a $\tau(X,Y)$–neighbourhood of 0. Then

$$S(V^{\circ}) \subset V^{\circ} \Rightarrow V_0 \subset S(V_0).$$

Part (a) of (24.11) is due to Jameson [1970] and Ng[1971] independently, while part (b) of (24.11) is due to Ng and Duhoux [1973].

There is the third duality theorem, due independently to Asimow [1968] and Ng [1969] (which is also generalized by Ng and Duhoux [1973]),is concerned with the directedness and additivity (also see Wong [1980], (3.5) and (3.8)).

A seminorm p on an OVS (X,X_+) is said to be:

(a) absolutely monotone if

$$-u \le x \le u \Rightarrow p(x) \le p(u);$$

(b) regular (or a Riesz seminorm) if

$$p(x) = \inf\{ p(u) : u \in X_+ \text{ and } -u \le x \le u \} \quad (x \in X).$$

As a consequence of (24.11), one can verify the following result, which should be compared with (24.10):

(24.12) **Corollary.** Let (X,X_+) be an OVS, let p be a seminorm on X, let

$$V_o = \{ x \in X : p(x) < 1 \} \quad \text{and} \quad V_c = \{ x \in X : p(x) \le 1 \},$$

and let V_c^π be the polar of V_c taken in X^*.

(a) p is absolutely monotone $\Leftrightarrow$ V_o is absolutely order–convex $\Leftrightarrow$

$$p(x) \le \inf\{ p(u) : -u \le x \le u \} \quad (x \in X).$$

(b) V_o is absolutely dominated if and only of for any $x \in X$ and $\epsilon > 0$ there exists a $u \in X_+$ such that

$$-u \le x \le u \text{ and } p(u) < p(x) + \epsilon.$$

(c) p is regular if and only if V_o is solid.

A LCS (resp. Hausdorff TVS), equipped with a cone, is called an ordered convex space, abbreviated ordered LCS (resp. Hausdorff ordered topological vector space, abbreviated ordered Hausdorff TVS).

(24.13) **Defintion.** An ordered LCS $(X, X_+, \mathscr{P})$ is said to be locally o–convex (resp. locally decomposable, locally solid) if $\mathscr{P}$ admits a neighbourhood base at 0 consisting of 0–convex and circled (resp. disked and decomposable, convex and solid) sets; in this case, the cone is said to be normal (resp. conormal, binormal) in $(X, \mathscr{P})$; while $\mathscr{P}$ is referred to as a locally o–convex (resp. locally decomposable, locally solid) topology on (X, X_+).

Remarks: (i) Characterizations of locally o–convex spaces are given by many authors (seee Schaefer [1966] or Wong [1980,p.98]). Criteria for locally decomposable (resp. locally solid) spaces can be found in Wong [1973 a] (resp. Ng [1971]). Moreover, an ordered LCS is locally solid if and only if it is locally o–convex and locally decomposable (see Wong [1973 a]).

(ii) The quotient space of a locally decomposable space by an order–convex, closed subspace is locally decomposable; the product space and the locally convex direct sum space of a family of locally decomposable (resp. locally o–convex) spaces are locally decomposable (resp.locally o–convex) (see Wong and Ng [1973]).

The notion of locally decomposable spaces was introduced by Bonsall [1955] on ordered normed spaces, and was extended by Jameson [1973], Wong [1973a, 1973b], Ng and Duhoux [1973] , and Walsh [1973] [he called conormal]. Krein [1940] appears to be the first one to consider locally o–convex B–space (see also Krein and Grosberg [1939]); the normality in general ordered LCS was investigated independently by Bonsall [1957], Namioka [1957], Schaefer [1958] (also see Schaefer [1966]), Kist [1958] and others; while the notion of locally solid topologies in general OVS without the lattice structure was introduced by Ng [1971] (it is a generalization of locally convex vector lattices).

Any conormal cone must be generating; the converse is true only for a complete, metrizable LCS with a closed cone as a special case of the following result shows.

(24.14) **Theorem** (Klee [1955]). Let X_+ be a cone in a metrizable TVS $(X, \mathscr{P})$, let $\{ V_n : n \geq 1 \}$ be a local base at 0 for $\mathscr{P}$ consisting of circled sets with $V_{n+1} + V_{n+1} \subset V_n$ $(n \geq 1)$, and suppose that $M = X_+ - X_+$. Then the family

$$\{ V_n \cap X_+ - V_n \cap X_+ : n \geq 1 \}$$

determines a uniquely metrizable vector topology on M, denoted by $\mathcal{P}_D$, which is finer than $\mathcal{P}_M$ (the relative topology on M). Furthermore, the following assertions hold:

(a) If $\mathcal{P}$ is locally convex then so is $\mathcal{P}_D$; in this case, $\{ D(V_n) ; n \geq 1 \}$ is also a local base at 0 for $\mathcal{P}_D$ (here each V_n is assumed to be a disk in X).

(b) If X_+ is $\mathcal{P}$-complete then $(M, \mathcal{P}_D)$ is complete.

(c) Supposed that X_+ is $\mathcal{P}$-complete. If $X = X_+ - X_+$ and $(X, \mathcal{P})$ is of the second category, then $\mathcal{P} = \mathcal{P}_D$, hence $(X, \mathcal{P})$ is complete.

(d) Suppose that X_+ is $\mathcal{P}$-complete. If each $\overline{V_n \cap X_+ - V_n \cap X_+}$ is a $\mathcal{P}$-neighbourhood of 0 (for any $n \geq 1$), then

$$X = X_+ - X_+ \quad \text{and} \quad \mathcal{P} = \mathcal{P}_D,$$

hence $(X, \mathcal{P})$ is complete.

In particular, an ordered Banach space $(E, E_+, \|\cdot\|)$ with a closed cone E_+ is locally decomposable if and only if E_+ is generating.

Proof. Part (a) is obvious since

$$D(V_n) \subset V_n \cap X_+ - V_n \cap X_+ \subset 2D(V_n) \qquad (\text{for all } n \geq 1)$$

whenever each V_n is disked. Parts (c) and (d) follow from part (b) by making use of Banach's open mapping theorem. [If $X = X_+ - X_+$ and $(X, \mathcal{P})$ is of the second category, then the embedding map $J : M \longrightarrow X$ is a bijection for which $J(M)$ is of the second category, hence J is open. If each $\overline{V_n \cap X_+ - V_n \cap X_+}$ is a $\mathcal{P}$-neighbourhood of 0, then J is almost open, hence J is open; consequently, (d) holds.]

Thus we complete the proof by showing that part (b) holds. Let $\{w_n\}$ be a $\mathcal{P}_D$-Cauchy sequence in M. As $\{ V_k \cap X_+ - V_k \cap X_+ : k \geq 1 \}$ is a local base at 0 for $\mathcal{P}_D$ there exists a subsequence $\{z_k\}$ of $\{w_n\}$ such that

$$z_{k+1} - z_k \in V_k \cap X_+ - V_k \cap X_+ \quad \text{(for all } k \geq 1).$$

For each $k \geq 1$, let $x_k, y_k \in V_k \cap X_+$ such that $z_{k+1} - z_k = x_k - y_k$. Then

$$z_{n+1} - z_1 = \Sigma_{k=1}^{n} (z_{k+1} - z_k) = \Sigma_{k=1}^{n} x_k - \Sigma_{k=1}^{n} y_k.$$

To show the $\mathscr{P}_D$–convergence of $\{w_n\}$, it suffices to verify the $\mathscr{P}_D$–convergence of the formal series Σx_n and Σy_n. Indeed, for each $n \geq 1$, let $u_n = \Sigma_{k=1}^{n} x_k$. Then $u_n \in X_+$ and

$$\text{(1)} \qquad u_{n+j} - u_n = x_{n+1} + \cdots + x_{n+j} \quad \in V_{n+1} \cap X_+ + \cdots + V_{n+j} \cap X_+ \subset V_n \cap X_+.$$

The sequence $\{ u_n \}$ is $\mathscr{P}$-Cauchy, hence $\mathscr{P}$-convergent to u, and thus we obtain, by passing to the limit as $j \longrightarrow \infty$ in (1), that

$$u - u_n \in V_n \cap X_+ \subset V_n \cap X_+ - V_n \cap X_+$$

[by the $\mathscr{P}$-closedness of V_n and X_+]. It turns out that the series $\Sigma_{k=1}^{\infty} x_k$ is $\mathscr{P}_D$–convergent to u. Similarly, one can show that the series $\Sigma_{k=1}^{\infty} y_k$ is $\mathscr{P}_D$–convergent. This completes the proof.

Let $(X, X_+, \mathscr{P})$ be an ordered LCS and $\mathscr{U}$ a local base at 0 for $\mathscr{P}$ consisting of disks in X. Setting

$$F(\mathscr{U}) = \{ F(V) : V \in \mathscr{U} \};$$

$$D(\mathscr{U}) = \{ D(V) : V \in \mathscr{U} \};$$

$$S(\mathscr{U}) = \{ S(V) : V \in \mathscr{U} \}.$$

It is easily seen that $F(\mathscr{U})$ determines a unique locally o–convex topology, denoted by $\mathscr{P}_F$ and called the <u>locally o–convex topology associated with</u> $\mathscr{P}$; moreover, $\mathscr{P}_F$ is the least upper bound of all locally o–convex topologies which are coarser than $\mathscr{P}$. Suppose that

$$X = X_+ - X_+.$$

Then $D(\mathscr{U})$ (resp. $S(\mathscr{U})$) determines a unique locally decomposable (resp. locally solid) topology, denoted by $\mathscr{P}_D$ and called the locally decomposable topology associated with $\mathscr{P}$ (resp. $\mathscr{P}_S$ is called the locally solid topology associated with $\mathscr{P}$); moreover, $\mathscr{P}_D$ is the greatest lower bound of all locally decomposable topologies which are finer than $\mathscr{P}$.

The construction of $\mathscr{P}_F$ is essentially due to Namioka [1957], that of $\mathscr{P}_D$ to Wong and Cheung [1971], and that of $\mathscr{P}_S$ to Wong [1969].

It should be noted that $\mathscr{P}_F$ and $\mathscr{P}_D$ are, in general, not consistent with the dual pair $<X,X'>$; but $(X,X_+,\mathscr{P}_F)'$ is the decomposable kernel of X' and $(X,X_+,\mathscr{P}_D)'$ is the order–convex hull in (X^*,X^*_+) of X' as parts of the following result shows.

(24.15) **Theorem.** Let $(X,X_+,\mathscr{P})$ be an ordered LCS with the dual cone X'_+, let $\mathscr{U}$ be a local base at 0 for $\mathscr{P}$, let $\mathscr{E}_{\mathscr{P}}$ be the equicontinuous bornology (i.e., the family of all $\mathscr{P}$-equicontinuous subsets of X'), let

$$D(\mathscr{U}^\pi) = \{\, D(V^o) = \Gamma(V^o \cap X'_+) : V \in \mathscr{U} \,\} \text{ and}$$

$$F^*(\mathscr{U}^\pi) = \{\, F^*(V^o) = (V^o \cap X^*_+) \cap (V^o - X^*_+) : V \in \mathscr{U} \,\}.$$

(a) $\mathscr{P}_F$ is the topology of uniform convergence on members in $D(\mathscr{U}^\pi)$, and

$$(X,X_+,\mathscr{P}_F)' = X'_+ - X'_+ \tag{1}$$

which is the decomposable kernel in (X^*,X^*_+) of X'. In particular, $\mathscr{P}$ is locally 0–convex if and only if $D(\mathscr{U}^\pi)$ is a fundamental family of $\mathscr{E}_{\mathscr{P}}$ (i.e, a base for the convex bornology $\mathscr{E}_{\mathscr{P}}$); and this is the case, if and only if X'_+ is a strict $\mathscr{E}_{\mathscr{P}}$–cone in the sense that the family

$$\{\, N \cap X'_+ - N \cap X'_+ : N \in \mathscr{E}_{\mathscr{P}} \}$$

is a fundamental family of $\mathscr{E}_{\mathscr{P}}$.

(b) Suppose that $X = X_+ - X_+$. Then $\mathscr{P}_D$ is the topology of uniform convergence on members in $\overset{*}{F}(\mathcal{U}^\pi)$, and

$$(X, X_+, \mathscr{P}_D)' = (X' + \overset{*}{X}_+) \cap (X' - \overset{*}{X}_+) = \overset{*}{F}(X'),$$

which is the order–convex hull in $(\overset{*}{X}, \overset{*}{X}_+)$ of $X' = (X, \mathscr{P})'$. Moreover, $\mathscr{P}$ is locally decomposable if and only if $\mathscr{E}_{\mathscr{P}}$ is a fundamental family of $\overset{*}{F}(\mathcal{U}^\pi)$; and this is the case, if and only if the order–convex hull in $(\overset{*}{X}, \overset{*}{X}_+)$ of any $\mathscr{P}$-equicontinuous subset of X' is $\mathscr{P}$-equicontinuous.

Proof. (a) As $F(\mathcal{U})$ is a local base at 0 for $\mathscr{P}_F$, it follows from (24.8.1) of (24.8) and $V^\pi = V^o$ $(V \in \mathcal{U})$ that $\mathscr{P}_F$ is the topology of uniform convergence on members in $D(\mathcal{U}^\pi)$ on account of

$$(F(V))^\pi = D(V^\pi) = D(V^o) \quad (\text{for all } V \in \mathcal{U}).$$

Therefore $(X, X_+, \mathscr{P}_F)' \subset X_+^{\prime} - X_+^{\prime}$. On the other hand, for any $f \in X_+^{\prime}$ there is a $V \in \mathcal{U}$ such that $f \in V^o \cap X_+^{\prime} \subset D(V^o)$, hence $X_+^{\prime} \subset (X, X_+, \mathscr{P}_F)'$. This proves (1).

For any disk $N \in \mathscr{E}_{\mathscr{P}}$, we have

$$\frac{1}{2}(N \cap X_+^{\prime} - N \cap X_+^{\prime}) \subset \Gamma(N \cap X_+^{\prime}) \subset N \cap X_+^{\prime} - N \cap X_+^{\prime};$$

thus $D(\mathcal{U}^\pi)$ is a fundamental family of $\mathscr{E}_{\mathscr{P}}$ if and only if $X_+^{\prime}$ is a strict $\mathscr{E}_{\mathscr{P}}$-cone; consequently the final part follows by making use of (24.8.1) of (24.8).

(b) Since $D(\mathcal{U})$ is a local base at 0 for $\mathscr{P}_D$, it follows from (24.8.1) of (24.8) and $V^\pi = V^o$ $(V \in \mathcal{U})$ that $\mathscr{P}_D$ is the topology of uniform convergence on members in $\overset{*}{F}(\mathcal{U}^\pi)$ on account of

$$(D(V))^\pi = \overset{*}{F}(V^\pi) = \overset{*}{F}(V^o) \quad (\text{for all } V \in \mathcal{U}).$$

To prove (2), let $f \in (X, X_+, \mathscr{P}_D)'$. Then there is a $V \in \mathcal{U}$ such that $f \in (D(V))^\pi = \overset{*}{F}(V^o) \subset \overset{*}{F}(X')$, hence $(X, X_+, \mathscr{P}_D)' \subset \overset{*}{F}(X')$. Conversely, it suffices to verify the following assertion

(3) $$0 \le f \le g \text{ with } g \in X' \text{ and } f \in X^* \Rightarrow f \in (X, X_+, \mathscr{P}_D)'.$$

In fact, there is a $V \in \mathscr{U}$ such that $g \in V^o$, hence

$$f \in F^*(V^o) = (D(V))^\pi \subset (X, X_+, \mathscr{P}_D)',$$

which obtains our assertion (3).

Clearly $\mathscr{E}_{\mathscr{P}}$ is a fundamental family of $F^*(\mathscr{U}^\pi)$ if and only if the order–convex hull in (X^*, X_+^*) of each $\mathscr{P}$-equicontinuous subset of X' is $\mathscr{P}$-equicontinuous. If $\mathscr{P}$ is locally decomposable, then $\mathscr{E}_{\mathscr{P}}$ is a fundamental family of $F^*(\mathscr{U}^\pi)$.

Conversely we assume that the order–convex hull in (X^*, X_+^*) of each $\mathscr{P}$-equicontinuous subset of X' is $\mathscr{P}$-equicontinuous. Then X' is order–convex in (X', X_+^*) [it is easily seen that X' is order–convex in (X^*, X_+^*) if and only if the order–convex hull in (X^*, X_+^*) of each $\mathscr{P}$-equicontinuous of X' is contained in X'], and $\{\overline{D(V)} : V \in \mathscr{U}\}$ is a local base at 0 for $\mathscr{P}$; consequently, $X = \overline{X_+ - X_+}$ or, equivalently, X_+' is proper. On the other hand, if

(4) $$X = X_+ - X_+,$$

then the assumption implies that $\mathscr{P}$ is the topology of uniform convergence on members in $F^*(\mathscr{U}^\pi)$ on account of $D(V) \subset V$ $(V \in \mathscr{U})$, so that $\mathscr{P} = \mathscr{P}_D$ (i.e., $\mathscr{P}$ is locally decomposable). Therefore we have to verify (4). To this end, it suffices to show that X_+^* is proper (see Ex.24–3).

In fact, let $f \in X_+^* \cap (-X_+^*)$. Then $f \in X_+' \cap (-X_+')$ by the order–convexity in (X^*, X_+^*) of X', and thus $f = 0$ since X_+' is proper.

<u>**Remark.**</u> In view of (24.15) (a), it can be shown that the completion of a locally o–convex space is locally o–convex. If $(X, X_+, \mathscr{P})$ is a metrizable locally decomposable space, in view of Klee's theorem (24.14), Banach's open mapping theorem and (24.15) (b), one can show that the closure $\tilde{X}_+$ of X_+ in the completion $(\tilde{X}, \tilde{\mathscr{P}})$ of $(X, \mathscr{P})$ is a

conormal cone in $(\tilde{X}, \tilde{\mathscr{P}})$ (see Wong [1976 b]).

The present form of (24.15) is taken from an article of Ng and Dahoux [1973]; in fact, part (a) contains the duality theorem of Schaefer and Bonsall concerned with normal cones and strict $\mathscr{B}$-cones (see Bonsall [1957], Schaefer [1966], Peressini [1967], Wong and Ng [1973]), while part (b) is a generalization of a result of Wong [1973 b].

For the special topologies $\sigma(X, X')$ and $\tau(X, X')$, we have the following interesting result.

(24.16) **Theorem.** Let $(X, X_+, \mathscr{P})$ be an ordered LCS whose topological dual and dual cone are denoted by X' and X'_+ respectively.

(a) (Bonsall [1957], Namioka [1957]). The weak topology $\sigma(X,X')$ is locally o-convex if and only if

$$X' = X'_+ - X'_+.$$

In particular, if $\mathscr{P}$ is locally o-convex, then $\sigma(X,X')$ is the coarsest locally o-convex topology on X consistent with $\langle X, X' \rangle$ and X' is a decomposable vector subspace of (X^*, X^*_+).

(b) (Wong [1973 b]). The following assertions are equivalent :

(i) The Mackey topology $\tau(X,X')$ is locally decomposable.

(ii) There exists a circled $\tau(X,X')$-equicontinuous subset N of X' such that either $F(N)$ (in (X',X'_+)) or $S(N)$ (w.r.t. X'_+) is $\tau(X,X')$-equicontinuous, and X' is order-convex in (X^*,X^*_+) i.e.,

$$X' = F^*(X') = (X' + X^*_+) \cap (X' - X^*_+);$$

(iii) $X = X_+ - X_+$ and X' is order-convex in (X^*,X^*_+).

In particular, if $\mathscr{P}$ is locally decomposable, then $\tau(X,X')$ is the finest locally decomposable topology on X consistent with $\langle X, X' \rangle$ and X' is order-convex in (X^*,X^*_+).

<u>**Proof.**</u> (a) Let $\sigma_F(X,X')$ be the locally o–convex topology associated with $\sigma(X,X')$. Then $\sigma_F(X,X') \le \sigma(X,X')$ and $(X,X_+,\sigma_F(X,X'))' = X'_+ - X'_+$ (by (24.15) (a)). Therefore, $X' = X'_+ - X'_+$ if and only if $\sigma_F(X,X') = \sigma(X,X')$ since $\sigma(X,X')$ is the coarsest topology consistent with $< X,X'>$.

(b) (i) ⇒ (ii) : For any disked, $\tau(X,X')$–neighbourhood V of 0, $D(V)$, and surely $S(V)$ (since $D(V) \subset S(V)$), is a $\tau(X,X')$–neighborhood of 0. Therefore, $F(V^o)$ (resp. $S(V^o)$) is $\mathcal{P}$–equicontinuous by (24.8.1) of (24.8) (resp. (24.11) (a)). On the other hand, (24.15) (b) shows that X' is order–convex in (X^*,X^*_+).

(ii) ⇒ (iii): Suppose that X_+ is not generating. Then there is an $0 \ne f \in X^*$ such that $f = 0$ on $X_+ - X_+$, hence $0 \le nf \le 0$ (for all $n \ge 1$), and thus $nf \in X'$; consequently,

$$nf \in F(N) \quad \text{or} \quad nf \in S(N) \quad (\text{for all } n \ge 1).$$

Let V be a disked $\tau(X, X')$–neighbourhood of 0 such that either $F(N) \subset V^o$ or $S(N) \subset V^o$. Then $f = 0$ on V [since $nf \in V^o$ for all $n \ge 1$], hence $f = 0$ on X [since V is absorbing], contrary to the fact that $f \ne 0$.

(iii) ⇒ (i): Let $\tau_D(X,X')$ be the locally decomposable topology on X associated with $\tau(X,X')$. Then $\tau_D(X,X')$ is consistent with $<X,X'>$ [by the order–convexity of X' in (X^*,X^*_+)], hence $\tau_D(X,X') = \tau(X,X')$ [by Mackey–Arens' theorem].

<u>**Remark.**</u> From the proof of part (b), we see that the following assertion holds. Let $(X, X_+, \mathcal{P})$ be an ordered LCS such that X' is order–convex in (X^*, X^*_+). Then $X_+ - X_+ = X$ if and only if there is a circled $\tau(X,X')$–equicontinous subset N of X' such that either $F(N) = (N + X'_+) \cap (N - X'_+)$ or $S(N)$(w.r.t. X'_+) is $\tau(X,X')$–equicontinuous.

Let (X,X_+) be an OVS, let $\mathcal{B}_{(ob)}$ be the family of all order–bounded subsets of X, and let $\mathcal{U}_{(0)}$ be the family of all disks in X each of which absorbs all members in $\mathcal{B}_{(ob)}$. It is easily seen that $\mathcal{U}_{(0)}$ determines a locally convex topology on X (not necessarily Hausdorff), which is called the <u>order topology</u> or the <u>order–bound topology</u>, and denoted by $\mathscr{P}_{(0)}$.

Clearly, $\mathscr{P}_{(0)}$ is the finest locally convex topology on X for which members in $\mathscr{B}_{(ob)}$ are topologically bounded and

$$X^b = (X,X_+,\mathscr{P}_{(0)})',$$

Consequently, $\mathscr{P}_{(0)}$ is Hausdorff if and only if X^b separates points of X.

If $X = X_+ - X_+$, then $\{ [-u,u] : u \in X_+ \}$ is a base for $\mathscr{B}_{(ob)}$, and $\mathscr{B}_{(ob)}$ is a convex bornology, hence $\mathscr{P}_{(0)}$ is the associated topology of $\mathscr{B}_{(ob)}$.

From now on, when we consider the order topology $\mathscr{P}_{(0)}$, we always assume that $\mathscr{P}_{(0)}$ is Hausdorff. Therefore $(X,X_+,\mathscr{P}_{(0)})$ is a bornological space, The converse is, in general, not true; namely, there exists a bornological topology on an OVS which is not the order topology (see Wong [1980,p.136]).

As a consequence of (24.16) (b), we have the following:

(24.17) **Lemma.** <u>The order topology</u> $\mathscr{P}_{(0)}$ <u>on</u> (X,X_+) <u>is locally decomposable if and only if</u> $X = X_+ - X_+$.

<u>Proof.</u> The necessity is obvious, while the sufficiency follows from (24.16) (b) since $X^b = (X,X_+,\mathscr{P}_{(0)})'$ is order–convex in (X^*,X^*_+) and $\mathscr{P}_{(0)} = \tau(X,X^b)$.

<u>Remark.</u> In general, $\mathscr{P}_{(0)}$ is not locally o–convex (see Namioka [1957]); but if (X,X_+) has the Riesz decomposition property in the sense that

$$[0,u] + [0,v] = [0,u + v] \quad (\text{for all } u, v \in X_+),$$

then $\mathscr{P}_{(0)}$ is the finest locally o–convex topology as shown by Schaefer [1966].

In terms of the order topology, we are able to give some sufficient conditions of continuity for positive linear maps. To do this, we require the following terminology: Let (X,X_+) be an OVS and $(Y,\mathscr{T})$ a LCS. An $T \in L^*(X,Y)$ is said to be (OT)–<u>bounded</u> if it sends order–bounded subsets of (X,X_+) into $\mathscr{T}$-bounded subsets of Y. The set consisting of all (OT)–bounded linear maps from (X,X_+) into $(Y,\mathscr{T})$, denoted by $L^{ot}(X,Y)$, is a vector subspace of $L^*(X,Y)$. It is clear that

$$X^b = L^{ot}(X,\mathbb{R})$$

and that if $(Y,Y_+,\mathscr{T})$ is an ordered LCS such that order–bounded subsets of Y are $\mathscr{T}$-bounded, then

$$L_+^*(X,Y) \subset L^{ot}(X,Y).$$

The following result, due to Wong [1972] , can be regarded as a general form for studying the continuity of positive linear mappings.

(24.18) **Theorem** (Wong [1972]). Let $(X,X_+,\mathscr{P})$ be an ordered LCS and $\mathscr{P}_{(0)}$ the order topology on X. Then $\mathscr{P}_{(0)} \leq \mathscr{P}$ if and only if

$$L^{ot}(X,Y) \subset L(X,Y) \quad \text{for any LCS } (Y,\mathscr{T}).$$

Proof. Necessity: Let $T \in L^{ot}(X,Y)$ and U a disked $\mathscr{T}$-neighbourhood of 0 in Y, Then $T^{-1}(U)$ is a disk in X that absorbs any order–bounded subset of (X,X_+), hence $T^{-1}(U)$ is a $\mathscr{P}_{(0)}$–neighbourhood of 0, thus $T \in L(X,Y)$ by the assumption that $\mathscr{P}_{(0)} \leq \mathscr{P}$.

Sufficiency. Since order–bounded subsets of X are $\mathscr{P}_{(0)}$–bounded, it follows that the identity map I_X is an (OT)–bounded map from $(X,X_+,\mathscr{P})$ onto $(X,\mathscr{P}_{(0)})$, and hence that I_X is continuous; consequently, $\mathscr{P}_{(0)} \leq \mathscr{P}$.

Remark. If $\mathscr{P}_{(0)} \leq \mathscr{P}$ and $(Y,Y_+,\mathscr{T})$ is an ordered LCS such that order–bounded sets are $\mathscr{T}$-bounded, it then follows from $X_+^*(X,Y) \subset L^{ot}(X,Y)$ that

$$L_+^*(X,Y) \subset L(X,Y);$$

in particular,

$$X_+^* \subset X^b = L^{ot}(X,\mathbb{R}) \subset X'.$$

(24.19) **Corollary**. (Namioka and Klee). Let $(X,X_+,\mathscr{P})$ be a metrizable ordered LCS and Y an LCS. Then each of the following conditions implies that

$$L^{ot}(X,Y) \subset L(X,Y) \text{ (in particular, } X_+^* \subset X^b \subset X'):$$

(i) $(X,X_+,\mathscr{P})$ is locally decomposable, and each increasing $\mathscr{P}$-Cauchy sequence in X has a supremum in X (in particular, X_+ is $\mathscr{P}$-complete);

(ii) X_+ is $\mathscr{P}$-complete and generating, and $(X,\mathscr{P})$ is of the second category.

In particular, positive linear functionals on an ordered, Fréchet space $(X,X_+,\mathscr{P})$ with a closed cone X_+ must be continuous.

Proof. It has only to verify the result for the condition (i) since if (ii) holds then (i) must be true by Klee's theorem (24.14).

Let $\{ V_n : n \geq 1 \}$ be a countable local base at 0 for $\mathscr{P}$ consisting of $\mathscr{P}$-closed, circled sets with $V_{n+1} + V_{n+1} \subset V_n$ $(n \geq 1)$. Then $\{ V_n \cap X_+ - V_n \cap X_+ : n \geq 1 \}$ is a local base at 0 for $\mathscr{P}$[since $\mathscr{P}$ is locally decomposable]. We now show that $\mathscr{P}_{(0)} \leq \mathscr{P}$.

Suppose, on the contrary, that there is a disked $\mathscr{P}_{(0)}$–neighbourhood W of 0 such that

$$V_n \cap X_+ - V_n \cap X_+ \not\subset 2^{2n}W \quad (\text{for all } n \geq 1).$$

For any $n \geq 1$, there exist $x_n, y_n \in V_n \cap X_+$ with $x_n - y_n \notin 2^{2n}W$. The sequences $\{ \Sigma_{k=1}^{n} 2^{-k}x_k , n \geq 1 \}$ and $\{ \Sigma_{k=1}^{n} 2^{-k}y_k , n \geq 1 \}$ are increasing $\mathscr{P}$-Cauchy sequences, hence there exist $u, w \in X_+$ such that

$$u = \sup_n \Sigma_{k=1}^{n} 2^{-k}x_k \quad \text{and} \quad w = \sup_n \Sigma_{k=1}^{n} 2^{-k}y_k .$$

As

$$-w \leq 2^{-n}(x_n - y_n) \leq u \quad \text{and} \quad 2^{-n}(x_n - y_n) \notin 2^{2n}W ,$$

it follows that $[-w,u]$ is not absorbed by the $\mathscr{P}_{(0)}$–neighbourhood W of 0, contrary to the construction of $\mathscr{P}_{(0)}$. This contradiction shows that $\mathscr{P}_{(0)} \leq \mathscr{P}$.

(24.20) Corollary (Schaefer). Let $(X,X_+,\mathscr{P})$ be a bornological ordered LCS. If X_+ is sequentially $\mathscr{P}$–complete and a strict $\mathscr{M}_{von}(\mathscr{P})$–cone (for definition see (24.15)(a)), then

$$L^{ot}(X,Y) \subset L(X,Y) \quad \text{(for any LCS } (Y,\mathscr{T}));$$

in particular, $X_+^* \subset X^b \subset X'$.

Proof. In view of (24.18) and the assumption that $(X,\mathscr{P})$ be bornological, it suffices to show that any disked $\mathscr{P}_{(0)}$–neighbourhood V of 0 absorbs any $\mathscr{P}$–bounded subset of X.

Suppose, on the contrary, that there is a $\mathscr{P}$–bounded set B in X which is not absorbed by V. Since X_+ is a strict $\mathscr{M}_{von}(\mathscr{P})$–cone, there is a disked, $\mathscr{P}$–bounded subset A of X such that $B \subset A \cap X_+ - A \cap X_+$. Hence $A \cap X_+$ is not absorbed by V. For any $n \geq 1$, there exists

$$x_n \in A_n \cap X_+ \quad \text{with} \quad x_n \notin 2^{2n}V.$$

The sequence $\{ \Sigma_{k=1}^{n} 2^{-k}x_k , n \geq 1 \}$ is an increasing $\mathscr{P}$–Cauchy sequence in X_+, hence $y = \sup_n \Sigma_{k=1}^{n} 2^{-k}x_k$ exists in X (by the sequential $\mathscr{P}$–completeness of X_+). As

$$0 \leq 2^{-n}x_n \leq y \quad \text{and} \quad 2^{-n}x_n \notin 2^{n}V_n \quad \text{(for all } n \geq 1),$$

we conclude that $[0,y]$ is not absorbed by the $\mathscr{P}_{(0)}$–neighbourhood V of 0, contrary to the construction of $\mathscr{P}_{(0)}$. This contradiction shows that V is a $\mathscr{P}$–neighbourhood of 0.

(24.21) **Corollary.** Let $(X,X_+,\mathscr{P})$ be an ordered LCS. If $\operatorname{Int} X_+ \neq \phi$, then $L^{ot}(X,Y) \subset L(X,Y)$ (for any LCS $(Y,\mathscr{T})$); consequently,

$$X_+^* \subset X^b \subset X'.$$

Proof. Let $e \in \text{Int}\, X_+$. The topology induced by the gauge of $[-e,e]$ is $\mathscr{P}_{(0)}$. As $e \in \text{Int}\, X_+$, there exists a circled $\mathscr{P}$-neighbourhood V of 0 such that $e + V \subset X_+$, and hence that $V \subset [-e,e]$. Therefore $\mathscr{P}_{(0)} \leq \mathscr{P}$, and the result follows from (24.18).

Other sufficient conditions for continuity of positive linear maps can be found in Ng [1973], Wong and Ng [1973], Peressini [1967] and Schaefer [1966,1974].

Exercises

24–1. Let $(X, \mathscr{P})$ be a TVS (not necessarily Hausdorff) and X_+ a cone in X. Show that:

(a) If X_+ is proper and closed, then $\mathscr{P}$ is Hausdorff.

(b) If there exists a $\mathscr{P}$–neighbourhood V of 0 such that $V \cap X_+$ is closed, then X_+ is closed.

24–2. Let $(X, \mathscr{P})$ be a LCS, let X_+ be a cone in X and X'_+ its dual cone in X'. Show that:

(a) $\overline{X}_+$ is proper if and only if X'_+ separates points of X.

(b) X'_+ is proper if and only if $X_+ - X_+$ is dense in $(X, \mathscr{P})$.

(c) $e \in X_+$ is a $\mathscr{P}$–interior point of X_+ if and only if $[-e, e]$ is a $\mathscr{P}$–neighbourhood of 0.

(d) Order–bounded subsets of X are $\mathscr{P}$–bounded if and only if $(X, \mathscr{M}_{\text{von}}(\mathscr{P}))^\times \subset X^b$.

(e) Suppose that X_+ is $\mathscr{P}$–closed. If $\{x_\tau, \tau \in D\}$ is a net in X such that

$$x_\tau \uparrow \quad \text{and} \quad x = \mathscr{P}\text{-}\lim_\tau x_\tau,$$

then $x = \sup_\tau x_\tau$.

(f) Suppose that $X = X_+ - X_+$. Then the family of all order–bounded subsets of X, denoted by $\mathscr{M}_{\text{ob}}$, is a convex bornology with $\{[-u, u] : u \in X_+\}$ as a basis.

24–3. Let (X, X_+) be an OVS with the dual cone X_+^*. Show that the following statements are equivalent :

(a) $X_+ - X_+ = X$ (i.e. X_+ is generating).

(b) X_+^* is proper.

(c) (Namioka [1957,p.17]). Each order–interval $[f,g]$ in $X^{\#}$ is $\sigma(X^{\#},X)$–compact.

(d) (Namioka [1957,p.17]). There is some order–interval $[\varphi, h]$ in $X^{\#}$ which is $\sigma(X^{\#},X)$–compact.

If, in addition, $\mathscr{P}$ is a Hausdorff locally convex topology on X such that its topological dual $X' = (X,\mathscr{P})'$ is order–convex in (X^*, X_+^*) (i.e., $X' = (X'+X_+^*) \cap (X'-X_+^*)$), then (a) is equivalent to the following :

(e) There is a circled $\tau(X,X')$–equicontinuous subset N of X' such that either $F(N) = (N+X_+')\cap(N-X_+')$ or $S(N)$ (w.r.t. X_+') is $\tau(X,X')$–equicontinuous.

24–4. Let (X,X_+) be an OVS, let M be a vector subspace of X and $M_+ = M \cap X_+$.

(a) If X_+ is proper then so is M_+.

(b) If M_+ is proper and if M is order–convex (i.e., $M = (M+X_+) \cap (M-X_+)$), then X_+ is proper.

24–5. Prove lemma (24.4).

24–6. (Namioka [1957], Bauer[1957]). Let (X,X_+) be an OVS, let M be a vector subspace of X and $f \in M^*$. Show that the following two statements are equivalent :

(a) There exists $g \in X_+^*$ which is an extension of f.

(b) There exists a convex absorbing subset V of X such that

$$f(y) \leq 1 \quad (\text{for all } y \in M \cap (V - X_+)).$$

24–7. Let (X,X_+) be an OVS, let V be a disked, absorbing subset of X, let p_V be the gauge of V and

$$V_0 = \{x \in X : p_V(x) < 1\}.$$

Show that V_0 is decomposable (resp. positive dominated) if and only if for any $\delta > 0$

$$V \subset (1+\delta)D(V) \quad (\text{resp. } V \subset (1+\delta)D_P(V)).$$

24–8. Prove (24.11).

24–9. Prove (24.12).

24–10. For a given $\alpha > 0$, we say that a cone E_+ in a normed space $(E,\|\cdot\|)$ is :

(i) α–normal (in $(E,\|\cdot\|)$) if $F(U_E) \subset \alpha U_E$;

(ii) α–generating (in $(E,\|\cdot\|)$) if $U_E \subset \alpha D(U_E)$;

(iii) almostly α–generating (in $(E,\|\cdot\|)$) if $U_E \subset \alpha \overline{D(U_E)}$;

(iv) approximately α–generating (in $(E, \|\cdot\|)$) if it is $(\alpha + \delta)$–generating for all $\delta > 0$.

Show that :

(a) E_+ is α–normal in $(E,\|\cdot\|)$

$\Leftrightarrow$ E'_+ is α–generating in $(E',\|\cdot\|)$. (Krein–Grosberg)

(b) E'_+ is α–normal in $(E',\|\cdot\|)$

$\Leftrightarrow$ E_+ is almostly α–generating in $(E,\|\cdot\|)$

$\Leftrightarrow$ E_+ is approximately α–generating in $(E,\|\cdot\|)$ provided that E_+ is complete (Ando [1962], Ellis [1964]).

24–11. For a given $\alpha > 0$, we say that a cone E_+ in a normed space $(E,\|\cdot\|)$ is :

(i) α–positively normal (resp. absolutely α–normal) in $(E,\|\cdot\|)$ if $F_P(U_E) \subset \alpha U_E$ (resp. $S(U_E) \subset \alpha U_E$);

(ii) α–positively generating (resp. absolutely α–generating) in $(E, \|\cdot\|)$ if $U_E \subset \alpha D_P(U_E)$ (resp. $U_E \subset \alpha S(U_E)$);

(iii) almostly α–positively generating (resp. almostly absolutely α–generating) in $(E,\|\cdot\|)$ if

$$U_E \subset \alpha\overline{D_P(U_E)} \quad (\text{resp. } U_E \subset \alpha\overline{S(U_E)});$$

(iv) approximately absolutely α–generating if it is absolutely $(\alpha+\delta)$–generating for all $\delta > 0$.

Show that :

(a) E_+ is α–positively normal in $(E,\|\cdot\|)$

$\Leftrightarrow$ E'_+ is α–positively generating in $(E',\|\cdot\|)$.

(b) E'_+ is α–positively normal in $(E',\|\cdot\|)$

$\Leftrightarrow$ E_+ is almostly α–positively generating in $(E,\|\cdot\|)$

(c) E_+ is absolutely α–normal in $(E,\|\cdot\|)$

$\Leftrightarrow$ E'_+ is absolutely α–generating in $(E',\|\cdot\|)$.

(d) E'_+ is absolutely α–normal in $(E',\|\cdot\|)$

$\Leftrightarrow$ E_+ is approximately absolutely α–generating in $(E,\|\cdot\|)$ provided that E_+ and $(E,\|\cdot\|)$ are complete. (Jameson [1970,(3.6.8)]).

(e) E' is absolutely α–normal in $(E',\|\cdot\|)$ and E' is order–convex in (E^*,E^*_+)

$\Leftrightarrow$ E_+ is approximately absolutely α–generating in $(E, \|\cdot\|)$ and the $\|\cdot\|$–top on E is locally decomposable.

(f) The norm $\|\cdot\|$ on (E, E_+) is regular if and only if its dual norm $\|\cdot\|$ is regular on (E',E'_+) provided that E_+ and $(E,\|\cdot\|)$ are complete (Davies [1968]).

24–12. A set B in an OVS (X,X_+) is said to be

(i) <u>directed upwards</u> if for any $b_1, b_2 \in B$, there is an $x \in B$ such that $b_i \leq x$ $(i=1, 2)$;

(ii) <u>additive</u> (on X_+) if B is convex, $0 \in B$ and its (extended) gauge p_B of B is additive on X_+ in the sense that

$$p_B(u_1+u_2) \geq p_B(u_1) + p_B(u_2) \quad (\text{for all } u_1, u_2 \in X_+).$$

Let (X, X_+) and (Y, Y_+) be OVS which form an ordered duality on the right, let V be a convex $\tau(X,Y)$–neighbourhood of 0, let p_V be the gauge of V and $V_0 = \{x \in X : p_V(x) < 1\}$. Show that :

(a) V is additive in X_+ if and only if V° is directed upwards.

(b) If V is directed upwards, then V° is additive on Y_+.

(c) Suppose that $D(V_o)$ is a $\tau(X,Y)$–neighbourhood of 0. If V^o is addiitive on Y_+ then V_o is directed upwards.

(d) V_o is additive on X_+ if and only if V_o is positively order–convex and $X_+ \backslash V_o$ is convex.

(e) V_o is directed upwards if and only if for any $x, y \in X$ and $\epsilon > 0$ there exists an $z \in X$ such that

$$x, y \leq z \quad \text{and} \quad p_V(z) < \max\{p_V(x), p_V(y)\} + \epsilon.$$

24–13. Let $(E, E_+, \|\cdot\|)$ be an ordered NS. Show that:

(a) The norm $\|\cdot\|$ is additive on E_+

$\Leftrightarrow$ U_E^o is directed upwards

$\Leftrightarrow$ Int U_E^o is directed upwards (Asimow [1968], Ng [1969])

(b) The dual norm $\|\cdot\|$ is additive on E'_+ and E' is order–convex in (E^*, E^*_+)

$\Leftrightarrow$ Int U_E is directed upwards. (Ng/Duhoux [1973])

24–14. Let $(X, X_+, \mathscr{P})$ be an ordered LCS. Show that:

(a) $(X, X_+, \mathscr{P})$ is locally decomposable.

$\Leftrightarrow$ $\mathscr{P}$ admits a local base at 0 consisting of disked, absolutely dominated sets

$\Leftrightarrow$ $\mathscr{P}$ is determined by a family of decomposable seminorms on X (Wong [1973.a]).

(b) $(X, X_+, \mathscr{P})$ is locally 0–convex

$\Leftrightarrow$ $\mathscr{P}$ is determined by a family of monotone seminorms

$\Leftrightarrow$ $\mathscr{P}$ is determined by a family of strongly monotone seminorms

$\Leftrightarrow$ $\mathscr{P}$ is determined by a family of absolutely monotone seminorms

$\Leftrightarrow$ if $\{x_\tau, \tau \in D\}$ and $\{y_\tau, \tau \in D\}$ are two nets in X such that $0 \leq x_\tau \leq y_\tau$ $(\tau \in D)$ and if $0 = \mathscr{P}\text{-}\lim_\tau y_\tau$ then $0 = \mathscr{P}\text{-}\lim_\tau x_\tau$.

(c) $(X, X_+, \mathscr{P})$ is locally solid

$\Leftrightarrow$ it is locally 0–convex and locally decomposable (Wong [1973.a])

$\Leftrightarrow$ $\mathscr{P}$ is determined by a family of Riesz seminorms (Ng [1971]).

24–15. Let $(X, X_+, \mathscr{P})$ be an ordered LCS such that $X' = (X' + X_+^*) \cap (X' - X_+^*)$ (order–convex in (X^*, X_+^*)).

Show that $X_+ - X_+ = X$ if and only if there exists a symmetric $\tau(X', X)$–equicontinuous subset N of X' such that either $F(N) = (N + X_+') \cap (N - X_+')$ or $S(N)$ (w.r.t. X_+') is $\tau(X', X)$–equicontinuous.

24–16. An OVS (X, X_+) is said to have the Riesz decomposition property if

$$[0, u] + [0, v] = [0, u+v] \quad \text{(for all } u, v \in X_+\text{)}.$$

Let (X, X_+) have the Riesz decomposition property and $X = X_+ - X_+$. Show that the order topology $\mathscr{P}_{(0)}$ in (X, X_+) is the finest locally solid topology. (Namioka [1957])

Special Symbols

$\mathcal{M}_{von}, \mathcal{M}_p, \mathcal{M}_c(\mathcal{P})$	Vector bornologies notations,182
$\mathcal{M}_{ind}$	The final vector bornology,192
$\mathcal{M}_{inf}(\mathcal{P})$	Infracomplete bornology,198
$\text{mix}\,\mathcal{U}_o$	327
$\mathcal{M}_{ob}$	The family of all order–bounded subsets of X,355
$\mathfrak{N}$	The class of all nuclear operators,314
$N_{\ell_1}(\cdot)$	314
O_E	The open unit ball in E,13
$Op(\mathbb{D}_o)$	The set of all $\mathbb{D}_o$–factorable operators,299
$p_V(\cdot)$	The gauge (or the Minkowski functional) of V,2
$\mathcal{P}$	A vector topology,110
$\mathcal{P}_{pro}$	The projective topology w.r.t.,149
$\mathcal{P}^x$	The bornological topology associated with $\mathcal{P}$,177
$\mathcal{P}^o$	The polar topology,220
$\pi_{\ell_1}(\cdot)$	316
$\mathcal{P}_F, \mathcal{P}_D, \mathcal{P}_S$	371,372
$\mathcal{P}_{(o)}$	The order (or order–bound) topology,377
Q_M	The quotient mapping,21
$q(T)$	The surjection modulus of T,98
$Q : F \twoheadrightarrow F_o$	The topological surjection,107
$R_\lambda(T)$	The inverse of T_λ,287
$\mathfrak{S}_{ini}$	The initial bornology,190

$\mathfrak{S}_t$	The associated topology of $\mathfrak{S}$ (vector bornology),202
$(\mathfrak{S}_t)_{von}$	The associated bornology of $\mathfrak{S}$,202
$\sigma(Y,X)$	The weak topology on Y,206
$\sigma(T)$	The spectrum of T,287
$\sigma_e(T)$	The point (or discrete) spectrum of T,287
$\sigma_c(T)$	The continuous spectrum of T,288
$\sigma_r(T)$	The residual spectrum of T,288
Space($\mathscr{U}$)	The space ideal associated with $\mathscr{U}$,299
$\mathfrak{G}$	The approximable operator ideal,301
S(V)	359
$S^*(N)$	360
S($\mathscr{U}$)	371
$sp\mathscr{U}_o$	327
$\check{T}$	The bijection associated with a linear mapping T,11
$\hat{T}$	The injection associated with a linear mapping T,11
$\|T\|$	Norm of T,47
T^*	The algebraic adjoint of T,54
T'	The dual operator of T,54
T''	The bidual operator of T,54
$\mathcal{T}_{ind}$	The inductive topology w.r.t. ,153
$\tau(X',X)$	The Mackey topology on X',220
$T_\lambda=\lambda I_E-T$	282

$tr(\cdot)$	The trace of ,315
U_E	The closed unit ball in E,13
$\mathcal{U}_X$	The local base at 0 for $\mathscr{P}$,110
$\mathcal{U}^x$	The family of all disked bornivorous,175
$\mathcal{U}\circ\mathcal{C}$	The product of operator ideals,302
$\mathcal{U}\circ\mathcal{C}^{-1}$	The right–hand quotient of operator ideals,302
$\mathcal{U}^{-1}\circ\mathcal{C}$	The left–hand quotient of operator ideals,302
$\mathcal{U}^{inj},\mathcal{U}^{sur},\mathcal{U}^{dual},\mathcal{U}^{reg}$	304,305
$\mathcal{U}^*$	The adjoint operator ideal of $\mathcal{U}$,317
$\mathcal{U}^\Delta$	The conjugate operator ideal of $\mathcal{U}$,317
$\mathcal{U}_o^{sur}(\mathcal{U}_o^{inf},\mathcal{U}_o^{rup})$	The superior (inferior, right–superior) extension of $\mathcal{U}_o$,320
$\mathcal{U}_{(o)}$	376
$v_o,v_1,\cdots,v_n$	A n–simplex,331
V_o,V_c	364
$x(M)$	Equivalence class notation,21
$[x,y]$	The inner product,28
$\langle x,f\rangle$	Linear functional notation,52
$X_{\mathscr{P}}$	A topological vector space,110
$X^{\mathfrak{S}}$	Bornological vector space notation,180
X^x	The bornological dual of $X^{\mathfrak{S}}$,202
$\langle X,Y\rangle$	A dual pair,205

$X''_\beta=(X'',\beta(X'',X'))$	The strong bidual of X,219
X'_{β_u}	The ultra–strong dual of X,221
$\mathfrak{X}$	The separable operator ideal,301
X_+	A (positive) cone,353
$X_+^*,X^\#,X^b$	355
X'_+	The dual cone of X_+,356

References

Andenaes, R.R.

[1970] Hahn–Banach extensions which are maximal on a given cone, Math. Ann. 188,90–96.

Ando, T.,

[1962] On fundamental properties of a Banach space with a cone, Pacific J. Math. 12,1163–1169.

Asimov, L.,

[1968] Universally well–capped cone, Pacific J. Math. 26(3), 421–431.

Banach, S.,

[1932] Opérations Linéaries (Monografje Matematyczue I, Warszawa).

Bauer, H.,

[1957] Sur le prolongement des formes linéair positives dans un espacs vectorial ordonné, C. R. Acad. Sci. Paris, 244, 289–92.

Bishop, E./Phelps, R.R.,

[1963] The support functionals of a convex set, Proc. Symp. in Pure Math. 7, Convexity, 27–35.

Bonsall, F.F.,

[1955] Endomorphisms of a partially ordered vector spaces without order unit, J. London Math. Soc. 30, 144–153.

[1957] The decomposition of continuous linear functional in non–negative components, Proc. Durham Phil Soc. A13, 6–11.

[1958] Linear operators in complete positive cones, Proc. London Math. Soc. (3) 8, 53–75.

Bourbaki, N.,

[1987] Topological vector spaces (Springer – Verlag, Berlin, New York, London, Paris).

Brézis, H./Nirenberg, L./and Stampacchia, G.,

[1972] A remark on Ky Fan's minimax principle, Boll. Un. Math. Ital., 6, 293–300.

Brodskii, M.S./Milman, D.P.,

[1948] On the center of a convex sets, Dokl. Akad. Nauk SSSR, 59, 837–840.

Browder, F.E.,

[1965a] Nonlinear monotone operators and convex sets in Banach spaces, Bull. Amer. Math. Soc. 71, 780–785.

[1965b] Fixed pointed theorems for non–compact mappings in Hilbert spaces, Proc. Nat. Acad. Sci. U.S.A. 53, 1272–1276.

Brown, A.B./Cairns, S.S.,

[1961] Strengthening of Sperner's lemma applied to homology theory, Proc. Nat. Acad. Sci. U.S.A. 47, 113–114.

Cohen, D.I.A.,

[1967] On the Sperner lemma, J. Combinatorial Theory, 2, 585–587.

Collins, H.S.,

[1955] Completeness and compactness in linear topological spaces, Trans. Amer. Math. Soc. 79, 256–280.

Cristescu, R.,

[1977] Topological vector spaces (Noordhoof, The Netherland).

Dasor, J.,

[1976] Factoring operators through c_0, Math. Ann. 220, 105–122.

Davie, A.M.,

[1973] The approximation problem for Banach spaces, Bull. London Math. Soc. 5, 261–266.

Davies, E.B.,

[1968] The structure and ideal theory of the predual of a Banach lattices. Trans. Amer. Math. Soc. 121, 544–555.

Davies, W.J./Figiel, T./Johnson, W.B./Pelczynski, A.,

[1974] Factoring weakly compact operators, J. Functional Analysis, 17, 311–327.

Day, M.M.,

[1973] Normed linear spaces (third edition, Springer–Verlag).

Dazord, J.,

[1976] Factoring operators through c_0, Math. Ann. 220, 105–122.

Deans, D.W.,

[1973] The equation $L(E,X^{**}) = L(E,X)^{**}$ and the principle of local reflexivity, Proc. Amer. Math. Soc. 40, 146–148.

De Grande–De Kimpe, N.,

[1977] Locally convex spaces for which $\Lambda(E) = \Lambda[e]$ and the Dvoretzky–Rogers theorem, Composition Math., 35(2), 139–145.

Diestel, J.,

[1984] Sequences and series in Banach spaces (Graduate Texts in Math. Springer–Verlag, Berlin).

Edwards, D.A.,

[1964] On a homemorphic affine embedding of locally compact cone into a Banach dual space endowed with the vague topology, Proc. London Math. Soc. 3(14), 399–414.

Ellis, A.J.,

[1964] The duality of partially ordered normed linear spaces, J. London Math. Soc. 39, 730–744.

Enflo, D.,

[1973] A counterexample to the approximation problem in Banach spaces, Acta Math. 130, 309–317.

Fan, K.,

[1952] Fixed–point and minimax theorems in locally convex topological linear spaces, Proc. Nat. Acad. Sci. U.S.A. 38, 121–126.

[1953] Minimax theorems, Proc. Nat. Acad. Sci. U.S.A. 39, 42–47.

[1957] Existence theorems and extreme solutions for inequalities concerning convex functions or linear transformations, Math. Z. 68, 205–216.

[1959] Convex sets and their applications (Lecture Notes, Argonne National Laboratory).

[1961] A generalization of Tychonoff's fixed point theorem, Math. Ann. 142, 305–310.

[1964] Sur un Théorèm minimax, C. R. Acad. Sci. Paris, 259, 3925–3928.

[1967] Simplicial maps from an orientable n–pseudomanifold into S^n with the octahedral triangulation, J. Combinatorial Theory, 2, 588–602.

[1970] A combinatorial property of pseudomanifolds and covering properties of simplexes, J. Math. Anal. Appl. 31, 68–80.

[1972] A minimax inequality and applications, inequalities III, Proceedings of the Third Symposium on Inequalities. Academic Press, 103–113.

[1981] A further generalization of Shapley's generalization of the Knaster–Kuratowski–Mazurkiewicz theorem, Game Theory and Mathematical Economics, North– Holland, 275–279.

[1984] Some properties of convex sets related to fixed point theorems, Math. Ann. 266, 519–537.

Figiel, T.,

[1973] Factorizations of compact operators and applications to the approximation problem. Studia Math. 45, 241–252.

Gale, D.,

[1984] Equilibrium in a discrete exchange economy with money, International J. of Game Theory, 13, 61–64.

Glicksberg, I.L.,

[1952] A further generalization of the Kakutani fixed point theorem, with applications to Nash equilibrium point, Proc. Amer. Math. Soc. 3, 170–174.

Grosberg, J. and Krein, M.,

[1939] Sur la décomposition des fonctionelles en composantes positive, C. R. (Doklady) Acad. Sci. URSS, 25(9), 723–726.

Grothendieck, A.,

[1955] Produits tensoriels topologiques et espaces nucléaires, Memoirs Amer. Math. Soc. 16.

[1956] Sur certaines classes de suites et le théorème de Dvoretzky–Rogers, Bol. Soc. Mat. Sao Paulo 8, 83–110.

[1973] Topological vector spaces (Gordon and Breach).

Hartman, P. and Stampacchia, G.,

[1966] On some nonlinear elliptic differential functional equations, Acta Math. 115, 271–310.

Hogbe–Nlend, H.,

[1977] Bornologies and functional analysis (North–Holland).

James, R.C.,

[1964] Characterizations of reflexivity, Studia Math. 23, 205–216.

Jameson, G.J.O.,

[1970] Ordered linear spaces (Lecture Notes in Math. 104, Springer–Verlag, Berlin).

Johnson, W.B./Rosenthal, H.P./Zipping, M.,

[1971] On bases, finite dimensional decompositions and weak structures in Banach spaces, Israel J. Math. 9, 488–506.

Junek, H.,

[1983] Locally convex spaces and operator ideals (Teubuer–Texte zur Mathematik, Band 56).

Kakutani, S.,

[1938] Two fixed–point theorems concerning bicompact convex sets, Proc. Imp. Acad. Tokyo, 14, 242–245.

[1941] A generalization of Brouwer's fixed point theorem, Duke Math. J. 8, 457–459.

Kalton, N. and N. Peck,

[1980] A re–examination of the Roberts example of a compact convex set without extreme points, Math. Ann. 253, 89–101.

Kirk, W.A.,

[1965] A fixed point theorem for mappings which do not increase distances, Amer. Math. Monthly, 72, 1004–1006.

Kist, J.,

[1958] Locally o–convex spaces, Duke Math. J. 25, 569–582.

Klee, V.L.,

[1955] Boundedness and continuity of linear functionals, Duke Math. J. 22, 263–269.

Knaster, B. Kuratowski, K. and Mazurkiewicz, S.,

[1929] Ein Beweis des Fixpunktsatzes für n–dimensionale Simplexe, Fund. Math. 14, 132–137.

Kneser, H.,

[1952] Sur un théoréme fondamental de la théone des jeux, C. R. Acad. Sci. Paris, 234, 2418–2420.

Komura, Y.,

[1964] Some examples on linear topological spaces, Math. Ann. 153, 150–162.

Köthe, G.,

[1969] Topological vector spaces I (Springer–Verlag, Berlin).

[1979] Topological vector spaces II (Springer–Verlag, Berlin).

Krein, M.,

[1940] Sur la décomposition minimale d'ure fonctionnele linéaire u_n composantes positives, Doklady Acad. Nauk SSSR 28, 18–22.

Larsen, R.,

[1973] Functional Analysis (Marcel Dekker, New York).

Leibowitz, G.M.,

[1970] Lectures on complex function algebras (Scott. Foresman & Co.).

Levin, M. and Saxon, S.,

[1971] A note on the inheriance of properties of locally convex spaces by subspaces of countable codimension, Proc. Amer. Math. Soc. 29, 97–102.

Lindenstrauss, J./Rosenthal, H.P.,

[1969] The L_p–spaces, Israel J. Math. 7, 325–349.

Lindenstrauss, J./Tzafriri, L.,

[1977] Classical Banach spaces I (Springer–Verlag, Berlin).

Lomonosov, V.,

[1973] On invariance subspaces of families of operators commuting with a completely continuous operators, Funkc. Anal. i Prilozenia, 7, 55–56.

Mackey, G.W.,

[1945] On infinite–dimensional linear spaces, Trans. Amer. Math. Soc. 57, 155–207.

[1946] On convex topological linear spaces, Trans. Amer. Math. Soc. 60, 519–537.

Markoff, K.K.,

[1936] Quelques théorémes sur les ensembles abéliens, C. R. Acad. Sci. URSS (N. S.), 1, 311–313.

Minty, G.J.,

[1962] Monotone (nonlinear) operators in Hilbert spaces, Duke Math. J. 29, 341–346.

Namioka, I.,

[1957] Partially ordered linear topological spaces, Memoirs Amer. Math. Soc. 24.

Narici, L. and E. Beckenstein,

[1985] Topological vector spaces (Marcel Dekker Inc. New York and Basel).

Nash. J.,

[1950] Equilibrium points in n–person games, Proc. Nat. Acad. Sci. U.S.A. 36, 48–49.

Ng, Kung–Fu (See also Wong, Yau–Chuen),

[1969] The duality of partially ordered Banach spaces, Proc. London Math. Soc. 19, 269–288.

[1971] Solid sets in ordered topological vector spaces, Proc. London math. Soc. 22, 106–120.

[1971b] On a theorem of Dixmier, Math. Scand. 26, 14–16.

[1973] A continuity theorem for order–bounded linear functionals, Math. Ann. 201, 127–131.

Ng, Kung–Fu/Duhoux, M.,

[1973] The duality of ordered locally convex spaces, J. London Math. Soc. (2)8, 201–208.

Peressini, A.L.,

[1967] Ordered topological vector spaces (Horper–Row, New York).

Pietsch, A.,

[1972] Nuclear locally convex spaces (Springer–Verlag, Berlin).

[1980] Operator ideals (VEB Deutscher Verlag der Wiss, Berlin).

Raebiger,

[1984/85] Some structure theoretical characterization of Grothendieck spaces.

Randtke, D.J.,

[1972] Characterization of precompact maps, Schwartz spaces and nuclear spaces, Trans. Amer. Math. Soc. 165, 87–101.

[1973] Representations for compact operators, Proc. Amer. Math. Soc. 37, 481–485.

[1973b] A simple example of a universal Schwartz spaces, Proc. Amer. Math. Soc. 37(1), 185–188.

[1974] A compact operator characterization of ℓ_1, Math. Ann. 208, 1–8.

Roberts, J.,

[1977] A compact convex set with no extreme points, Studia Math. 60, 255–266.

Robertson, A.P./Robertson. W.J.,

[1973] Topological vector spaces (2nd ed. Cambridge Tracts no. 53, Cambridge Univ. Press).

Royden, H.L.,

[1968] Real analysis (Collier Macmillan, London).

Rudin, W.,

[1966] Real and complex analysis (McGraw–Hill Book Comp. New York).

Schaefer, H.H.,

[1966] Topological vector spaces (Macmillan, New York).

[1974] Banach lattices and positive operators (Springer–Verlag, Berlin).

Schatten, R.,

[1950] A theory of cross–spaces (Princeton Univ. Press).

Schauder, J.,

[1930] Der Fixpunktsatz in Funktionalräumen, Studia Math. 1, 123–139.

Schwartz, L.,

[1981] Geometry and probability in Banach spaces (Lecture Notes in Math. 852, Springer–Verlag, Berlin).

Shapley, L.S.,

[1973] On balanced games without side payments, Mathematical Programming (Eds, Hu, T.C. and Robinson, S.M.) New York, Academic Press, 261–290.

Shih, M.H.,

[1986a] Covering properties of convex sets, Bull. London Math. Soc. 18,57–59.

[1986b] On certain matching properties of convex sets, Bull. London Math. Soc. 18, 192–194.

Shih, M.H./Tan, K.K.,

[1985] Generalized quasi–variational inequalities in locally convex topological vector spaces, J. Math. Anal. Appl. 108, 333–343.

Shih, M.H./Lee, S.N.,

[1992] A combinatorial Lefschetz fixed–point formula, J. Combinatorial Theory, Ser. A (in press).

Sperner, E.,

[1928] Neuer Beweis für die Invarianz der Dimensionszahl und des Gebietes. Abh. Math. Sem. Ham. Univ. 6, 265–272.

Stephani, I.,

[1976] Operator ideals with the factorization property and ideal classes of Banach spaces, Math. Nachr. 74, 201–210.

Terzioglu, T.,

[1970] On compact and infinite–nuclear mappings, Bull. Soc. Math. Roumaine 14, 93–99.

[1971] A characterization of compact linear mappings, Arch. Math. 22, 76–78.

[1972] Remarks on compact and infinite–nuclear mappings, Math. Balkanika 2, 251–255.

Treves, F.,

[1967] Topological vector spaces, distributions and kernel (Acad. Press, New York).

Tychonoff, A.,

[1935] Ein Fixpunktsatz, Math. Ann. 111, 767–776.

Valdivia, M.,

[1971] Absolutely convex sets in barrelled spaces, Ann. Inst. Fourier Grenoble 21(2), 3–13.

[1972] Some examples on quasi–barrelled spaces, Ann. Inst. Fourier Grenoble 22(2), 21–26.

Walsh, B.,

[1973] Ordered vector sequence spaces and related classes of linear operators, Math. Ann. 206, 89–138.

Wong, Yau–Chuen,

[1969] Locally o–convex Riesz spaces, Proc. London Math. Soc. 19, 289–309.

[1970] A short proof of Schaefer's theorem. Math. Ann. 184, 155–156.

[1972] The order–bound topology, Proc. Cambridge Phil. Soc. 71, 321–327.

[1973a] Open decomposition on ordered convex spaces, Proc. Cambridge Phil. Soc. 74, 49–59.

[1973b] A note on open decomposition, J. London Math. Soc. (2)6, 419–420.

[1973c] On the continuity of order–bounded linear functionals. Nata Mathematica (2)6, 40–45.

[1976a] The topology of uniform convergence on order–bounded sets (Lecture Notes in Math. 531, Springer–Verlag, Berlin).

[1976b] Characterizations of the topology of uniform convergence on order–intervals, Hokkaido Math. J., (2)5, 164–200.

[1979] Schwartz spaces, nuclear spaces and tensor products (Lecture Notes in Math. 726, Springer–Verlag, Berlin).

[1980] An introduction to ordered vector spaces (Lectures delivered at Institue of Mathematics, Academia Sinica at Taiwan).

[1980b] A factorization theorem for Schwartz linear mappings, Math. Japonica, 25(6), 655–659.

[1982a] A characterization of quasi–Schwartz linear mappings, Indag. Math. 85(3), 359–363.

[1982b] On a formula of the regular hull of operator ideals, Mathematies Seminar Notes 10, 797–802.

[1987] On a factorization of operators through a subspace of c_0, Nonlinear and Convex Analysis, Lecture Notes in Pure & Appl. Math. 107 (Dekker), 295–298.

Wong, Yau–Chuen/Cheung, Wai–Lok,

[1971] Locally absolute–dominated spaces, United College J. (Hong Kong) 9, 241–249.

Wong, Yau–Chuen/Ng, Kung–Fu,

[1973] Partially ordered topological vector spaces (Oxford Math. Monographs, Clarendon Press Oxford).

Index